COMBINE HARVESTERS

Theory, Modeling, and Design

COMBINE HARVESTERS
Theory, Modeling, and Design

PETRE MIU

CRC Press
Taylor & Francis Group
Boca Raton London New York

CRC Press is an imprint of the
Taylor & Francis Group, an **informa** business

CRC Press
Taylor & Francis Group
6000 Broken Sound Parkway NW, Suite 300
Boca Raton, FL 33487-2742

First issued in paperback 2017

© 2016 by Taylor & Francis Group, LLC
CRC Press is an imprint of Taylor & Francis Group, an Informa business

No claim to original U.S. Government works

ISBN 13: 978-1-1387-4827-9 (pbk)
ISBN 13: 978-1-4665-0512-4 (hbk)

To Professor Dr. Heinz Dieter Kutzbach.

Special dedication: To my daughter, Iulia.

To all my students, research colleagues, and followers.

The greatest enemy of knowledge is not ignorance, it is the illusion of knowledge.

—Stephen Hawking

Truth is what stands the test of experience.

—Albert Einstein

Contents

Preface

Combine Harvesters: Theory, Modeling, and Design was designed to be both introductory and comprehensive for a two-course sequence for undergraduate and graduate students, respectively. In regard to the theory, analytical techniques, engineering design expertise, and software use, this book also serves as a useful reference tool for scientists and practicing engineering professionals in academia and industry around the world.

The main impetus for this work was my realization that no similar textbook exists to provide a systematic, unified, and in-depth approach to combine harvester interdisciplinary engineering knowledge, from fundamentals to design applications. Given its engineering content based on the author's practice, this book is also recommended for other courses in biomechatronics and biosystems engineering that emphasize functional analysis, design, and research in field machinery, crop harvesting dynamics, and modeling and controls for agricultural systems. Considering *green engineering* in the evolving design strategy, as outlined in this book, results in minimizing environment pollution and risks to human health and a better quality of life, while improving the environment when the knowledge is accordingly applied in practice by engineers, technologists, and farmers—the ultimate beneficiaries.

Approach

The author's objective was to provide the users with very well-organized engineering knowledge that is progressive, qualitative, and quantitative. The claim to be progressive and qualitative is based on (1) the coverage of the latest stochastic and deterministic modeling methods and evolutionary computational techniques that lead to calculation, simulation, and optimization of combine harvester processes; (2) its unitary approach to combine harvester modeling, control, and design topics; and (3) the depth of engineering and mathematics, and the use of simulation software (including MATLAB®) throughout the text. The claim to be quantitative is proved by the book content in terms of the comprehensiveness of topics and the interdisciplinary approach, ranging from specific properties of processed materials to immune systems for fault diagnosis and autonomous guidance of vehicles. The book may also be used by specialists from other emerging technologies, such as biosystems robotics and applied biomechatronics.

The manuscript of this book evolved mainly from lecture notes, research outcomes, and papers developed by the author in the modeling, simulation, evolutionary optimization, and design of combine processes. A deliberate attempt was made to round this knowledge with additional material from other contributors' experiences and industry practice. The material in this book ranges from theoretical developments to practical applications through the author's research experience as an Alexander von Humboldt fellow, my teaching and design expertise in harvesting machinery, and design and control of large computer numerical control machining centers, and my reliability engineering experience for aerospace electronics.

Target

The students, engineers, researchers, and other colleagues who will be using this book are offered a progressive exposure to the knowledge of combine harvester engineering, starting with the combine systems construction basics, then going through key concepts in theory, and ending with advanced modeling concepts and design optimization techniques. Since process modeling and simulation necessarily entail mathematical representations and computer programs, the users shall feel confident in handling statistics, derivatives, and integrals as well as MATLAB programming. Basic knowledge of engineering control and machine elements design is desired as well. The examples provided throughout the text are intended to not only facilitate the learning process, but also be useful handy tools for related applications.

Main Features of the Book

- Book organization from technical specifications and basics of combine harvester construction through comprehensive modeling and simulation theory development to optimal control and design optimization of machine processes: cutting, husking, conveying, threshing, and separation, including specific power requirements for each unit.
- An integrated approach of modeling, simulation, optimization, and control that is uniformly applicable to any structure of combine systems.
- A top-down mechatronic design approach integrating kinematics and dynamics of processed materials with machine systems mechanics, control, and design techniques.
- Use of MATLAB software and Simulink, and associated toolboxes. These are primarily intended not for students' software training, but for learning the engineering fundamentals and problem-solving techniques outlined in the book.
- Examples of current combine systems/elements design throughout the book.
- Applications/exercises, many of them inspired by the author's engineering and research experience.
- Both SI (metric) and imperial/U.S. measuring units used throughout.
- Exhaustive material that cannot be conveniently integrated into the book presented in separate appendices at the end.

Organization of the Book

In this 15-chapter book, the author identifies, organizes, and develops key concepts of combine harvester process theory and mathematical models for equipment and machine process simulation and optimization. This most up-to-date knowledge is accompanied by practical applications adopting a top-down method for mechatronic design of combine harvesters, starting from the general technical specifications, developing machine layout as defined by engineering calculations, and finishing with design considerations of major subassembly processes.

The theory, calculations, and examples have been developed to serve for:

- Combine process modeling and simulation
- Development of combine process and driving task-based control systems
- Sizing and top-to-bottom design of combine assembly and components

The author strongly encourages direct communication with users whose qualified suggestions, comments, and questions may help clarify some concepts and topic approaches and may also help improve this work in the next edition.

MATLAB® is a registered trademark of The MathWorks, Inc. For product information, please contact:

The MathWorks, Inc.
3 Apple Hill Drive
Natick, MA 01760-2098 USA
Tel: 508 647 7000
Fax: 508-647-7001
E-mail: info@mathworks.com
Web: www.mathworks.com

Acknowledgments

First, I would like to acknowledge the suggestions of all the professors who have unconditionally considered and supported my publishing proposal for this book. The following faculty are due recognition for their support: Heinz-Dieter Kutzbach (Hohenheim University, Stuttgart, Germany), Alvin Womac (University of Tennessee, Knoxville, Tennessee), Bernie Engel (Purdue University, West Lafayette, Indiana), and Paul McNulty (University College Dublin, Dublin, Ireland).

I would like to express my high appreciation to Professor Marcel Segarceanu, who mentored me as an MS student and PhD by inoculating the passion for high-end research and teaching at the Politehnica University of Bucharest, Romania.

Writing this book was mainly possible due to my research work as an Alexander von Humboldt fellow at Hohenheim University, Institute of Agricultural Engineering, Stuttgart, Germany. I am deeply indebted to my German mentor, Professor Dr. Heinz-Dieter Kutzbach, as well as my colleague Dr. Peter Wacker for helping me to obtain full access to an uncountable amount of ready-to-use, high-quality, rigorous German-style organized experimental data, which has been acquired through the substantial effort of the research team. My research activity was positively influenced by my German colleague Folker Beck as well. As part of such a team, I am honored to have been able to write this book. However, it was not easy!

My work could not have reached this level of presentation without special support with very high-quality pictures from CLAAS (Jörg Huthmann) and Case IH (Gerald E. Salzman), for which I express many kind thanks.

The fruition of this book became possible due to the interest and strong support, throughout the project, of Jonathan Plant, executive editor, and Joselyn Banks-Kyle, project coordinator, at Taylor & Francis/CRC Press.

The typesetting, proofreading, and art processing were done through a great collaborative effort of the team at Deanta Global Publishing. Thus, for many hours of sharp attention and prompt responses, I am grateful to the entire team, particularly to Sheyanne Armstrong (copy editor) and Michelle van Kampen (project manager).

I also express my thanks to Cynthia Klivecka, the Taylor and Francis project editor, for overseeing the entire production of this book.

My lovely wife, Dana, deserves a written acknowledgment of my gratitude for encouraging me while enduring my hard work over the three-year period of writing this book.

It is my fervent hope that this book will serve not only as a useful reference, but also as a challenge for creativeness and further development in the field. I invite the professional engineers and academia to give me their input on the book topics as well as on other topics that should be included in the next printing. I appreciate your dedication to our profession! Enjoy it and keep in touch!

Petre Miu
Richmond Hill, Ontario, Canada

Author

Petre Miu, an Alexander von Humboldt fellow, is the owner of Projenics (www.projenics.net), Canada, a firm that offers multidisciplinary engineering high-end services for product design and scientific services for process/business modeling, simulation and optimization, data analysis, and statistics.

He worked as a research associate at the Biosystems Engineering Department, University of Tennessee, Knoxville, Tennessee (2005–2006), and as a professor at the Politehnica University of Bucharest, Romania (12 years), where he developed and taught the following courses: Calculation and Design of Harvesting Machinery, Introduction to Biomechatronics, and Robots in Biosystems Engineering.

He earned a PhD degree in engineering (1995) from the Politehnica University of Bucharest. Dr. Miu began his most prolific and prestigious scientific work in Stuttgart, Germany, at Hohenheim University, Institute of Agricultural Engineering, where the German government awarded him two consecutive postdoctoral Alexander von Humboldt research grants (1997–1998 and 1999–2000).

Dr. Miu has extensive experience in the design of harvesting machinery and equipment, as well as in other areas, such as computer numerical control machining centers, robotics, and electronic packaging of radio frequency radar systems.

Other awards: Copernicus research grant (1993) and Tempus research grant (1991) at Hohenheim University, Stuttgart, Germany. Editorial duties: Reviewer for the Elsevier journal *Computers and Electronics in Agriculture*. Publications: Two technical books (*Introduction to Mechatronics* and *Robots ASK (Applications, Structure, Kinematics)*), numerous scientific articles published in prestigious international journals, and conference papers. Research and development areas: Process modeling, simulation, optimization, and control; biomechatronics; robotics; and biomass-based biofuels and bioproducts.

1

Introduction to Combine Harvesters

1.1 Introduction

A modern *combine harvester*, or simply *combine*, is a versatile machine designed to efficiently harvest a variety of grain crops from the field to deliver clean *grains*, usually collected in the machine tank and discharged periodically for transportation and further processing or storage.

The following main crops are harvested using combine harvesters: wheat, rice, barley, oats, rye, triticale (hybrid of wheat and rye), soybeans, flax (linseed), sunflower, and corn (maize). Actually, modern combines can harvest more than 80 types of grain crops, from canola seeds to beans, and from clover to corn. To harvest grain crops, a combine harvester is self-propelled and controlled (by a human operator or an automated pilot) on certain paths in the field; combine harvesters are also driven on public roads or transported with a special trailer to different fields when travel distances are long.

While clean grains are collected in the combine tank, *material other than grain* (MOG) that enters the combine is left behind the machine, on the field, in a continuous windrow (later to be baled), or further chopped and spread continuously on the entire width of the harvesting area. The MOG is composed of fragmented, dried stems and leaves of the crop plants. Although the MOG has limited nutrients, it enriches the soil and changes the soil texture through chemical decomposition, or it may be used for livestock feeding and bedding.

The combine harvester combines all technological operations of grain crop harvesting: *cutting* and *gathering* of the plants, grain *threshing* and *separating*, and grain *cleaning* and *collecting* in the combine tank. Using such technology requires certain growing conditions for the grain crops, but it ensures a rapid collection of clean grains, minimizes grain losses, and clears the field of plants at a relatively reduced cost.

In the following, we briefly define and describe the above-mentioned processes. The processes of *cutting* and *gathering* the plants are performed by removable heads (called headers) that are designed for particular crops, harvesting technologies (e.g., corn grains vs. corn ears), or both. The header is mounted in front of the combine, usually in symmetry with the combine width. The main types of headers are standard header (platform header), draper header (for wheat, rice, barley, oats, rye, triticale, and soybeans), corn header, stripper header (for rice), sunflower header, and pick-up header (for beans).

The *threshing* and *separating* processes are the detachment of the grains from the flowery cover (panicle, ear, etc.) and the separation of grain from the MOG. These processes are performed by a threshing system that can consist of one module or a sequence of threshing modules. Since not all grains are separated by the threshing system, a conventional combine is equipped with straw walkers to shake the straw and recover the rest of the threshed grains.

In the threshing unit, the MOG becomes a mixture of long stalks, chaff, and small fragments of spikes, stalks, leaves, and husks. The combine *cleaning* shoe separates the grain from the MOG that is released with the soil at the back of the machine. The clean grains are collected into the combine tank and discharged when necessary.

Because the crop material follows a certain path into the combine, conventional combine harvesters work on fields with a maximum hillside slope of 6% (side-to-side or transverse direction) and 8% for the longitudinal direction (front to back). Hillsides can have slopes as steep as 45%. Working on high slopes involves leveling the machine to prevent grain and chaff from sliding to the lower side of the separating and cleaning units, while the header remains at all times parallel to the soil to cut and collect the crop properly. Leveling the body of the combine allows the straw walkers to operate more efficiently, maintaining the same machine feedrate as when working on flat fields. Leveling the separating units repositions the machine's center of gravity relative to the base area defined by the wheels' position; this improves the machine's stability against tipping on the hill. The best hillside combines have leveling on both the transversal and longitudinal planes with the following inclinations: downhill slopes of up to 12%, uphill slopes of up to 30%, and transversal slopes of up to 42%. This manner of combine harvesting is on the decline; however, big modern combine harvesters, which have axial rotary grain threshing and cleaning units with improved design allow the machine to work on transversal slopes of up to 20%; besides, these combines are much more stable on hillsides.

Grain crops may be harvested using different technologies. This book organizes, generates, and integrates knowledge for modern, self-propelled combines, including hillside combines, which perform in the field all the operations mentioned above. The theory, calculations, and design are very useful in designing pulled combines or their components, such as threshing and separating units.

1.2 Technological Requirements for Combine Harvesters

Harvesting of grain crops with combine harvesters constitutes a very important task among agricultural activities. Therefore, combine harvesters must satisfy certain technological requirements regarding the crop type, field conditions, weather, and postharvesting technologies, as well as quality performance indices of combine processing of grain crops. The importance of combine harvesting derives from many considerations:

- It is the main harvesting technology of cereals in the world.
- It is seasonal, though it has to be done at the right moment, over a short period of time (e.g., 4–6 days for wheat harvesting), and with minimal grain losses. Optimally, harvesting of a crop starts when the quantity of useful organic substances (proteins, lipids, amino acids, etc.) contained by grains reaches a maximum.
- When the weather is uncooperative, the plants fall on the ground; the machine is able to lift the plants and cut and properly process them inside the machine. Not getting the grains inside the combine may determine a major portion of grain losses.
- For a better quality and quantity of grains, sometimes it is better to modify the harvesting and storage technology, for example, first cutting the plants and placing

them in windrows or swaths, followed later by threshing, separating, drying, and so forth.

- The combine has to have modular construction that allows the connection of different equipment or subassemblies, such as different headers, concaves, and so forth, to be flexibly equipped and operated for a variety of grain crops. Consequently, the combine and appropriate equipment are adaptable and flexible, that is, versatile.

- The combine process parameters are adjustable in relatively large ranges to accommodate technological requirements when processing a variety of grains and MOG that vary in terms of shape, size, moisture, mechanical resistance, aerodynamic properties, and so forth.

- The speed of the machine varies according to the crop conditions, while the process parameters are maintained at optimum values. That implies at least two split, independent channels of power transmission from the same engine.

- Operating the machine properly results in an efficient harvesting operation. This could be substantially improved by monitoring and controlling the machine processes.

- Combine operator comfort (less dust, noise, and vibrations, coupled with proper temperature and humidity) in an optimally controlled cab is of great importance for machine design, manufacturing, and operation.

- Increasing combine feedrate is driven by two basic requirements: increasing the production of grain crops and the necessity of grain harvesting within optimal harvesting time periods.

- Self-propelled combine harvesters have already reached the limiting width of roads, so cost-effective continuous improvement of combine harvester performance will certainly be obtained through modeling, simulation, and optimization of both processes and component design, coupled with implementing a high degree of process automation, control, and improvement of grain transportation and storage logistics (Miu and Kutzbach, 2000).

- The above conditions require highly trained and skilled operators, possible implementation of autonomous vehicle guidance, and very well-planned harvesting operations involving precision farming.

- The combine operating rate can additionally be improved by working in a field until the crop is fully harvested (avoiding switching between different fields), providing enough trailers for unloading harvested grain on the go and for transportation, scheduling operators in shifts with proper breaks, and performing good preventive maintenance. A proper capacity of grain storage, drying, or postharvest processing is very desirable.

1.3 History of Combine Harvester Development

Wheat harvesting technology goes back to 75 AD. At that time, Roman Gaius Plinius Secundus (23–79 AD), known as Pliny the Elder, described in his encyclopedia, *Naturalis Historia*, a reaping device pushed by oxen (Book XVIII, Ch. 72) to gather the ears of wheat and barley from the plant stems. This device was mainly composed of a primitive chassis with two wheels, a collecting box mounted on the chassis, and a front row of sharp-pointed teeth for ear reaping. Although this device was quite wasteful, destroying the straws as well, it remained in use for more than 300 years. During the Dark Ages (derived from the Latin *saeculum obscurum*, which denotes the medieval period), scythes and sickles were used to gather grain crops, as depicted on a later Roman bas-relief in Augusta Trevororum (today's Trier town, Germany).

In 1799, the English inventor Joseph Boyce was granted the first recorded patent for a *mechanical reaper*. The reaper cut an approximately 0.6 m (2 ft) wide path and laid grain to the lateral side. Eight years later, Plucknett was the designer of the first *horse-pulled reaper* (Maw and Dredge, 1879).

Figure 1.1 shows the 1811 reaper implementation of James Smith, of Deanston (England), which was pushed by two horses. It consisted of a vertical shaft with a cutting disc and a cone that was rotated through a long, horizontal shaft. During a trial, it cut more than 0.405 ha (1 acre) per hour, but the cutting disc required sharpening four times. Except for the irregularity of cut material lying on the field, the machine showed a great deal of original ingenuity (Maw and Dredge, 1879).

In 1826 in Scotland, the Reverend Patrick Bell designed (but did not patent) a *reaping machine*, which used the scissors principle of plant cutting—a principle that is still used today (Figure 1.2a and b). The Bell machine was pushed by horses. A few Bell machines were available in the United States.

In 1831, during the first public demonstrations, Cyrus McCormick claimed that he developed in 18 months a *pulled reaper* that cut and guided the plants laterally. Using a revolving reel, the reaper swept the cut plants onto a platform, from which they were raked into piles by a man walking alongside the machine. Consequently, McCormick was granted a patent for the reaper on June 21, 1834 (Patent X8277 (U.S.)). In 1835, McCormick received a second patent for an improved reaper (Wikipedia, 2011). The improvements referred to cutting the grain by the serrated edge of a straight and vibrating cutter, operated by a crank, with the plants being sustained by fingers. The blade was serrated like a sickle, except that the angle of the teeth was reversed for every alternate tooth (Maw and Dredge, 1879). By 1847, Cyrus McCormick began the mass manufacture of his improved reaper (Figure 1.3) in a Chicago factory. In England, the production of McCormick's "Virginia Reaper" started in 1855.

In 1835, in the United States, Hiram Moore built and patented the *first combine harvester*, which was capable of reaping, threshing, and winnowing cereal grains. With a length of 5.2 m (17 ft) and cut width of 4.57 m (15 ft), this combine harvester was pulled by 20 horses fully handled by farmhands.

FIGURE 1.1
Smith's reaper, 1811.

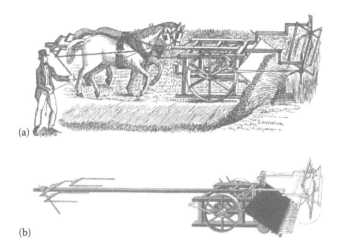

(a)

(b)

FIGURE 1.2
Bell's reaper, 1826. ((a) Maw, W.H. and Dredge, J., *Engineering: An Illustrated Weekly Journal*, Vol. XXVII, 1879, figure 24, p. 544 (663); (b) Crochet, B., *150 ans de machinisme agricole*, Editions de Lodi, 2006.)

FIGURE 1.3
McCormick's improved reaper, 1847. (From Crochet, B., *150 ans de machinisme agricole*, Editions de Lodi, 2006.)

In 1841, 2 years after a truly portable steam-powered engine had been developed, the manufacturer Alexander Dean (Birmingham, England) built the first *threshing machine powered by steam*. In the same year, in Liverpool, Ransomes and Co. exhibited a threshing machine powered by a disc engine (Figure 1.4). This machine could be drawn by horses or made self-propelled by a pitched chain passing over a pulley on the main axle. A similar machine was produced in 1842 by Messrs. Tuxford in Boston. Both the engine and machine were mounted on a single frame.

In 1857, John F. Appleby (Wisconsin) invented the *twine binder* (Maw and Dredge, 1879).

In the United States, by 1860, combine harvesters had been built with a cutting width of several meters; they were drawn by up to 30 horses. In 1872, William Deering, with his Chicago-based Deering Co., built the *first reaper-binder* (Cornways, 2011). The machine was able to cut small cereal plants and tie them into small bundles, or *sheaves*. Such sheaves are usually arranged into conical teepees to allow the grain to dry before threshing.

The 1879 Kilburn Show of Royal Agricultural Society of England debuted a few stationary threshing machines; one of them was built by Clayton and Shuttlework (Figure 1.5).

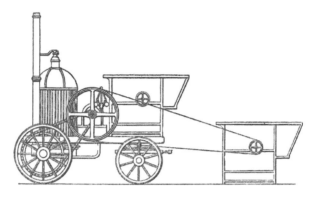

FIGURE 1.4
Ransome's threshing machine powered by steam. (Maw, W.H. and Dredge, J., *Engineering: An Illustrated Weekly Journal*, Vol. XXVII, 1879, figure 38, p. 549 (668).)

FIGURE 1.5
Threshing machine of Clayton and Shuttleworth, 1879. (Maw, W.H. and Dredge, J., *Engineering: An Illustrated Weekly Journal*, Vol. XXVII, Office for Advertisements and Publication, London, 1879, figure, p. 556 (675), description at page.)

The machine was fed by a worker who stood on a platform on the side. The width of the threshing drum was 1600 mm (63 in.), and the fan blew through the threshed material mixture as it fell through the exit to the sack. The machine was equipped with leveling devices. Similar machines were made by Nalder and Nalder of Wantage, England (Maw and Dredge, 1879).

By 1890, several companies, such as Best, Houser and Haines, and Young and Berry, were manufacturing horse-drawn combines. They were quite large, with header widths up to 30 ft. The crew team included up to 10 people (driver, separator man, header tender, sack jig, and sewers).

Such heavy combines created problems with tipping on hillside fields, sometimes injuring people or igniting fires due to coals falling out of the boiler heaters. Then, there was the problem of setting everything back. In 1891, the Californian Stockton-based Holt Manufacturing Company invented a *tilting mechanism* to level the combine thresher on slopes up to 30° and improve its balance. The leveling was allowed by two separate wooden frames that enabled the drive wheels to be raised or lowered independently. The combines incorporating this mechanism were called hillside combines, and that was considered a major technical breakthrough that improved harvesting work safety, reduced fire hazards, and prevented repair costs due to the machine tipping over.

In 1911, in Illinois, the Holt Manufacturing Company acquired the facility of a farm implement maker and founded the Holt Caterpillar Company. The Holt Company used the first combustion engines to drive a threshing machine.

The first tractor-pulled combine was made in 1925 by International Harvester from Chicago (Wiley, 2010). It was also possible to adapt the existing horse-drawn combines by replacing the horse hitch with a tractor one. In 1939, the company Massey Harris from Wisconsin created the Model 21, the first self-propelled combine with a combustion engine. It was small compared with a horse-pulled combine. With the header placed in the front and equipped with a grain tank, this combine was more efficient and able to harvest around 12 ha (30 acres) a day. The Model 21 combine was a big milestone in combine development history.

In 1927, the German company Krupp developed the first *reaper-binder* equipped with a power take-off drive. Then, the first *reaper-thresher-binder* was developed in 1940 by the German manufacturer CLAAS. It was drawn and driven by a tractor using a power take-off drive (Figure 1.6).

In 1951, the German company Fahr built the first self-propelled combine MD-1, with a free-cutting width of 2.1 m. In 1953, CLAAS launched Herkules, a self-propelled combine that was able to harvest up to 5 metric tons of wheat a day.

The technology of self-propelled combine harvesters had spread and evolved so that by 1960, they outnumbered the tractor-drawn machines. Besides being operated by a single operator, self-propelled combines used their engine for both propulsion and harvesting processes. They became more and more efficient due to using increased-power engines. At the same time, the engines were equipped with *self-cleaning rotary screens* to prevent overheating of the engine due to the chaff and dust that clog the engine radiator, blocking the cooling airflow. Newer combine harvesters are equipped with diesel engines. A schematic of a conventional combine is pictured in Figure 1.7a, and a cutaway model of a modern conventional combine is shown in Figure 1.7b.

Then, engineers concentrated on improving the threshing unit, straw walkers, and cleaning shoe separation processes. Also, the width of the headers was simultaneously

FIGURE 1.6
CLAAS' first reaper-thresher-binder, 1940.

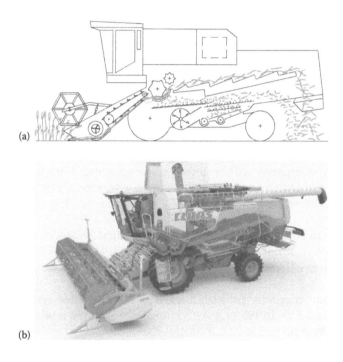

FIGURE 1.7
(See color insert.) (a) Schematic of a conventional combine. (b) Cutaway of a modern conventional combine 3D model. (Courtesy of CLAAS.)

increased with the cutting speed. The introduction of axial-flow rotary combines in 1977 by International Harvester represented one of the most significant technical advances in modern combine development (Figure 1.8).

In 1963, International Harvester introduced the combine model 403 with *transversal (side-to-side)* and *longitudinal (front-to-back) leveling*. This machine could operate on a 36% side slope and 34% uphill grade or 12% downhill grade. The leveling mechanism was based on an oil-cushioned pendulum that activated two hydraulic distributors, allowing leveling of the thresher by two independent hydraulic systems (Ganzel, 2007).

Self-propelled combines started with standard manual transmissions. The machine speed varied based on engine revolutions per minute set up through the acceleration pedal.

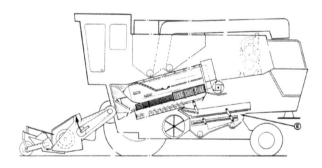

FIGURE 1.8
Axial-flow IH combine schematic. (Kydd, H.D. Combine types. PAMI Gleanings, Prairie Agricultural Machinery Institute, Humboldt, Canada, 1980.)

In the early 1950s, combines were equipped with a *variable-speed drive* that was based on a V-belt transmission with a variable-width sheave controlled by a spring and a hydraulic actuator or manual adjusting mechanism. Varying the sheave width allows the V-belt to ride on different sheave diameters, thus modifying the transmission ratio. This drive was coupled with the input shaft of the gearbox. When *hydrostatic transmission* (a system of hydraulic pumps and motors) was introduced by Versatile Manufacturing, this technology was also transferred to combine harvesters. The hydrostatic transmission has been coupled with the input shaft of a regular four-speed gearbox, forming a hydrostatic drive system. During harvesting (e.g., when the third or second gear is used), the operator can continuously modify the machine speed from zero to the maximum ground speed allowed by the selected gear. Thus, the standard mechanical clutch is no longer needed.

While the basic combine design works for all grain crops, *corn harvesting* follows two approaches. In both of them, a specialized *corn head* is used. For every corn row, such head is equipped with snap rolls that strip and pull down the corn stalk and leaves, while the ears are collected and enter the feeder throat. Then, in a regular combine, the corn ears are threshed, and after cleaning, the corn grains are collected into the grain tank. The other technology, developed in Russia and Eastern Europe, uses a *corn ear husker* to remove the husks and preserve the ear as a whole. This technology has a great advantage for corn drying and storage because the airflow exchange through the ears is much greater than that through the bulk grains. Later, the corn ears are threshed with stationary threshing machines or delivered to livestock without further processing.

Today's combine harvesters can cut swaths of more than 12 m (39 ft) and are equipped with global positioning system devices to track their position, as well as to assess or measure the grain production. Combines have *air conditioning, heaters, cushioned seats,* and *adjustable steering wheel height* for the comfort of the combine operator (Figure 1.9). Details about the construction of modern combine harvesters are given in Section 1.4.

The history of combine harvesters allows us to understand and learn about their technical evolution: the operating crew has been reduced and the overall efficiency has been increased countless times, while preserving grain quality at high purity. An increased modern grain production coupled with combine harvesting keeps grain prices low. This saves money and time for the harvest workers, who can carry on different activities in other fields or use the saved time and money for their hobbies. More about the history of grain harvesting and the development of grain harvesters has been written by Quick and Buchele (1978) and Crochet (2006).

1.4 Construction of Modern Combine Harvesters: Specifications and Performance

1.4.1 Conventional Combine Harvesters

Modeling, simulation, control, design, and development of modern combine harvesters require a thorough understanding and suitable representation of combine system construction and processing of various grain crop properties. The objective is to optimize the combine harvesting process outputs—minimized grain loss, harvesting time, that is, harvesting costs—while preserving grain quality (less damaged grain, clean grain) and the environment (less soil compaction and soil fertilization with fragmented MOG).

FIGURE 1.9
Cab of a modern combine. (Courtesy of SAME Deutz-Fahr.)

A *combine harvester* is a mixed and multifunctional system that consists of different types of processing units and their drives, which process a large variety of crops. Such large, self-propelled machines may cost more than US$500,000.

A *self-propelled combine* consists of specific processing units, their drive systems, and a tank mounted on a chassis along with the ground driving system (engine, mechanical/ electrical power drives, wheels, and steering mechanism). The cab with seat, control console, all driving and process control instrumentation, and climate control system gives the operator maximum convenience in controlling the machine with minimum effort.

Two main types of combine harvesters are prevalent: *conventional combines* and *rotary (axial-flow) combines*. Both combine types have a T-shape longitudinal flow of processed material, beginning with collecting the material from the field with the header to discharging the MOG on the soil, behind the machine. A conventional combine is defined by its *tangential threshing system* and *straw walkers*, while a rotary combine has an *axial-flow threshing system*. Combine harvesters are usually manufactured in series of different dimensions to accommodate various farm production needs. Although today there are many rotary combines on the market, manufacturers have returned to carrying conventional combines alongside their rotary models.

FIGURE 1.10
(See color insert.) Longitudinal section view of a conventional combine harvester. (Courtesy of Same Deutz-Fahr.)

The longitudinal section view of a conventional combine harvester is shown in Figure 1.10. The main parts of a self-propelled *conventional combine* are as follows:

- Chassis with engine and fuel tank, ground power train, wheels, and steering mechanism (and chassis leveling system for hillside/sidehill combines)
- Header (auger header/draper/pick-up head/corn head/chopping corn head/sunflower head)
- Threshing–separating system (one, two, or more modules)
- Separating straw walkers
- Cleaning system (with leveling system for hillside combines)
- Conveying, storage, and discharge system
- Cab with seat, driving control, process control console and instrumentation, and climate control system

The crop is cut and collected by the header; then this material is carried up the feeder throat by a *chain-and-flight elevator*, and then fed into the threshing unit/system. A description of headers and their process can be found in Section 1.5.

The combine *threshing unit/system* is the most important assembly from the point of view of working processes and the power requirement (Campbell, 1980; Kim and Gregory, 1989a, 1989b). A *tangential threshing unit* (Figure 1.11a and b) is mainly composed of a rotary threshing drum (commonly known as cylinder) and a concave (a meshed grill) whose position can be adjusted relative to the cylinder by a manually or automatically driven mechanism. The cylinder speed must be continuously adjustable as required by the threshing process of different crops.

In a tangential threshing unit, the crop material passes the narrow space bordered by rotational rasp bars of the cylinder and concave surface. As high-speed movies show, the threshing process is accomplished mainly due to the impact of the rasp bars or other active elements on material, accompanied by friction with and drawn by the rasp bars and concave. The action of the grooved rasp bars and concave bars in separating the grains and chaff from the straw is controlled by an adjustable rotation of the cylinder against the concave, through which the threshed mixture (grain and fragmented MOG) separates and falls onto the cleaning shoe. The concave could be provided with *dis-awning* plates (Quick

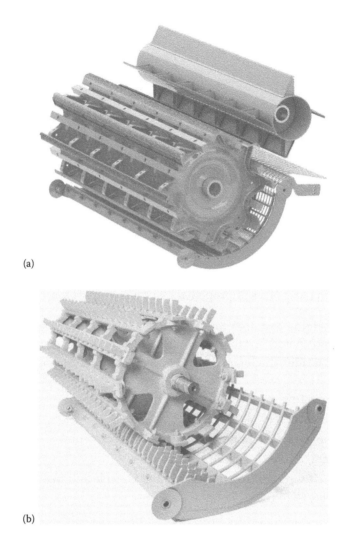

(a)

(b)

FIGURE 1.11
(a) Conventional tangential threshing unit. (Courtesy of Laverda.) (b) Threshing unit for rice. (Courtesy of Laverda.)

and Buchele, 1978), which remove the awns (bristle-like appendages, e.g., in wheat and barley) through generating additional friction with the moving crop material. The concave segments must be interchangeable, as required by the harvesting of different crops whose stem and grain size, as well as their mechanical and aerodynamic properties, vary. The threshed but unseparated grains that exit the threshing unit are recovered from the straw by the straw walkers, whose oscillatory movement shakes the straw and leads the grains to the cleaning shoe system.

A small fraction of unthreshed ears or panicles are conveyed back to the threshing unit and rethreshed. The clean grains are conveyed with a paddle elevator or auger for temporary storage in the combine tank. A flapping auger discharges the grain from the tank into a trailer for later processing or storage.

Increasing the throughput capacity of conventional combines has been achieved through the following changes:

- Increasing the width of the tangential threshing unit from 0.7 to 1.7 m or even more.
- Increasing the concave wrap angle from 85°–90° to medium values of 110°–120° up to 145°.
- Increasing the cylinder diameter up to 800 mm and maintaining the wrap angle.
- Decreasing the size of concave rods from 8–9 mm to 3.5–4.5 mm. This increases the proportion of the active separation surface of the concave relative to the entire surface.

In certain crops and favorable harvesting conditions, grain separation through the concave can be as high as 90%, though a proportion of 80% grain separation in a tangential threshing unit is common at the nominal feedrate.

A more efficient way of increasing the throughput capacity of conventional combines is using a series of tangential threshing units with two or more cylinder-concave units (Figure 1.12a and b). A beater can be placed between two cylinders or an impeller at the end of the threshing module.

A *two-cylinder threshing system* has a 65%–75% higher threshing–separating area, while the separating area of the straw walkers is diminished by 20%–25% within the same size of the machine. Overall, the capacity throughput of a two-cylinder conventional combine is 15%–25% greater than the capacity of a one-cylinder combine.

This technical solution of a two-cylinder threshing system has two advantages: increasing the separating area of the grains through concaves and using a less intensive processing action (smaller rpm) with the benefit of less grain damage. These advantages explain the replacement of the straw walkers with a separating system made of multiple rotary separators mounted in a series in previous combine models developed by CLAAS and New Holland.

The tangential threshing/separating system of Fendt's L-Series combines (Figure 1.13) is made of a regular tangential unit, an active beater, and a *multicrop separator*, that is, a full cylinder with welded in-line brackets and plates. The concave is swung under the separator when harvesting crops with relatively long straw in wet conditions, or the concave of this separator can be rotated above the cylinder when processing crops that require less agitation of the material for grain separation or in dry conditions of brittle crops (Figure 1.14). An arrangement of two or more cylinders, beaters, and so forth, determines the change of material direction when passing from one cylinder to the next; that increases the impact of the rasp bars on the material by changing the beating angle and grain path for better separation.

Depending on crop properties, harvesting conditions, and threshing unit construction and settings, the material that exits the threshing system is composed of certain proportions of unthreshed grain, threshed but unseparated grain, chaff, and straw of different sizes. In conventional combines, this material falls on the *straw walker system*, which performs the task of separation of the remaining grains from the straw. A conventional combine may have four to eight straw walkers with a stepped separation surface (Figure 1.15). Each straw walker is usually made of a U-shaped, long sheet metal channel for collecting separated grain and chaff. The upper edge of the lateral plates of the channel is cut into a sawtooth configuration. On top of the channel are mounted separation grates a few steps along. The number of steps, their height, and the inclination angle of straw walkers differ among manufacturers. The straw walker parallel motion is generated by a pair of spaced-apart crankshafts, of which one is the driver. The crankshaft rotation has an angular offset

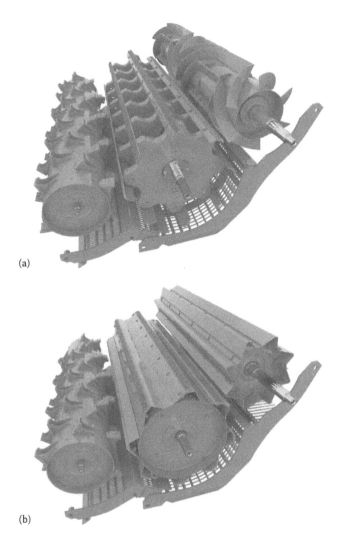

(a)

(b)

FIGURE 1.12
(a) Two-cylinder tangential threshing module with final beater. (Courtesy of CLAAS.) (b) Two-cylinder tangential threshing module. (Courtesy of CLAAS.)

phase that generates an orbital, parallel motion. The crank throw is about 0.05–0.08 m. Adjacent straw walkers are arranged to be substantially in opposite phase, so that as one moves forward, the other moves rearward. In this way, the straw is lifted and then moved rearward by its engagement with the sawtooth edges. Due to this parallel motion, the recovered grains through the separation web slide backward through the walker channel to the cleaning shoe.

To improve the separation efficiency, on top of the straw walkers an auxiliary system has been transversely mounted, such as a cross-shaker (John Deere), agitation tines (CLAAS), a swinging shaker (Laverda), or a transversely mounted *multifinger separation intensifier system* (Figure 1.16). This system may consist of a rotary drum having an eccentric axle mounted herein. Rigid, retractable fingers are mounted on the axle, whose position can be modified. During rotation of the drum, the fingers, projecting to the greatest (adjustable) length at the bottom of the drum, tear apart the mat of straw, increasing grain separation.

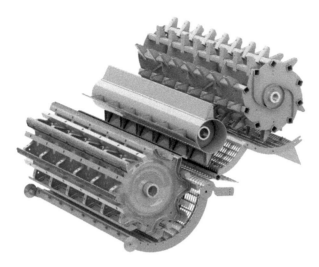

FIGURE 1.13
Tangential threshing module with a middle beater and a multicrop separator. (Courtesy of Fendt.)

FIGURE 1.14
Tangential threshing module with a rotated concave above the cylinder. (Courtesy of Fendt.)

The breaking-up action performed by the fingers on the straw can be optimized by adjusting the following parameters: the elevation of the drum relative to the straw walkers, drum rotation, rotation direction, inclination, and length of retractable fingers outside of the drum.

Drum rotation is usually varied within a 100–200 rpm range. By changing the fingers' inclination angle, the material is conveyed more or less aggressively, which may avoid possible entanglement problems.

Straw walkers limit the throughput capacity of a combine due to relatively reduced separation intensity compared with that of a threshing unit. The most intensive separation of

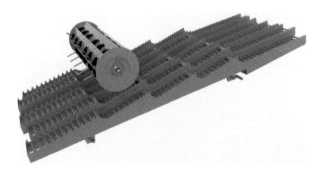

FIGURE 1.15
Straw walkers with multifinger separation intensifier system. (Courtesy of CLAAS.)

FIGURE 1.16
Retractable, multifinger rotary drum. (Courtesy of CLAAS.)

the grain takes place when the straw falls down from a step onto the successive section. The inclination angle of the straw walker sections is also very important for optimization of the grain separation process (Steponavicius et al., 2011).

The *cleaning shoe*, necessary to both conventional and rotary combines, is usually composed of a receiving element (preparation floor), a grain pan, a chaffer, a sieve, and a blower system (Figures 1.17 and 1.18). The preparation floor receives the crop material directly from the threshing concave (or rotor grates), straw walkers, or possibly a return pan under the straw walkers. The preparation floor's delivery edge extends a series of conventional fingers that break up the straw mat during its step fall to the grain pan. The blower or fan system directs a portion of airflow through this walker-type step, partially blowing the chaff downstream to the end of the cleaning shoe system.

FIGURE 1.17
Conventional cleaning shoe system. (Courtesy of Laverda.)

FIGURE 1.18
Cleaning shoe system. (Courtesy of CLAAS.)

The grain pan is provided with an underair blast, and the grain pan thus does double duty: as a conveyor due to its reciprocating movement and as a precleaner. The grain pan also has conventional fingers at its end, where the material falls down on a second step. At this stage, the material already stratified with the grains segregated by their passage along the grain pan is delivered to a chaffer.

Beneath the chaffer, a sieve assembly includes a conventional sieve portion and a chaffer extension coplanar with the sieve portion and spaced from it by a transverse tailing slot, located above the tailing auger of the cleaning shoe. The chaffer and sieve assembly are provided with a conventional air blast from the cleaning shoe fan system. The clean grains are collected over the cleaning shoe width by an auger that transports them to one side of the machine, and then they are loaded into the grain tank by a paddle conveyor. *Tailings*, including unthreshed ears, or panicles, and some heavy MOG components are recovered through the tailing slot, and then collected by a tailing auger and returned to the threshing system for rethreshing. Light trash, including chaff and grain loss, is discharged to the ground from the rear of the cleaning shoe by the air blast from the blower.

When the straw reaches the end of the straw walkers, it falls out the rear of the combine. The straw can be baled for livestock bedding or spread by rotating spreaders at the back of the combine. Figure 1.19 shows the process of a straw chopper followed by a straw/chaff spreader

FIGURE 1.19
Straw chopper and MOG spreader. (Courtesy of CLAAS.)

FIGURE 1.20
Straw chopper and adjustable-blade spreader. (Courtesy of Laverda.)

with a swinging discharge to spread the chaff and chopped straw at the desired width from the rear of the machine, for a good mulching of straw and chaff in the soil. A different MOG chopper/spreader is shown in Figure 1.20; the chaff and chopped straw are guided by a system of metal shields with an adjustable angle of orientation for the desired width of material spreading. Proper positioning of these shields allows a certain compensation for crosswinds.

Table 1.1 shows a comparison of technical specifications of a few modern conventional combine harvesters made by very well-known manufacturers.

A major objective in combine harvester design is balancing the machine throughput capacity with the process efficiency of the main systems: crop gathering, threshing, separating, and cleaning. The wide range of crop types, harvesting conditions, and machine operations and controls make this task very difficult.

Today, conventional combines are used less due to their relatively limited material throughput (up to 9 kg/s), which is dictated by the size of threshing–separating and straw walker systems. The limited performance perspectives of this separating system have eventually led to an integral transfer of the grain separation task to the axial-flow threshing system that is discussed below.

Clean grains are temporarily stored in the combine tank, and then they are discharged by an auger whose swing angle from the combine longitudinal position is higher than 90°.

The other main components of conventional combines are also common to rotary combines; they are described in the next section.

1.4.2 Rotary Combine Harvesters

In 1975, New Holland released the first rotary combine, the model TR70 Twin-Rotor combine with two longitudinally mounted axial threshing and separating rotors (Figure 1.21).

Figure 1.22 shows a longitudinal cutaway of a rotary (Case IH Axial-Flow AFX Series) combine harvester. The tangential threshing module and straw walkers have been completely replaced by a larger, rotary threshing–separating system. In this system, a smooth, continuous flow of material moves on a helical path within the space bordered by the rotor and a 360° wrapping system of concave and cage. The AFX axial-flow threshing unit has three zones: feeding, threshing, and separating. The feeding transition cone draws high volumes of crop material into the threshing space. Inside the concentric cage of the rotor there are adjustable vanes to control the crop flow. The recently developed ST rotor

TABLE 1.1

Technical Specifications of Modern, Conventional Combine Harvesters

Manufacturer	CLAAS	John Deere	New Holland	Fendt	Deutz-Fahr
Model	Lexion 560	T670	CX860	6335C	6090
Dimensions					
Length without header to auger end, mm (in.)	9,200 (362.2)	—	9,969 (392.5)	8,910 (351)	9,240 (364)
Height in transport, mm (in.)	3,870 (152.4)	—	3,960 (156)	4,000 (157.5)	3,990 (157)
Weight with tires, kg (lb)	14,000 (30,865)	13,640 (30,070)	15,500 (34,171)	13,360 (29,453)	11,070 (24,405)
Engine					
Rated power, kW (hp)	216 (290)	268.5 (360)	246 (330)	246 (336)	228 (310)
Maximum power, kW (hp)	227 (305)	293 (393)	266 (357)	265 (360)	—
Number of cylinders	6	6	6	6	6
Piston displacement, L (in.^{3})	7.2 (442)	9 (549)	7.8 (476)	7.4 (451.5)	7.15 (436)
Rated speed, rpm	2,100	2,200	2,100	—	—
Fuel tank capacity, L (gal)	605.6 (160)	800 (211.3)	750 (198.2))	620 (163.8)	—
Header					
Header widths, m (ft)	—	—	—	4.8–7.6 (15.7–24.9)	5.4–7.2 (17.7–23.6)
Cutting frequency, strokes/min	—	—	—	1,254	1,220
Feeding System					
Number of chains	3	4	4	4	—
Chain size	557	—	CA550HD	—	—
Slat design	Serrated	Serrated T-slat	Serrated	—	—
Reverser drive type	Electrohydraulic	Mechanical-hydraulic	Electrohydraulic	—	—
Torque-sensing drive available	—	No	Standard	—	—
Housing lateral float available	Optional	Yes	Standard	—	—
Threshing and Separating System					
Number of threshing cylinders	Preseparation + 1	1	6	1 + 1 separation	1 + 1 separation
Cylinder width, mm (in.)	1,420 (56)	1,676 (66)	1,575 (62)	1,600 (63)	1,521 (60)
Cylinder diameter, mm (in.)	600 (24)	660 (26)	750 (29.5)	600 (23.6)	600 (23.6)
Cylinder speed, rpm	—	—	—	430–1,310	420–1,250
Cylinder speed control	Electrohydraulic	Electrical	Electrohydraulic	—	Electrical
Concave wrap angle, °	—	—	—	120	121

(*Continued*)

TABLE 1.1 (Continued)

Technical Specifications of Modern, Conventional Combine Harvesters

Manufacturer	CLAAS	John Deere	New Holland	Fendt	Deutz-Fahr
Model	Lexion 560	T670	CX860	6335C	6090
Concave area, m^2 (ft^2)	—	—	1.18 (12.7)	0.99 (10.66)	1.13 (12.16)
Beater diameter, mm (in.)	—	—	—	380 (14.96)	—
Beater speed, rpm	—	—	—	800	—
Separating cylinder width, mm (in.)	N/A	—	1,575 (62)	1,600 (63)	1,521 (60)
Separating cylinder diameter, mm (in.)	N/A	—	750 (29.5)	600 (23.6)	Separation 590 (23.2)
Total separating area, m^2 (ft^2)	1.71 (18.4)	1.75 (18.84)	7.15 (76.97)	2.25 (24.22)	2.1 (22.6)
Separating Straw Walkers					
Number of straw walkers	—	—	N/A	6	6
Number of walker steps	—	—	N/A	4	4
Straw walker length, m (ft)	—	—	N/A	4.26 (13.98)	—
Straw walker area, m^2 (ft^2)	6.25 (67.27)	8.76 (94.3)	N/A	6.81 (73.3)	6.7 (72.12)
Cleaning System					
Leveling system	Optional	No	Optional	—	Optional 20%/6%
Total sieve area, m^2 (ft^2)	4.93 (53.07)	4.98 (53.6)	4.57 (49.17)	5.58 (60.06)	6.32 (68.03)
Total cleaning area, m^2 (ft^2)	6.25 (67.27)	6.79 (73.09)	6.5 (69.96)	8.64 (93)	6.32 (68.03)
Fan type	4-turbine fans	4-fan motors	6-blade fan	Radial fan	—
Fan speed, rpm	—	—	—	350–1,050	—
Grain Handling System					
Tailing elevator type	Paddle	Paddle	Dual auger	—	—
Clean grain elevator type	Paddle	Paddle	Paddle	—	—
Grain tank capacity, L (gal)	9,867 (2,607)	11,000 (2,906)	11,630 (3,073)	9,000 (2,378)	8,500 (2,246)
Tank unloading rate, L/s (gal/s)	99 (26.16)	78 (20.6)	105 (27.24)	105 (27.24)	90 (23.78)
Unloading auger length, m (ft)	6.6 (21.65)	6.553 (21.5)	5.49 (18)	5 (16.4)	Optional 5.6 (18.4)
Standard unloading height, m (ft)	3.96 (13)	4.29 (14.08)	3.79 (12.42)	4.45 (14.76)	4.3 (14.1)
Crop Residue Disposal					
Straw chopper	2-speed	2-speed	Optional	64 knives	Optional
Straw spreader	With adjustable vanes	Yes	—	—	Optional
Chaff spreader	Variable speed	Yes	Optional	—	Optional
Quick switch chopping/swathing	—	—	—	Yes	—

TABLE 1.1 (Continued)

Technical Specifications of Modern, Conventional Combine Harvesters

Manufacturer	CLAAS	John Deere	New Holland	Fendt	Deutz-Fahr
Model	Lexion 560	T670	CX860	6335C	6090
Power Train					
Drive type/number of gears	Hydrostatic/3	Hydrostatic/3	Hydrostatic/4	Hydrostatic/4	Hydrostatic/4
Ground speed, km/h (mph)	—	—	—	0–20 (0–12.43)	0–30 (0–18.65)
Transport speed at 1,550 rpm, km/h (mph)	—	—	—	20 (12.43)	30 (18.65)
Brakes, turning against	Disc	Drum	—	—	—
Brakes parking	Drum	Drum	Disc	—	—
Final drive type	Double reduction	—	Planetary	—	—
Steering					
Tread width, adjustable axle, mm (in.)	2,990 (117.7)/3,300 (129.9)	—	—	—	—
Tread width, rear wheel assistance, mm (in.)	3,150 (124)/3,675 (144.7)	—	—	—	—
Standard steering type	Dual cylinder	Dual cylinder	Dual cylinder	Dual cylinder	—
Turning radius, mm (in.)	9,370 (369)	—	—	—	—
Tires					
Drive tire size	800/65 R32	800/65 R32	900/60 R32	800/65 R32	650/75 R32
Steering tire size	500/82 R28	480/80 R26	14.9 × 24	460/70 R24	405/70 R20
Cab					
Operator seat suspension	Luxury/air ride	Yes	Air ride	Yes	Yes
Instructor/passenger seat	Standard	RH console	Standard	Yes	—
Control, position	Tilt/telescope	RH console	RH console	Yes	Yes
Monitor	Yes	Yes	Yes	Yes	Yes
Heating	—	—	—	Yes	Yes
Automatic air conditioning	—	—	—	Yes	Yes

Note: — Data not provided or feature not available.

features a smaller diameter of the longitudinal rotor tube and taller threshing elements mounted on the tube. This configuration allows higher throughputs of material and a lower power requirement.

Compared with tangential threshing systems, the rotary ones offer the following advantages:

- Reduced damage to fragile grains.
- Better adaptability of all crop processing.
- Low level of grain losses.

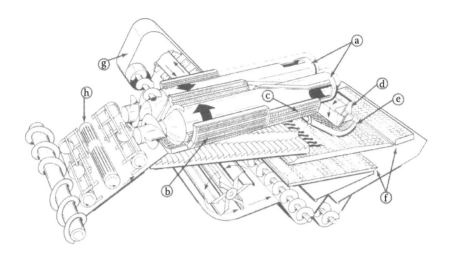

FIGURE 1.21
Sperry New Holland TR70: (a) rotors, (b) threshing concave, (c) separating concave, (d) back beater, (e) beater grate, (f) cleaning shoe, (g) tailing return, and (h) stone ejection roller.

FIGURE 1.22
(See color insert.) Section view of Case IH Axial-Flow AFX combine harvester. (Courtesy of Case IH.)

- Reduced sensitivity to the variation of volumetric feedrate.
- Small variation in grain loss with increasing MOG/grain mass ratio.
- Very efficient at threshing crops with relatively high moisture.
- Although the specific power requirement (kW/(kg/s)) of the axial system is 16%–20% higher, the throughput capacity is 50%–90% higher in axial than in tangential threshing units. That is due to a higher separating intensity in a rotary threshing–separating system than in straw walkers.
- Fewer adjustments, low vibration level, and less maintenance when compared with the straw walkers.

The disadvantages of rotary combines are an increased power requirement and a higher degree of MOG fragmentation and separation through concaves and grates that lead to

FIGURE 1.23
Cutaway of a Gleaner R-Series rotary combine. (Courtesy of Gleaner.)

overcharging the cleaning shoe system. Because rotary combines do not preserve the quality of the straw, it is more difficult to bale or remove it from the field, though it is easier to incorporate the crop residue that results from rotary combines into the soil furrows during plowing.

Later, each major combine manufacturer adopted quite different rotary combine constructions. The Gleaner combines (Figure 1.23) follow the design of Allis Chalmers manufacturer; the rotary threshing/separating unit is mounted crossways so that the material is fed tangentially into the left end of the rotor. To intensify the after-threshing chaff separation process, a pair of rollers accelerates the material movement to the cleaning shoe across from the airstream blown by a fan.

New technical solutions for a threshing–separating system combine a tangential threshing system with an axial threshing–separating system, forming a hybrid threshing–separating system. Figure 1.24 shows this arrangement (APS threshing system + Roto Plus separation system) in a modern CLAAS Lexion combine (Figure 1.25). The first cylinder is an accelerator of material movement that is followed by the tangential threshing cylinder and main concave; the impeller behind them divides the material into two parts for feeding two axial separating rotors that separate the remaining grains in the fragmented straw.

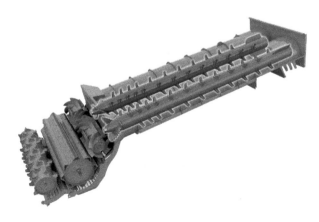

FIGURE 1.24
Hybrid threshing–separating system. (Courtesy of CLAAS.)

FIGURE 1.25
(See color insert.) Cutaway of a Lexion CLAAS combine with a hybrid threshing–separating system. (Courtesy of CLAAS.)

A similar threshing–separating system is used in John Deere's C-Series combines (Figure 1.26). The material is fed into a large tangential threshing cylinder, and then an intermediate beater prevents material wrapping and slugging. The following overshot beater further processes the material and pushes it to two counterrotating tine separators.

The hybrid threshing–separating systems allow the setting of a lower peripheral speed of the cylinders and beaters, whose action will be later complemented by the rotors' effect. Such technical solutions favor a higher material throughput, lower mechanical damage of the grains, and minimal threshing and separation losses.

The separated grain, chaff, and some fragmented straw are directed to the cleaning shoe, using different technical solutions: an assembly of augers (Figure 1.26), a backward inclined floor (Figure 1.21), or a pair of rollers, as used in the Gleaner combines (Figure 1.23).

The mechanical resistance of grains' detachment from their flowery cover varies within relatively large ranges (Jbail, 1994). Thus, developing a constant threshing intensity, coupled with an increased acceleration for grain separation along the threshing space, an axial threshing unit shall develop a variable peripheral speed along the rotor. At smaller speeds, the grains from the ear center will be initially detached, and then the other grains will be detached at progressively higher speeds. That means the rotor/concave radius shall increase from the feeding zone to the separating zone. An exponential variation law of the radius of such a

FIGURE 1.26
Cutaway of a John Deere rotary combine. (Courtesy of John Deere.)

threshing/separating unit shall be close enough to the law that describes the grain detachment from the cereal or corn ears. At this time, although patents have been granted, no combine harvester has been equipped with an axial threshing unit with a variable rotor radius.

The other main components of rotary combines (cleaning shoe, grain handling system, and the tank) are also common to conventional combines, as described in Section 1.4.1. Table 1.2 shows a comparison of technical specifications of several modern rotary combine harvesters.

1.5 Equipment for Combine Harvesters

To harvest a large variety of crops, combine harvesters need special equipment or attachments, such as front headers, closed threshing cylinders, cleaning sieves, huskers, stalk choppers, and corresponding material conveyors. Additional equipment/features may be necessary when the machine operates on hillside fields, or for rice harvesting that requires high underframe clearance and traction aids, half-tracks, or even full crawler ground drive.

Combines that belong to one or more of a series developed by a manufacturer are equipped with interchangeable front headers for gathering and cutting the plants of particular crops. We distinguish the following headers: grain header, draper header, stripper header, corn header, chopping corn header, sunflower header, and pick-up header.

1.5.1 Grain Headers

Grain headers (Figure 1.27a and b) are used for harvesting cereals with small grains: wheat, barley, oats, rye, triticale, rice, soybean, flax, milo, and other cereal crops. The headers are built in series of four to six sizes (e.g., 6–12 m, 20–40 ft) to be coupled with combines of different sizes for a variety of crops as needed.

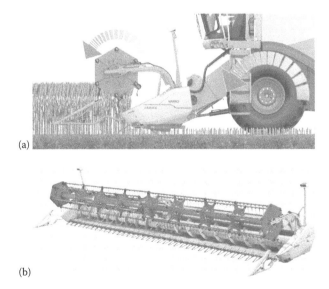

(a)

(b)

FIGURE 1.27
(a) Lateral view of a grain head. (Courtesy of CLAAS.) (b) Grain head. (Courtesy of CLAAS.)

TABLE 1.2

Technical Specifications of Modern, Rotary Combine Harvesters

Manufacturer	Case IH	Gleaner AGCO	John Deere	Massey Fergusson	New Holland	CLAAS
Model	AF 7120	R76	9770 STS	9795	CR9060	740/740 Terra Trac
Dimensions						
Length without header to auger end, mm (in.)	9,398 (370)	8,611 (339)	—	9,957 (392)	9,430 (371.3)	—
Height in transport, mm (in.)	4,089 (161)	3,581 (141)	4,521 (178)	3,734 (147)	3,950 (155.5)	—
Weight with tires, kg (lb)	15,912 (35,080)	12,800 (28,500)	14,862 (32,765)	13,425 (29,597)	15,944 (35,150)	16,785 (37,000)
Engine						
Rated power, kW (hp)	268 (360)	261 (350)	268 (360)	261 (350)	268 (360)	268 (360)
Maximum power, kW (hp)	309 (415)	280 (375)	293 (393)	280 (375)	310 (415)	294 (394)
Number of cylinders	6	6	6	6	6	6
Piston displacement, L (in.^{3})	8.7 (531)	8.4 (513)	9 (548)	8.4 (513)	8.7 (531)	9.3 (567.5)
Rated speed, rpm	2,100	2,100	2,200	2,100	2,100	1,900
Fuel tank capacity, L (gal)	1,000 (264.2)	568 (150)	945 (250)	605 (160)	757 (200)	800 (210)
Feeding System						
Number of chains	4	4	3	4	3	—
Chain size	557 HD	557	557	557	CA557	—
Slat design	Serrated	Serrated	Undershot	Nonserrated	S-slat	—
Reverser drive control	Power plus CVT drive	Electrohydraulic	Electrohydraulic	Electrohydraulic	In cab/hydraulic	—
Torque-sensing drive available	—	Yes	Yes	Yes	Standard	—
Housing lateral float available	Optional	Yes	Yes	Yes	Standard	—
Threshing and Separating System						
Number of threshing cylinders/rotors	0/1	0/1	0/1	0/1	0/2	2/2
Rotor/cylinder type	Longitudinal rotor	Transverse rotor	Rotary	Rotary	Twin rotary	Cylinder/rotor

Rotor description	Staggered rasp bars	Chromed reversible bars	Single tine separator	Rasp bar	Bar staggered, spiraled	Dis-awning plates
Rotor length, mm (in.)	2,612 (102.8)	2,235 (88)	3,124 (123)	3,556 (140)	2,642 (104)	4,200 (165.4)
Rotor diameter, mm (in.)	762 (300)	635 (25)	762 (30)	700 (27.5)	432 917	445 (17.5)
Rotor speed control	Electronic, cab	—	Electrohydraulic	Hydrostatic	Electrohydraulic	360–1,050 rpm
Concave and grate area, m² (ft²)	1.07 (11.52)	0.54 (5.81)	1.1 (11.84)	1.42 (15.28)	—	3 (32.3)
Cylinder diameter, mm (in.)	—	—	—	—	600 (24)	600 (24)
Cylinder width, mm (in.)	—	—	—	—	1,420 (56)	1,420 (56)
Separating area, m² (ft²)	1.91 (20.56)	2.97 (31.97)	1.5 (16.15)	1.45 (15.6)	1.78 (19.16)	1.44 (15.5)
Cleaning System						
Leveling system	Self-leveling	—	—	—	Standard	Standard
Precleaner area, m² (ft²)	1.02 (10.98)	—	0.98 (10.55)	—	1.02 (10.98)	—
Chaffer area, m² (ft²)	1.73 (18.62)	2.51 (27.02)	1.89 (20.34)	2.856 (30.74)	1.73 (18.62)	—
Sieve area, m² (ft²)	1.73 (18.62)	2.19 (23.57)	1.62 (17.44)	2.441 (26.27)	1.73 (18.62)	—
Total separating area, m² (ft²)	5.4 (58.13)	4.99 (53.71)	4.49 (48.33)	5.35 (57.6)	5.4 (58.13)	5.1 (54.85)
Fan type	Cross-flow plus	Transverse	12-blade scroll	Transverse	6-blade	6-turbine
Fan diameter, mm (in.)	381 (15)	279 (11)	500 (19.7)	279 (11)	—	—
Grain Handling System						
Tailing elevator type	Trisweep	Paddle	Paddle	Paddle	Paddle	—
Clean grain elevator type	Paddle	Paddle	Paddle	Paddle	Paddle	—
Grain tank capacity, L (gal)	11,100 (2,932.3)	11,629 (3,072)	10,572 (2,793)	10,570 (2,792)	11,100 (2,932.3)	10,572 (2,793)
Tank unloading rate, L/s (gal/s)	113 (29.85)	141 (37.25)	116 (30.64)	159 (42)	112 (29.6)	116 (30.7)
Unloading auger length, m (ft)	6.7 (264)	6.99 (275)	6.55 (258)	7.44 (293)	6.4 (252)	—
Standard unloading height, m (ft)	4.39 (173)	4.71 (185.5)	4.37 (172)	4.34 (171)	4.41 (173.8)	—
Crop Residue Disposal						
2-speed straw chopper	Standard	Standard	Standard	Standard	Standard	—
Straw spreader	Variable speed	2-speed	No	2-speed	Standard	—
Chaff spreader	—	Yes	Integrated with chopper	Yes	Standard	Standard

(*Continued*)

TABLE 1.2 (Continued)

Technical Specifications of Modern, Rotary Combine Harvesters

Manufacturer	Case IH	Gleaner AGCO	John Deere	Massey Fergusson	New Holland	CLAAS
Model	AF 7120	R76	9770 STS	9795	CR9060	740/740 Terra Trac
Power Train						
Drive type/number of gears	Hydrostatic/4	Hydrostatic/4	Hydrostatic/3	Hydrostatic/4	Hydrostatic/4	—
Ground speed, km/h (mph)	—	—	—	—	—	0–25 (0–15.5)
Transport speed, km/h (mph)	—	—	—	—	—	25 (15.5)
Brakes, turning against	Dual-caliper disc	Drum—hydraulic	Drum	Drum—hydraulic	Dual-caliper disc	—
Brakes parking	Disc	Mechanical drum	Drum	Mechanical drum	Disc	—
Final drive type	Bull gear	Spur gear S-41	Single reduction	Spur gear S-41	Heavy duty	—
Steering						
Tread width, adjustable axle, mm (in.)	265 (10.4) axle extension	3,023 (119)/3,630 (143)	—	3,023 (119)/3,630 (143)	3,213 (126.5)/3,663 (144.2)	—
Tread width, rear wheel assistance, mm (in.)	2,994 (117.9)	3,200 (126)/3,658 (144)	—	3,073 (121)/3,683 (145)	3,213 (126.5)/3,717 (146.3)	—
Standard steering type	Single cylinder	Dual cylinder	Single cylinder	Dual cylinder	Dual cylinder	—
Turning radius, mm (in.)	4,000 (157)	6,858 (270)	—	6,426 (253)	4,394 (173)	—
Tires						
Drive tire size	520/85 R42	30.5 L32	3.5 LR32	30.5 L32	900/60 R32	620/70 R42
Steering tire size	540/65 R30	16.9 × 24	18.4 R26	16.9 × 24	600/65 R28	28 LR 26
Cab						
Operator seat suspension	Air suspension	Luxury/air ride	Adjustable air ride	Luxury/air ride	Air ride	Yes
Instructor/passenger seat	Standard	Yes	Yes	Yes	Standard padded	—
Control, position	Tilt/telescope	Tilt/telescope	Tilt/telescope	Tilt/telescope	Tilt/telescope	—
Monitor	Standard	Standard	Yes	Standard	Standard	Standard

Note: —, Data not provided or feature not available.

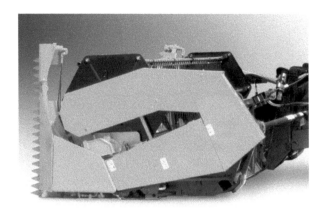

FIGURE 1.28
Grain head with side cutter bars. (Courtesy of SDF.)

A grain header is made of a platform with a frontal, reciprocating knife cutter bar; two lateral dividers of the plants; cross-mounted, equidistant crop-lifting devices; a reel with slates and metal or plastic fingers; and an intake auger combined with a retractable-finger drum that pushes the gathered material into a combine feeder-house. A grain header may have some improved or additional attachments as follows: extendable platform, flexible floating cutter bar, full-fingered intake augers, full-width retracting fingers, side cutter bars (Figure 1.28), different knife-guard combinations, headlights, or an extended platform for rapeseed harvesting. All headers have a single-lever hydraulic hookup.

Due to reel rotation, each bar/slate with fingers enters vertically among the plants, which are then pushed toward the cutter bar, falling on the platform, and gathered to the header center by the intake auger. The cutting height of the plants and vertical and horizontal position of the reel relative to the cutter bar, as well as reel rotation relative to the machine speed and crop type, affect shattering of the grains and collecting of grain losses. A flexible cutter bar can flex over field contours and ridges to better cut the plants that have pods close to the ground (e.g., soybean). An extendable platform (e.g., CLAAS VARIO header) is hydraulically adjusted by extending its cutter bar position under the crop prior to contact with the reel bar to improve gathering tall plants or for straight-cutting canola in a variety of crop conditions.

1.5.2 Draper Headers

Draper headers differ from grain headers in that they use two fabric or rubber conveyors and a center feeding belt conveyor apron instead of a cross auger (Figure 1.29). This combination allows faster collecting of the material toward the header center than cross-augers, that is, higher throughputs. Due to its increased efficiency, a draper header requires less power, though the draper header needs more maintenance/repair time than the grain header, whose cross-augers have a higher mean-time-between-failure characteristic. Draper headers are offered in a wide range of cutting widths with flexible or rigid cutter bars.

An interesting draper header was built by CLAAS; the MAXFLO header (Figure 1.30a) features an intake system composed of two compression augers mounted laterally onto the intake auger with retractable fingers (Figure 1.30b). The speed of the auger can be adjusted to three settings (150, 200, and 250 rpm), and its moving direction can be reversed by the operator from inside the cab. The speed of the conveyor belts is monitored by the machine

FIGURE 1.29
Draper header. (Courtesy of Case IH.)

control system. In areas where crop swathing is required, this draper header can be converted into a swathe platform by removing the central auger, sliding one of two belts over (to close the feeder-house section), and having both belt conveyors move in the same direction (both left or right) as needed.

1.5.3 Corn Headers

Combine harvesters are equipped with row crop *corn headers* for harvesting 6–24 rows at a time (Figure 1.31). These corn headers are built in series of 6 rows, 8 rows, 12 rows, 16 rows, and so forth. The number of harvested rows is determined by the combine class following capacity and engine power requirement, for example, 165–215 kW (220–290 hp) for 6–8 rows, 225–345 kW (300–460 hp) for 8–12 rows, and so forth. The distance between cornrows can be as tight as 0.38 m (15 in.), although 0.76 m (30 in.) is the most common.

Corn headers can be classified based on their main functions as follows:

- *Corn headers* for collecting the ears only, while the stalks are pulled down
- *Chopping corn headers* that collect the ears and chop the stalks whose fragments are spread on the field
- *Integral corn headers* that collect the ears, and chop and collect the stalk fragments to be left on the soil as a continuous swath or to be collected into a trailer that moves in parallel with the combine

Most of the corn headers have common features, such as:

- Crop/row dividers, whose spacing accurately matches the row width (some models may have an adjustable row spacing design)
- Durable poly snouts or leading edges of the cornrow units, and ear saver flaps
- Gathering chains with hardened pins, gearbox-driven row units, and a transversal auger for gathering the ears
- Automatic header height control, terrain contour–following capability, and flexible sensing wand–based row steer function, as well as single-unit coupling systems for hydraulic and electric connections with the machine

(a)

(b)

FIGURE 1.30
(a) MAXFLO draper header. (Courtesy of CLAAS.) (b) Detail of CLAAS draper header.

FIGURE 1.31
Corn header view. (Courtesy of Case IH.)

Using whisker-based sensors (Figure 1.32) attached to the center row snouts, the row's path is better tracked and the combine direction is corrected through the automatic steering control; that allows a reduction of ear loss due to operator fatigue, lack of attention, or distractibility of any nature.

FIGURE 1.32
Snouts with ear saver flaps and whisker-like sensors. (Courtesy of Case IH.)

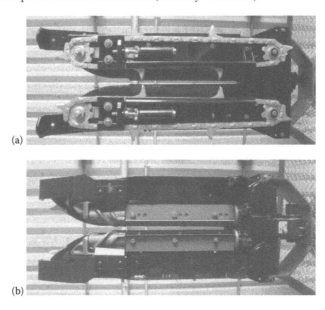

FIGURE 1.33
(a) Gathering chains with tapered fingers. (Courtesy of Case IH). (b) Cantilevered stalk rolls and plates. (Courtesy of Case IH.)

Feeding each row unit of a *corn header* is ensured by poly snouts and by tapered-finger gathering chains (Figure 1.33a) that drive the plant stalks between a pair of cantilevered stalk rolls (Figure 1.33b) placed underneath the deck plates. The ear-detaching process consists of pulling down the stalks by the stalk rolls (Figure 1.34), while the ears are retained on top of the snapping plates and pushed further backward by the chains toward the gathering auger. The detaching process of the ears from the stalks is based on the fact that the tensile strength of the stalk is greater than the tensile strength of the ear peduncle, though this difference in tensile strength may be affected by the way the stalks are handled. That depends on the settings of the rolls and plates, as well as the chain speed versus combine

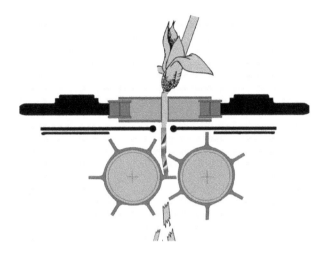

FIGURE 1.34
Schematic of ear-detaching process. (Courtesy of Geringhoff.)

speed. The opening between the plates needs to be adjusted just enough to accommodate a certain range of ear peduncles and stalk sizes—wide enough to avoid butt shelling, but not too wide to let the ears escape.

Some corn headers have the deck plate position hydraulically adjusted. The chains and stalk rolls of each row unit are driven through a row unit gearbox.

The flight-to-trough clearance of the ear auger shall be tight enough to prevent small ear shelling. Auger speed shall be correlated with the machine speed.

Each row unit of a *chopping corn header* is fitted with a two-blade or three-blade rotating cutter (Figure 1.35), which is driven right from the row unit gearbox. This cutter chops the stalks that are pulled down by the stalk rolls. The chopped material is spread on the field, and later it is incorporated in soil during plowing or tillage operations. The Geringhoff rolls sever the stalks while being pulled down. This innovative design eliminates a subsequent stalk-shredding operation, though some stalks break and enter the machine with the ears.

Instead of rotating cutters at each row unit, an *integral corn header* has underneath it a transversally positioned rotary cutter with two to four longitudinal blades. This rotary cutter chops the stalks and simultaneously throws the mixture of chopped stalks, leaves, and possibly weeds back to an auger that collects them and leaves them on the field as a continuous swath. The cut material may also be collected into a trailer that moves in parallel with the combine.

Special row crop headers have been developed for sunflowers, sorghum, and soybeans.

1.5.4 Sunflower Headers

Sunflowers headers need to be collected by the shuttles. The detaching process of the sunflower heads aims to cut the stalk underneath the flower head by a cutting bar (Figure 1.36).

The hydraulically adjustable front plate keeps the flower heads in a forward position, while the snapping roller prevents their cutting too soon (Figure 1.37). To collect all flower heads and prevent seed shaking, the seed pans are quite long, and they can be adjusted relative to the stalk diameter via adjustment rails. The gradient of the seed pans, reel position, and rotation speed can also be adjusted to accommodate the most diverse crop conditions during harvesting and to maximize the throughput of the machine. The auger collects all sunflower heads and pushes them into the feeder-house.

FIGURE 1.35
Three-blade cutter for stalk chopping. (Courtesy of CLAAS.)

FIGURE 1.36
View of a sunflower header. (Courtesy of SDF.)

Some grain headers can be relatively easily modified to harvest sunflowers as follows: the metal pans are fitted to the cutter bar and a modified reel with wide slates will handle the plants toward the platform and cutter bar (Figure 1.38).

Instead of a reel, other sunflower headers are equipped with poly snouts and a pair of guiding chains with fingers for each row. Due to plant shaking during handling, such headers have higher sunflower seed losses on the field.

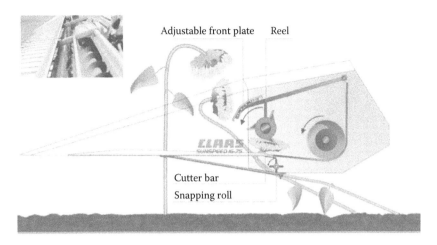

FIGURE 1.37
(See color insert.) Schematic of a sunflower head working process. (Courtesy of CLAAS.)

FIGURE 1.38
Modified grain header for sunflower harvesting. (Courtesy of CLAAS.)

1.5.5 Pick-Up Headers

The *pick-up (rake-up) header* enables the swath harvesting of all crops, especially beans, peas, rapeseed, grass seed, canola, and lentils. A pick-up header (Figure 1.39) consists of a wide conveyor belt with equidistant bars that carry a lot of metal tines (fingers). The tines go deep into the stubble and lift up the swath on the belt, and then a large auger collects the material and pushes it into the feeder-house of the combine.

The belt rollers are vulcanized to provide an increased traction that means reduced slipping of the belt. The belt and auger are hydraulically driven, and that allows a variable speed based on harvesting conditions and working capacity of the combine. The pick-up header floats in front of the machine following the rough terrain. A mechanical or hydraulic suspension changes the pick-up flotation sensitivity, and two or four individual rollers help maintain the belt at a constant height from the soil. The pick-up has a single-point, hydraulic multicoupler that allows direct fitting to the combine hydraulic system.

1.5.6 Stripper Headers

In certain regions, *stripper headers* have become increasingly popular for harvesting rice and small grain crops. Up to 25% of rice in the United States, Australia, and South

FIGURE 1.39
View of a pick-up header. (Courtesy of Case IH.)

America is harvested with combines equipped with stripper headers. Although the collecting grain loss of a stripper is higher than that of a header, it is generally of an acceptable level in rice harvesting. This is because stripper harvesting has some advantages, as follows: high throughput (up to 17 kg/s), high harvesting speeds (as high as 2.2 m/s or 5 m/h), reduced fuel consumption, and fewer maintenance operations (Kutzbach and Quick, 1999).

The stripper header (Figure 1.40) was originally developed by Keith Shelbourne (Shelbourne Reynolds, UK) in 1989. The latest design of Shelbourne, RVS and CVS stripper headers (Figure 1.41), are stripping rotors that rotate under an adjustable hood and a small platform, followed by an intake auger. The stripping rotor that spins backward is fitted with rows of stripping teeth (Figure 1.42) that strip the grain from the ears or panicles as the combine moves the header forward. The auger and rotor are placed closer together, and most grains are moved directly from the rotor to the deep-flight auger. Thus, only the grain and some leaves are the primary plant components that pass through the threshing unit of the combine. The stems of the plants are left standing in the field.

The latest stripper models feature more stainless steel than previous rice models. The stripper is equipped with a variable-speed belt drive that enables the operator to adjust the rotor speed from the cab.

FIGURE 1.40
View of a Shelbourne stripper header. (Courtesy of Shelbourne Reynolds.)

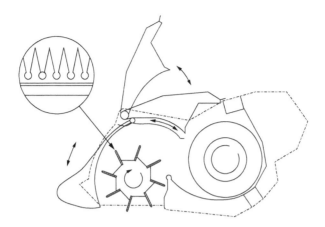

FIGURE 1.41
Schematic of a stripper header. (Courtesy of Shelbourne Reynolds.)

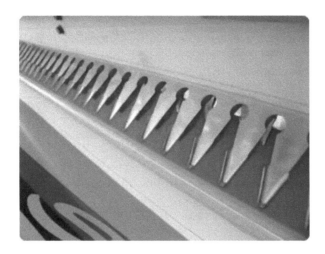

FIGURE 1.42
Teeth of a Shelbourne stripper header. (Courtesy of Shelbourne Reynolds.)

1.6 Power Systems of Combine Harvesters

Modern harvesters have a high capacity for material processing, while the ground speed of the combine in the field can be easily controlled as a function of crop type, crop variety, and field conditions. Processing a high material throughput requires maintaining the engine at constant speed, a threshing–separating cylinder/rotor, a straw walker crankshaft, and a cleaning shoe driving shaft. At the same time, the ground speed of the machine has to be varied from zero to the required harvesting speed or to the transport speed on the road.

A modern combine harvester requires different power systems (mechanical, hydraulic, pneumatic, electric, and electronic), respectively, for propulsion, driving, actuating, and control of its processing equipment and systems. Some of the engine power is used for combine process control as well as cab environment control (heating, air conditioning, etc.).

The propulsion system of a combine harvester includes a diesel engine and a power transmission system. Ideally, the engine operates at a constant rotation that is controlled by the engine control unit. The classical power transmission system includes a master clutch on the input shaft, a manual stepped transmission, a differential, drive axles, final reduction drives, and brakes.

The driving system is composed of the master clutch, parking brake, gearshift mechanism, steering system, and engine control unit.

1.6.1 Engine

A typical modern combine harvester is equipped with a diesel engine (Figure 1.43) to power all functional systems of the machine, in addition to propelling it in the normal course of operation or down the road. The engine typically has six to eight cylinders and develops a nominal power within the 150–300 kW (200–400 hp) range. The load on the engine varies according to the loads applied by various assemblies that are activated, operated, and deactivated during routine use of the machine. Hydrostatic drives use more power than conventional stepped drives, especially on slopes with the grain tank fully loaded.

The combine usually operates at a set engine speed such as low speed (about 1200 rpm), medium speed (about 1600 rpm), and high speed (more than 2000 rpm). Once the speed has been selected by the operator, the engine control unit controls the speed by dynamically adjusting the amount of fuel injected into the engine cylinders. The engine control unit may use a power curve or an algorithm for a power curve to proactively adjust fuel flow rate, thereby adjusting the engine power. The engine control unit can also combine input from a proactive algorithm with input from a reactive algorithm.

Generally, the process power requirement can be roughly calculated using a specific power index of 9 kWh/t of MOG throughput. Conventional combines need an engine with a total power of about 155 kW/m of cylinder width.

FIGURE 1.43
Cutaway of a 8-7L engine. (Courtesy of Case IH.)

1.6.2 Power Transmission System

Self-propelled combines were initially equipped with *conventional stepped transmissions* whose input shaft rotation was modified using a mechanical continuous variable transmission (variable-diameter, pulley-based V-belt transmission) that was mechanically or hydraulically controlled to allow the operator to speed up or slow down. A *clutch* was still needed for stopping the machine or changing transmission gears. This type of transmission has a very good efficiency: above 90% for the gearbox and 85%–87% for the total transmission system. However, this technical solution does not allow an efficient split of the mechanical power flow required for crop processing and that required for combine propulsion.

The latest trend in combine propulsion is using a *hydrostatic continuous variable transmission* (CVT) that allows a continuous variation of the ground speed of the machine while maintaining the constant engine speed required for steady operation of material processing systems.

Hydrostatic continuously variable transmissions with automatic control were introduced after 1995 for standard tractors (Renius and Resch, 2005). A CVT allows an infinite transmission ratio of nominal speed down to zero, while the engine speed remains constant. The most basic hydrostatic CVT is a system composed of at least one hydrostatic pump and a motor. At least one of these units must have a continuously variable displacement.

Figure 1.44 shows a drive transmission with an axial piston variable-displacement motor and two-stage shift gearbox.

The engine turns the variable-displacement pump, which is capable of high flow rates at very high pressure of transmission fluid. The pressurized fluid is then directed to the variable-displacement motor for high efficiency. The motor displacement is usually higher than the pump displacement. A very well-designed and -manufactured hydrostatic CVT with swash plate units can have about 8% speed loss and 10% torque loss at a typical working fluid pressure between 200 and 300 bar (2900 and 4350 psi) (Renius and Resch, 2005). The driving shaft of a CVT can have a constant angular velocity over a range of output velocities. This translates into better fuel economy than other transmissions by enabling the engine to run at its most efficient rotational speed.

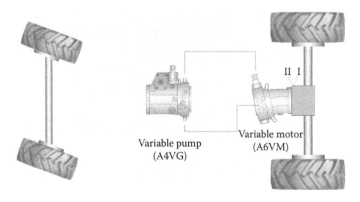

FIGURE 1.44
Continuous variable transmission with axial piston variable-displacement motor and two-stage shift gearbox. (Courtesy of Bosch-Rexroth.)

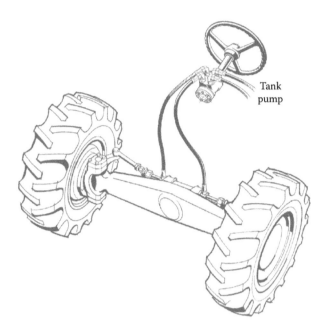

Tank
pump

FIGURE 1.45
Components of an open-center hydrostatic steering system. (Courtesy of Sauer-Danfoss.)

1.6.3 Steering Mechanism

The basic hydrostatic steering mechanism used in combine harvester construction is made of the following components: steering wheel, steering unit, gear pump, tank with oil, and steering cylinder (Figure 1.45). The steering unit is supplied with oil by a fixed-displacement pump. When the steering wheel is turned, the steering unit directs an oil flow rate to the corresponding side of the steering cylinder, while the oil from the opposite side flows toward the tank. The oil flow rate is proportional to the rotation angle of the steering wheel. When the vehicle is not being steered, the oil is pumped at a low pressure toward the steering unit.

The steering mechanism may also have improved features for load sensing, oil flow amplifying, and electrical control. The power steering, electrohydraulic power steering, and steering sensors can be integrated to form an *autoguided steering system*, which can be coupled with a global positioning system (GPS) receiver, enabling autoguidance of combines in the field, monitoring of machine position in the field, and collection of harvest yield data.

1.6.4 Wheels

Combine harvesters are equipped with rubber wheels that are configured for dry or paddy fields (Figure 1.46). Their size, which is different for the front axle and rear axle, depends on combine weight and load, as well as on maximum allowable pressure on the soil. Paddy-use tires can be classified as *high-lug tires*, *wide-lug tires*, and *paddy-lug tires* (Sakai, 1999).

Research trials show significant differences in the level of soil compaction from a combine harvester, depending on the tire make and type of tire fitted. This can have a significant impact on plant root growth and crop yield even within crop rotation. If very wet soil conditions are encountered, the tire inflation pressure should be lowered to the

FIGURE 1.46
Rubber tires. (Courtesy of SDF.)

minimum recommended for the load being carried. To decrease the compaction of the soil, the weight is distributed evenly over a larger footprint; that means wider tires, dual tires, or a self-leveling track system (Figure 1.47) with tandem wheels should be used. The tracks give better grip in wet and muddy conditions and also spread the weight of the combine, reducing soil compression, while the powered rear axle easily turns the machine. The Mobile-Track System (MTS) of CLAAS Lexion combines allows half the ground pressure (average 0.72 bar or 10.5 psi), as well as faster ground speeds—as high as 29 km/h or 18 mph (Wehrspann, 2010).

Provided that driving the combine on the road is not extensive at moderate speeds, somewhat lower pressure in the tires can be acceptable, but only if the manufacturer specifications are met.

Tables 1.1 and 1.2 compare the specifications of modern combines by manufacturer; they show the tire size versus combine weight with tires.

FIGURE 1.47
Self-leveling track system with tandem wheels. (Courtesy of CLAAS.)

FIGURE 1.48
Header inclination on nonhillside combine harvesters. (Courtesy of CLAAS.)

1.6.5 Chassis/Cleaning Leveling Systems

The header of a modern combine can be inclined up to a certain degree for harvesting the crops on hillside fields (Figure 1.48). However, in a combine without a leveling system, the threshed material moves to the lower side of the straw walkers and cleaning shoe; this results in losing the air provided by the fan through the sieve's clear openings. Thus, the straw walkers and cleaning shoe will work in a less efficient way and the grain separation losses will increase.

Each leveling system has a design limitation in terms of the slope, which quantifies the terrain inclination to the horizontal. The slope is usually expressed as a percentage (tangent of the angle of inclination × 100). The tangent of the angle is the ratio between the *rise* (distance measured on the vertical) and the *run* (distance measured on the horizontal).

There are a few methods of leveling systems used in combine harvester construction, as follows:

- *Hillside transverse leveling* (side to side, perpendicular to the moving direction) of the chassis and header. In this case, as the combine moves on the elevation contour lines (across the hill slope), the chassis, with all processing systems and cab, is continuously maintained in a horizontal position (Figure 1.49); the wheels may or may not be vertical. The chassis rotates around two coaxial pivots mounted on the wheel undercarriages due to the lateral lift cylinder action that is mechanically or electronically controlled.

  The feeder-house tilt frame rotates around a pivot pin located at the bottom or top of the feeder-house, and is leveled synchronously with the chassis by a single double-action cylinder. The inclination angle is between 20° and 30° with a maximum of 44°.

- *Hillside longitudinal leveling* (front to back). In this case, as the combine moves along the hill slope, two hydraulic cylinders that are joined with the chassis and rear wheel undercarriage lift the back side of the combine according to the hill inclination angle (Figure 1.50). In this case, the best leveling performance reaches 35° when the combine travels uphill and 12° for downhill traveling.

- *Hillside full leveling* (both transverse and longitudinal directions), as shown in Figure 1.51.

- Hillside transverse leveling of the cleaning shoe system, up to 20° in rotary combines (Figure 1.52).

Although hillside leveling requires a complicated design, it offers several advantages in terms of increased processing efficiency on sidehill and normal threshing, separation,

FIGURE 1.49
Combine with transverse leveling of the chassis. (Courtesy of SDF.)

FIGURE 1.50
Combine harvester with longitudinal leveling system. (Courtesy of SDF.)

and cleaning operations. Thus, the straw walkers are able to operate at the highest MOG throughput, while the grain loss is maintained at a low level. Then, leveling improves combine stability on the hill by maintaining the machine's center of gravity within the area defined by the points of wheel contact with the soil.

1.6.6 Hydraulic System

Modern combine harvesters must meet the requirements of high processing capacity, efficiency, easy operation, and maximum reliability during the harvesting season while preserving the quality of harvested grains.

FIGURE 1.51
Combine harvester with full leveling system. (Courtesy of Laverda.)

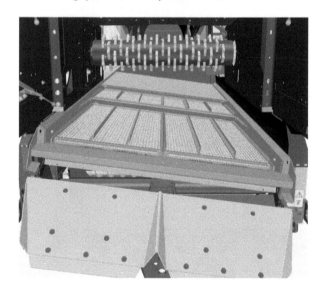

FIGURE 1.52
Tilted cleaning shoe system. (Courtesy of Case IH.)

High-performance hydraulic drives and control systems are essential components for generating and controlling the movement (speed, speed direction, acceleration, and damping), as well as the position of different mechanical processing units of the machine. Thus, in a combine harvester, the hydraulic or electrohydraulic systems can be used for the following:

- Hydrostatic continuous variable transmission of power transmission systems (discussed above)
- Gearbox with dynamic brakes
- Steering system (discussed above)
- Remotely powered brake system
- Hillside machine transverse or longitudinal autoleveling

- Positioning, driving, and control of the header and its components: reel, auger, or deck plates of a corn header
- Control of variable-speed V-belt-based drive
- Fan drives with constant or variable-displacement pumps
- Grain tank extensions
- In/out positions of the tank discharge auger

Hydraulic system schematics are represented by graphic symbols standardized by ISO 1219-1:2006, which focuses on general design rules of the symbols. The second part of the standard, ISO 1219-2, describes typical circuit diagrams through examples. In connection with these standards, ISO 5598 describes the specific terminology in English, French, and German. An extract of ISO 1219-1's graphic symbols is given in Table 1.3.

A general schematic of closed-circuit hydraulics equipment used in combine harvesters is shown in Figure 1.53. The open or closed hydraulic system (Figure 1.54) is composed of the following components: oil reservoir, filter, hydrostatic pump, hydrostatic motor, pipes, fittings, and heat exchanger. A closed-circuit piston pump is normally provided with a charge pump, charge pressure relief valve, and control (direct displacement control or servo control). It may also be equipped with relief valves or pressure limiters for high-pressure protection. The hydraulic actuator converts hydraulic pressure into force and linear displacement (hydraulic cylinder) or into torque and angular displacement (hydraulic motor).

The hydraulic system of the header fulfills its functions for two modes with the combine: transport mode and working mode. When the combine is down the road, due to the road irregularities, coupled with rubber tire elasticity, the header vibrates accordingly. The hydraulic active damping of header vibrations prevents the machine from becoming uncontrollable, allows a higher driving speed, and increases the driver comfort. In working mode, the hydraulic system serves to

- Lift the header at a certain height over the soil when harvesting tall crops (corn, sunflower). The pitching movement is reduced by active vibration compensation.
- Tilt the header for harvesting crops on hillside fields.
- Lift the header to ensure constant pressure on the ground (e.g., for picking up crop swaths). This also prevents possible damage to the header, cutter bar, and crop separators.
- Lift the reel.
- Horizontally position the reel relative to the cutter bar.
- Rotate the reel.
- Reverse the rotation system for the header auger and feeding conveyor.

When using mechanical-shift gearboxes coupled with a hydraulically controlled variable-speed drive, it is possible to shift the gears while the machine is in motion. Shifting through a frictional connection makes the gearbox less prone to failure due to improper shifting (Bosch-Rexroth, 2011).

The hydraulic systems and electrohydraulic systems are controlled automatically or by the combine operator according to the harvest conditions.

TABLE 1.3

Extract of ISO 1219-1:2006: "Fluid Power Systems and Components—Graphic Symbols and Circuit Diagrams—Part 1: Graphic Symbols"

Component Description	Symbol	Component Description	Symbol
Shutoff valve		Nonreturn valve	
Diaphragm accumulator		Hose line	
Pressure sensor		Flow sensor	
Pressure gauge		Adjustable restriction valve	
Pressure regulator		Pressure relief valve	
Pump, one direction of flow, fixed displacement, one direction of rotation		Hydraulic motor, one direction of flow, fixed displacement	
Pump, two directions of flow, variable displacement, one direction of rotation		Hydraulic motor, two directions of flow, fixed displacement	
Directional 2/2-way valve (two ports, two positions, muscular control)		Directional 3/2-way valve (three ports, two positions, muscular control)	
Directional 4/2-way valve (four ports, two positions, muscular control)		4/3-way valve, four ports, three positions, closed in midposition	
4/2-way solenoid actuated valve		4/3-way valve, four ports, three positions, relieving in midposition	
Single-acting cylinder		2/2-way stem actuated valve	
Double-acting cylinder		Filter	
Heater		Cooler	

1.6.7 Electrical, Electronic, and Control Systems

Electrical, electronic, and control systems have multiple roles, as follows:

- Increasing machine/operator productivity
- Maximizing machine availability and adaptability to various crops and harvesting conditions

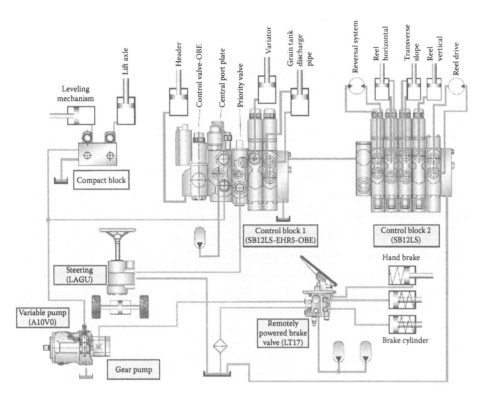

FIGURE 1.53
General schematic of hydraulics implementation in a combine harvester. (Courtesy of Bosch-Rexroth.)

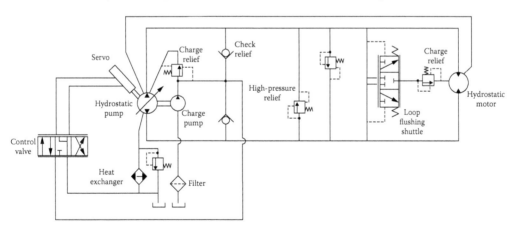

FIGURE 1.54
Schematic of typical closed-circuit hydraulic transmission. (Courtesy of Sauer-Danfoss.)

- Preserving quality of harvested grains
- Minimizing harvesting grain losses and costs
- Reducing operator stress while improving overall machine operability
- Acquiring data about the crop (yield mapping, moisture, maturity)
- Integrating harvest management

These systems are connected with mechanical components, a hydraulic system, multiple sensors, GPS (where proven economically), an information and control center in the cab, and possibly the harvest management control center.

The electronic and control system may perform control functions of the following systems and working units of the combine:

- Engine power control
- Axle hydrostatic drives with fixed- and variable-displacement motors and auxiliary drives for steering axles
- Single-wheel drives with electronic anti-wheel-slip control
- Electronic drive management with control units
- Implementation of hydraulics with load-sensing systems
- Leveling of hillside combine or cleaning shoe
- Hydraulic steering and remotely powered braking systems
- Automatic steering down the crop row for corn and sunflower harvesters
- Electrohydraulic header/component control (cutter bar position, platform height, vertical and horizontal position of the reel)
- Fan drives

The load on the combine engine varies in accord with the power requirement from the driving system on different road and field conditions and from the operating units, which are activated, adjusted during operation, and deactivated during routine use of the combine harvester. Thus, the *engine power control system* adjusts the engine power to respond to engine speed changes, and weight changes due to grain collecting in the tank, and in anticipation of changes in loads being imposed on the engine. Based on a power curve (algorithm) stored in its memory, the engine control unit dynamically maintains a constant engine speed in the face of different load requirements, by changing the fuel flow into engine cylinders. Although this is typically a reactive response, there are other functional options, such as proactive change inputs, at predetermined rates of change, in the rate of fuel delivery due to anticipation of load requirements.

Other systems of *power management control* monitor the available engine power to additionally power subsequent processing units, such as a chopper or straw agitation system that may be selectively configured and disengaged.

The introduction of microcontrollers to agricultural machinery led to the development of bus configurations and data structures to support continuing machinery evolution. The *controller area network (CAN) data bus* used in *combine harvester electronic drive management* is an acknowledged standard that integrates the

- Control of the drive transmission and hydraulic leveling mechanism of the machine
- Control of ground speed, reel speed, fan speed, and implemented hydraulic drives
- Sensors, control units, display, and multifunction joysticks, which are interconnected and communicate simultaneously
- Stations with the same rights and connected via a serial bus
- Diagnostics parameter settings and process monitoring

The transmission errors due to electromagnetic interference are automatically corrected by retransmitting the data.

The next protocol for combine harvesters will be *FlexRay*, a distributed network protocol that has been developed to improve existing CAN technology. It allows the transfer of the data at higher frequencies (10 Mbit/s) than CAN protocols (250 kbit/s). The FlexRay protocol allows data transmission and reception at predetermined time frames, which helps to eliminate the errors occurring when multiple messages are sent out (Shearer et al., 2010). Additionally, this protocol allows for various bus technologies: point-to-point connection, multidrop bus, passive star, cascaded active star, or hybrid (both multidrop and star) network (Shaw and Jackman, 2008).

1.7 Cab, Information, and Control Center

During the harvesting period, a combine harvester can be operated for an extended number of hours a day, by one or alternate operators. The operator drives the combine and observes most of the combine processing functions. Thus, to increase machine-operator productivity, decrease harvest costs, preserve harvest quality, and reduce operator stress while offering protection against dust and sun radiation, modern combines are equipped with a cab with seating accommodation for machine driving (Figure 1.9), and a control center for setting the tasks, acquiring process data, supervising process parameters, and making adjustments. The cab is environmentally controlled by a heating, ventilating, and air conditioning system to maintain a comfortable working space. Current cab design focuses on many aspects:

- Operator's seating accommodation
- Good view of the header, crop, and road during transportation, as well as the inside control panel
- Ergonomic use of a multifunction control lever or joystick
- Increased size to accommodate the air exchange, and around the operator's seating
- Optimized cab environment control system with easily replaceable components
- Good access for maintenance and repair of the threshing unit and its drive

A series of international standards regulate many of the above-mentioned design aspects, while addressing other design features, depending on the manufacturer's design tradition, expertise, and own standards; economic reasons; and ultimately designer experience and esthetic faculties.

The control center comprises devices for process parameter control, work quality index supervision and adjustment, data registration and management, and communications with the operation manager. Adjustment settings of the main functional parameters for specific crop and harvest conditions are updated by the operator or electrohydraulic controllers in the memory of a microprocessor.

Most modern combine harvesters are equipped with devices for real-time supervision of and warning on the functional regime and dynamics of work quality indices. Besides checking functional parameters of the engine (rotational speed, oil pressure, and temperature), fuel level, and machine speed, the control panel displays information and allows setting of the functional parameters of combine working units, such as rotational speed of different shafts, grain level in the tank, and proportion of grain loss and damaged grain. As an example, in Figure 1.55, a monitor displays a series of such settings and the status of process parameters from a Case IH crop simulator.

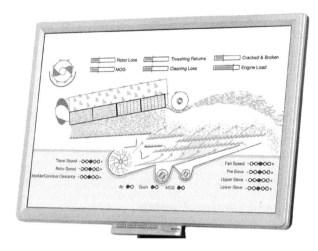

FIGURE 1.55
Crop flow simulator: monitor displaying process parameter status. (Courtesy of Case IH.)

Such information is gathered by specific sensors mounted in different locations in the combine. Through different methods, these sensors convert instantaneous values of the material (grains and MOG) weight, volume, throughput, impact force on a surface, frequency of grain impacts, and so forth, into electric signals. Then, a microprocessor converts all signals into corresponding values of measured parameters, which are further used to control the machine processes.

Soils and crop yields are not uniform; that is, they vary according to terrain location. The throughput of the grains gathered from a certain field surface can be converted into crop yield associated with a certain map area. When coupled with information from a GPS, the data can be stored, plotted, and used in the future for precision farming.

The practice of precision farming is based on mapping systems and applies farming concepts that refer to

- Crop yield monitoring and control of plant growth
- Precise fertilizing based on soil nutrients, moisture, and crop needs
- Controlling weeds and pests
- Spatial information acquisition, database structuring, and integration and management of all data

Combine harvesting using a GPS can be very well integrated into precision farming practice as an essential functional component. Further details about using GPS are described in Chapter 15: "Guidance and Control of Autonomous Combine Harvesters."

1.8 Combine Harvester Performance

Combine harvesters must harvest a wide range of crop types of different sorts whose size of grains or seeds, plant height, and plant density are very diverse—not to mention that the crop condition and field/terrain configuration are also important influencing factors on combine crop processing. Consequently, the combine working capacity and quality of work are significantly influenced by such factors. By design and operation settings, a modern combine must achieve a proper level of performance indices, as follows:

- Throughput of material (grains and MOG): q, kg/s (lb/s)
- Proportion of grain separation from MOG: s_G, %
- Overall proportion of grain losses: s_L, %
- Grain damage: s_D, %
- Grain cleanliness: s_C, %
- Rate of grain tank unloading: s_U, kg/s (lb/s)
- Straw/stalk fragmentation mean size: μ_{pf}, mm (in.)
- Straw/stalk fragmentation standard deviation: σ_{pf}, mm (in.)
- MOG spreading uniformity (variance): s_p^2, %
- Specific processing power requirements: P_{sp}, kW/(kg/s) (hp/(lb/s))
- Overall power requirements: P_{sw}, kW/(kg/s) (hp/(lb/s))

Apart from these working performance indices, a combine harvester is characterized by a series of technical specifications, such as transport speed, turning radius, accessible terrain slope during harvesting operations, operator seating and cab environment conditions, operation autonomy, and so forth.

The combine *throughput (feedrate) of material* is the instantaneous value of material (grains and MOG) mass (kg/s or lb/s) that can be processed by the machine in current crop and field conditions, providing that the proportions of grain losses and damaged grains reach acceptable values consistent with a conventional standard or regulation. Accordingly, one can also define the *grain throughput* q_G and *MOG throughput* q_p. For a given class of combines, the general tendency is to maximize the material throughput by decreasing the length of the crop harvesting period, as favorable weather conditions may change.

As the grains are removed from the ears in a combine, the flows of grains and MOG get separated during different processing stages performed by certain working units. Thus, one of the basic design requirements refers to correct sizing of consecutive working units due to varying specific loads of grains and MOG.

The *proportion of grain separation* (%) depends on the overall performance of the threshing unit as well as the ability of the cleaning shoe to recover the unthreshed ears to be sent back for threshing.

The overall *proportion of grain losses* is related to the total mass of grains that should be harvested over a certain field area. Thus, the proportion of grain losses is composed of a series of losses corresponding to successive processing operations of working units, as follows: *gathering grain losses* s_{gL} (header/pick-up header, stripper), *threshing grain loss* s_{tL}, *separating losses* (*threshing unit separating loss* s_{sL}, *straw walker separating loss* s_{wL}, *cleaning shoe separating loss* s_{cL}), and *grain handling losses* (conveyors) s_{hL}, all measured as percentages.

The *grain damage* means the percentage of damaged grain mass relative to the gathered grain mass. A damaged grain means a broken grain, damaged grain coat, or internal grain damage not visible with the naked eye. Most grain damage takes place in the threshing unit, while less damage occurs during grain handling by different conveyors (header conveyor, feeding conveyor, and other conveyors).

A high level of damaged grains and a low level of grain cleanliness may require additional processing operations and influence grain germination, storage, use, and price.

The other combine performance indices are self-explanatory; however, they will be clearly defined in the corresponding chapters with theory and modeling of combine processes.

Table 1.4 classifies the dependence of combine performance indices on the type of actions taken by working units and their physical components. Thus, the performance indices depend on crop properties and harvesting conditions, desired work quality

TABLE 1.4

Combine Performance Indices and Influence Factors

Processing Unit	Functional Component	Action	Influence Factor/Parameter	Effect
Grain header and draper header	Plant separator Reel Cutter bar (flexibility for field contour) Draper conveyor(s) Collecting auger(s)	Plants shattering, that is, grain removal Improper plant approach Plants shattering, grain removal Uncut plants Uncollected, cut plants Damaged grain by friction	Working height above the soil Machine ground speed Reel bar entering direction Reel rotational speed Cutter bar versus machine speed Auger rotational speed Conveyor versus ground speed Clearance to platform	Grain losses Grain damage
Corn header and sunflower header	Row unit/row snouts Stalk rolls Gathering chains Snapping plates Collecting auger	Improper plant guiding Stalks breaking Stalks improperly pulled down Stalks breaking Not detaching ears Small ear shelling	Steering direction Height above the ground Center-to-center distance Roll rotational speed Chain speed Plate clearance Flight-tip-to-trough clearance	Grain losses Grain damage and loss Grain damage
Feeding conveyor	Chain conveyor	Grains shattering	Peripheral speed Clearance to the feeder-house	Grain damage
Threshing system	Cylinder(s)/ concave(s) Rotor(s)/concave/ cage	Incomplete grain threshing Low level of grain separating Aggressive action on material	Peripheral speed, concave clearance, design of rotor, concave and cage, throughput	Grain losses and grain damage
Straw walkers	Straw walkers	Low level of separation	Design, crankshaft rpm and phase angular offset	Grain losses
Cleaning shoe	Sieve Fan	Low level of separation	Design, dynamics, inclination angle Speed, airflow angle	Grain losses
Grain handling system	Pivoted auger/ paddle conveyors/tank unloading auger	Grain squeezing	Flight-tip-to-trough clearance	Grain damage

(grain damage and grain cleanliness), acceptable level of overall losses, desired through-put, combine design, level of process automation and control, and harvest operations management.

References

Bosch-Rexroth Group. 2011. Drive and control systems for combine harvesters and forage harvest-ers. http://www.boschrexroth.pl/country_units/europe/united_kingdom/en/products_solutions/mobile_hydraulics/branches/harvesters/hydrostatic_drive_transmission/ (accessed January 15, 2012).

Campbell, D.W. 1980. Modeling the combine harvesting. PhD dissertation, University of Saskatchewan.

Cornways. 2011. Mähdrescher: Historisches. http://www.cornways.de/hi_maeh.html (accessed October 17, 2011).

Crochet, B. 2006. *150 ans de machinisme agricole*. Paris: Editions de Lodi.

Ganzel, B. 2007. Farming in the 1950s and 60s. York, NE: Wessels Living History Farm. http://livin-ghistoryfarm.org/farminginthe50s/machines_14.html (accessed September 12, 2011).

ISO 1219-1:2006. Fluid power systems and components—Graphic symbols and circuit diagrams—Part 1: Graphic symbols for conventional use and data processing applications. Geneva: ISO.

Jbail, F.A.M. 1994. Study of the resistance of grains detachment from cereal ears. PhD dissertation, Politehnica University of Bucharest.

Kim, S.H. and J.M. Gregory. 1989a. Power requirement model for combine cylinders. ASAE Paper 891592. St. Joseph, MI: American Society of Agricultural Engineers.

Kim, S.H. and J.M. Gregory. 1989b. Straw to grain ratio for combine simulation. ASAE Paper 891593. St. Joseph, MI: American Society of Agricultural Engineers.

Kutzbach, H.D. and G.R. Quick. 1999. Harvesters and threshers. In *CIGR Handbook of Agricultural Engineering*. Vol. III. *Plant Production Engineering*, ed. B.A. Stout, 311–347. St. Joseph, MI: American Society of Agricultural Engineers.

Kydd, H.D. 1980. Combine types. *PAMI Gleanings*. Humboldt, Canada: Prairie Agricultural Machinery Institute.

Maw, W.H. and J. Dredge. 1879. *Engineering: An Illustrated Weekly Journal*. Vol. XXVII.

Miu, P.I. and H.D. Kutzbach. 2000. Simulation of threshing and separation processes in threshing units. *Agrartechnische Forschung, Sonderheft* 6:1–7.

Quick, G.R. and W.F. Buchele. 1978. *The Grain Harvesters*. St. Joseph, MI: American Society of Agricultural Engineers.

Renius, Th.K. and R. Resch. 2005. Continuously variable tractor transmissions. ASAE Distinguished Lecture 29, 1–37. St. Joseph, MI: American Society of Agricultural Engineers.

Sakai, J. 1999. Tractors: Two-wheel tractors for wet land farming. In *CIGR Handbook of Agricultural Engineering*. Vol. III. *Plant Production Engineering*, ed. B.A. Stout, 54–94. St. Joseph, MI: American Society of Agricultural Engineers.

Shaw, R. and B. Jackman. 2008. An introduction to FlexRay as an industrial network. IEEE 978-1-4244-1666-0/08:1849–1854.

Shearer, A.S., S.K. Pitla, and J.D. Luck. 2010. Trends in the automation of agricultural field machinery. Presented at Club of Bologna EIMA 2010, Bologna, Italy.

Steponavicius, D., S. Petkecius, and R. Domeika. 2011. Determination of rational values of inclination angles of straw-walker sections. *Engineering for Rural Development* 26.–27.05:76–81.

Wehrspann, J. 2010. Newest combine technologies from Claas, Case IH, New Holland, AFCO, and John Deere. *Farm Industry News* 43(7):8.

Wikipedia. 2011. Cyrus McCormick. http://en.wikipedia.org/wiki/Cyrus_McCormick (accessed October 15, 2011).
Wiley, C. 2010. Combine harvester—Innovating modern wheat farming, impacting the way the world thinks about bread. Seattle, WA: HistoryLink.org. http://www.historylink.org/index.cfm?DisplayPage=output.cfm&file_id=9483 (accessed September 12, 2011).

Bibliography

Chase, H. 1921. *The Wonder Book of Knowledge*. Philadelphia: John C. Winston Company.
John Deere Co. 2011. John Deere combines. http://www.deere.co.uk/common/docs/products/equipment/combines/YY0914252_E.pdf (accessed October 10, 2011).
Kirchheimer, S. 2011. The history of farm equipment. eHow. http://www.ehow.com/about_5399510_history-farm-equipment.html (accessed September 12, 2011).
Miu, P. 1995. Modeling of the threshing process in cereal combine harvesters. PhD dissertation, University of Bucharest.
Sauer-Danfoss. 2011. General steering components—Technical information. 520L0468—Rev. BE. www.sauer-danfoss.com (accessed October 2011).
Smith, C.W., J. Betran, and E.C.A. Runge. 2004. *Corn: Origin, History, Technology, and Production*. Hoboken, NJ: John Wiley & Sons.

2

System Modeling, Simulation, and Control

2.1 Introduction

System modeling and *simulation* are extensive techniques for emulating, investigating, evaluating, and improving the *process*, design, and functional parameters of the system that performs the process. The system receives *input* and provides *output* for some purpose. A system/process associated with a proper *controller* forms a *control system* to deliver the desired output (response) at a desired level of performance, given a certain input.

Most of the time, a system consists of interacting subsystems that perform specific processes in a mixed synchronous or successive manner. When the system generates time-dependent response variables whose rates of change are significant, it is called a *dynamic system*. A combine harvester is a mixed, multifunctional, and dynamic system that consists of different interacting working units, which process a very large type and variety of crops. To monitor and automatically control the combine subsystem processes, in most parts of this book we concentrate on process theory and modeling in reference to process/system optimization, and design of physical components, as applied for the design of combine control.

The objective of this chapter is to provide students and researchers with relatively advanced knowledge of process/system function analysis, mathematical modeling, computer simulation, and automatic control that may be applied in the research and mechatronic design of self-propelled machines, particularly combine harvesters. The topic is both exciting and challenging, and requires a multidisciplinary approach to master. Although this knowledge is not exhaustive, it is strongly recommended that the practitioners whose training is dated read, understand, and take notes on the concepts, approaches, and methods used or developed to enhance their comprehension and skills in applying them. Readers are also referred to other excellent books, such as those in the References and Bibliography at the end of this chapter.

2.2 System/Process Modeling

System definition is the first step in an overall methodology for achieving a set of objectives (Ayyub and Klir, 2006). According to Albert Einstein, "the mere formulation of a problem is often far more essential than its solution" (Einstein and Infield, 1938). In engineering, the formulation of a system is a critical task in the process of analysis, modeling, simulation, and design.

A *system* is a set of interacting *components*, which have certain *attributes*, forming an integrated whole that receives an *input*, performs a *function/process* or a series of processes, and generates an *output* for a certain purpose. A common function of a system is to alter material, convert energy, or process information.

The components' type and the way the components are arranged define the system *structure*.

Because the system components interact, they are interdependent through a set of specific physical and *functional relationships* defined during the (conceptual) design process of that system. Then, the system may interact with other systems to get the input and deliver the output, which characterizes the *interconnectivity* of the system. This interaction, coupled with internal functional relationships, defines the *behavior* of the system. A system may have one or more functions that are performed by its component(s); that means a system may contain *multifunctional components* as well. If a system consists of different types of components, it is termed a *mixed* or *multidomain system*.

Given a system, the following expression can be formally stated:

$$S = (C, A, R) \tag{2.1}$$

where S, C, A, and R denote, respectively, a *system*, a *set of components*, a *set of attributes*, and a *set of relationships* defined on C. Although Equation 2.1 has an overly simple form, its symbols may contain different sets of elements with distinguished properties and subjected to certain relationship requirements.

Each nondivisible component of the system must meet the following two requirements (Blanchard and Fabrycky, 2011):

- Its attributes and behavior influence the attributes and behavior of at least one adjacent component.
- Its attributes and behavior influence the properties and behavior of the system.

When designing a system, the objectives and purpose must be clearly and comprehensively defined so that the system performs the necessary function to provide a desired output for each given input.

Redesigning a system assumes that an end user—the customer—exists; thus, the project is financed. To redesign an existent or new system with its associated processes, the following techniques have to be applied:

- Problem identification
- System/process definition
- Input data collection and analysis
- System/process modeling
- Process simulation/model validation
- Experimental design
- System/process analysis
- System/process simulation-based optimization
- System control development
- Final design and launch in production

In the next sections, we discuss the classification of the systems and the corresponding models that describe them. This chapter only refers to physical, human-made, and computer-based information systems that can be associated with the book subject.

Application 2.1: Combine Harvester System

As a physical, dynamic system, a *combine harvester* interacts in terms of input with the crop and operator, and then after material processing, it delivers an output consisting of clean grain and material other than grain (MOG) as a swath, spread MOG, or chopped MOG that is loaded onto a trailer. The combine system can be further divided into subsystems that may be considered independent systems by themselves. It is up to the researcher how this division is done; that means which components are inside parts of a certain system within its boundary and which ones are left outside. For the purpose of this example, one can mention that the combine harvester system converts the input crop feedrate to an output feedrate of clean grain based on a global processing function that is, generally speaking, controlled by the operator, who, based on the sensor's feedback, makes decisions for setting of the machine's functional parameters during the harvesting process. The combine interacts with the crop and the trailer using specific interface subsystems, such as the header and tank-unloading auger. The environment represented by terrain and weather conditions influences the combine processes as a disturbance.

2.2.1 System/Process Definition and Data Analysis

For a practitioner in the field, *problem identification* means enumeration of the problems with an existing system and associated process to help develop requirements for a new proposed system. Thus, the problem formulation process consists of the formal problem statement and formulation of the specific objectives of the project and specific issues to be addressed. At this stage, one must define the *performance measures/quantitative criteria* as the basis on which different system configurations will be assessed.

Formulation of specific objectives of a project is mainly related to the optimization of the level of performance indices of the system/process, and minimization of the cost of product redesign or new product development for a given team's knowledge and expertise, as well as the technical and financial resources of the organization.

To *define a new system/process* means to determine the type, function, location, and integration of a physical system, and its functional and performance specifications that derive from requirements generated at a higher level, as well as requirements from interaction with adjacent systems. The *system specifications* are statements that explicitly state what the system/process is to be and do. At this stage, the system configuration and hypotheses about system performance must be defined as precisely as possible.

Phenomena occurring in physical processes, which integrate the subjects of this book, are described with quantities measured by physical units. Each physical unit is characterized by a dimension and a numerical value. There are seven *independent units*: length (meter, m), mass (kilogram, kg), time (second, s), electric current (ampere, A), temperature (Kelvin, K), amount of substance (mole, mol), and luminous intensity (candela, cd). Other units can be expressed in the seven base units. For reader convenience, this book uses both SI and imperial system units.

Collecting input data and *data analysis* are steps in modeling the system to be developed. Data consists of unconnected numbers or symbols representing entities with appropriate levels of reliability or belief. In terms of availability, the sources for input data are historical records, scientific and engineering literature, competitor models (manufacturer specifications, reverse engineering), organization expertise (in-house research, practitioners' and management estimates), and direct observation of previously developed models. In developing a model, the data shall be obtained from all sources and correlated accordingly since some data may be outdated, not relevant to the project, or not reliable enough. This is, of course, provided that time and resources (financial, personnel, and equipment) allow such an approach.

There is a risk associated with collecting and using input data. The risk of using historical data comes from data referring to quite old models or data that was collected at a now outmoded technical level. The data organized in Tables 1.1 and 1.2 is an example of the starting process of collecting input data. Some specifications may be provided based on lab measurements or an educated extrapolation and have yet to be proven in a real environment.

Reverse engineering is an activity of gathering information and data connected to technological principles of an existing device or equipment by analysis of its structure, function, and operation; physical measuring of the component sizes and material properties; direct experimenting in real conditions; and cost calculation. Reverse engineering is an expensive activity consisting of comprehensive market research, product specification analysis, acquisition of models of interest, and the gathering of information of interest. The risks associated with using the acquired data and information include patent infringement and creation of a similar product that is already on the market. However, this way of direct learning, coupled with a strong technical expertise, may lead to developing a better product and avoid the repetition of possible conceptual or even design mistakes. When used only for learning how to improve the new product, this practice, although not recommended due to professional ethics and law regulations, may prove economically efficient by saving time and investments costs of fundamental or applied research activities.

The data obtained by an organization through *in-house research* represents a reliable source of knowledge to build on. This data and knowledge accumulate from the research and direct analysis of previously developed products. This knowledge may be limited by a series of factors, such as management policy, staffing capacity, research budget, lab equipment availability, schedule constraints, financial risk, and even lack of special expertise and innovation.

Analysis of input data starts with organizing the available data, classifying it based on specific criteria, and comparing it on a certain basis. This shall be accompanied by further mathematical processing, engineering calculations, and estimations.

Data abundance does not necessarily grant us certainty, and sometimes it can lead to error in decision making, with undesirable outcomes due to either overwhelming and confusing situations or overconfidence (Ayyub and Klir, 2006).

It is a common error to fit a mathematical equation to a set of data for given experimental limits and then apply it by extrapolation beyond those limits (Grimvall, 2008a). Such a process, unless cautious, ceases to be a proper interpretation of available data and may later prove to be unsafe to use. Among other subjects, in Section 2.3, data-fitting procedures and techniques are proposed and discussed while drawing attention to certain instabilities that can follow such improperly applied procedures.

2.2.2 System/Process Modeling Concepts

Formulating a system allows engineers and scientists to develop a comprehensive understanding of the nature of the problem, underlying physical phenomena, and processes. But, as physicist Werner Karl Heisenberg said, "What we observe is not nature itself, but nature exposed to our method of questioning" (Heisenberg, 1958). Thus, scientists and engineers need to define an image or a model of the system that performs a process.

System modeling is the process of developing a model of the system. The generic term *model* refers to a conceptual representation or physical entity that resembles, describes, or conveys information about the system behavior in performing the process. For the purpose of this book, we define a *model* as a mathematical representation of a system and its afferent process, while a *three-dimensional (3D) or two-dimensional (2D) model* is the design or image representation of the system assembly with its components.

System modeling *purpose and significance* is outlined as follows:

- Modeling is a combination of science and art; it relies on intuition, application of scientific laws, empirical.
- Modeling allows exploration of the intrinsic behavior of a system/process.
- Modeling allows computer simulation of a system/process and prediction of the effects of changing functional and constructional parameters of the system.
- Modeling, followed by simulation and optimization, may decrease the overall cost of system development.
- Modeling, simulation, and optimization of a system rely almost exclusively on high-speed computational techniques and generated imagery.

Here are some *features of the models* and their development circumstances:

- A model is developed when the system being modeled is not accessible or does not exist, or when a scaled-down version of the system was not previously developed.
- A model of a system is not that system; the system being modeled is unique, while the model may assume different forms. There might be many models representing the same system under different sets of system operating restrictions.
- A model is primarily developed to emulate known data, explain experimental results, and fill in desired information by interpolation or extrapolation.
- A model reproduces the data in a form that is suitable for further use, or extracts information about a specific physical quantity.
- A model is similar to but simpler than the system it describes.
- A model incorporates most of the significant features of the system and most relevant effects in the studied phenomena.
- A model embodies both scientific principles and previously obtained experimental data related to the system.
- Model complexity is increased iteratively while maintaining or improving its accuracy.
- A good model is a judicious trade-off between realism and simplicity and between usefulness and complexity, unless economical, military, or other reasons prove otherwise.

No matter what the purpose is, every mathematical model can be characterized by a set of criteria such as sensitivity, robustness, and accuracy.

The *sensitivity of a model* to a parameter is quantified by the magnitude of change in the system output with variation of that parameter. A large sensitivity of the model to a parameter suggests that the system output can change substantially with relatively small variation in that parameter, and vice versa. Thus, the sensitivity of a model can be understood as the persistence of a modeled system to the perturbations in a single parameter.

Methods for optimal experimental design use model sensitivity analysis to determine the conditions under which an experiment is to be conducted to maximize the information content of the data (Emery and Nenarokomov, 1998).

The *robustness of a model* is associated with the sensitivity of the model to errors in the data the model is based on. A model is robust if its results remain true, even though the model may not be very accurate. The robustness of a model can be understood as the *stability of the system behavior* under simultaneous changes in model parameters. Thus, the robustness of a model can be interpreted as the persistence of the modeled system to simultaneous perturbations in many model parameters.

The *accuracy of a model* is a measure of the model's ability to accurately describe the phenomena involved in the process that is performed by the modeled system.

The model accuracy depends on how well the model incorporates the key features of the system and relevant aspects of the process, and how robust the model is with respect to its mathematical form and the numerical values of input parameters (Grimvall, 2008a).

Mathematical models can be categorized based on the following main *classification criteria*:

- *System nature*: Physical system, management system, and meta-model (a system of rules that evaluate different models of the same system).

- *Dynamics*: Steady-state or dynamic models. These are models whose outputs show no variation over time and space; that is, they do not account for the element of time while the outputs vary over time and across space. Dynamic models are typically composed of differential equations.

- *Probability*: Deterministic or stochastic models. A deterministic model uses mathematical relationships to determine a unique set of variable states of the system based on model parameters and previous states of considered variables. Due to random events that are characteristic to the processing of phenomena, a stochastic model is based on probability distributions of variable states, rather than on unique variable states.

- *Reasoning*: Deductive, inductive, and floating. A deductive model is developed as a logical structure based on theory. An inductive model is developed based on observations (experimental data) and generalization from them. A floating model is not based on theory or experimental findings, but on an expected structure, for example, a model of phenomena that are based on a possible catastrophe theory.

Modeling of combine harvester physical systems and processes will employ steady-state and dynamic models, as well as deterministic and stochastic models. The next section discusses this topic in detail, along with different mathematical modeling methods.

2.2.3 Parameters and Variables

A mathematical model of a system consists of one equation or an assembly of equations, inequalities, and data (quantities) employed to describe the behavior of the system in quantitative values of its process performance indices. Mathematical models are essentially composed of relationships between parameters and variables. Model equations can be solved mathematically or translated into a computer code to obtain numerical solutions for state variables.

Parameters define characteristics, features, or measurable factors of the modeled system. A parameter is normally a *constant*, which stands for a certain property (e.g., properties of component material and size, such as the length of a grain sieve), at least in a single simulation, and it is changed only when adjusting the system behavior, that is, model response, is desired.

The quantities that can be measured independently in an experiment are called *variables*. There are independent (input) variables, state variables, and dependent (output) variables. An *independent variable* is does not depend on the process; it varies by default independently, or it is varied by the researcher. In the mathematical models of dynamic systems, time is an example of an independent variable. There are also independent variables that vary with spatial dimensions. Depending on the context, an independent variable is also known as a predictor variable, controlled variable, or explanatory variable.

A *state variable* represents the state of the system/process. The value of a state variable at an instant gives information about the process state at that instant; for example, the threshed and segregated grain proportion on a certain location of the concave in the threshing space (i.e., at a certain moment) is an example of a state variable of threshing-separating processes. The number of state variables for a dynamic system is not unique, but is limited to the minimum number of variables, which satisfies the requirement of characterizing future states given the current state of the system and future inputs.

A *dependent variable* is the response that is measured. The inequalities in a model constrain the variation of one or more dependent variables. A dependent variable is also known as a response variable, measured variable, observed variable, or outcome variable.

The equation models of dynamic systems are called *lumped parameter models* because the spatial variation of the system parameters is negligible, or else is being approximated by lumped sections with a constant parameter value (Klee and Allen, 2011).

2.3 Deterministic Models

In a deterministic model, the output is precisely quantified through known mathematical relationships derived from physical laws among input, state, and output variables. Basically, a given input will always determine the same output. One may think that mathematical relationships are often deterministic, though this is not the case because of the difficulty to correctly express a physical law.

A mathematical model of a dynamic system where the single independent variable is time is composed of ordinary differential equations (ODEs). The differential equations are usually developed by applying known fundamental principles of physics (e.g., Newton's law of motion) that govern the kinematics and dynamics of a modeled system. The mathematical models may be linear or nonlinear, with single, double, or multiple inputs and outputs. The inputs are known or unknown excitations applied to the system. The outputs are the system responses.

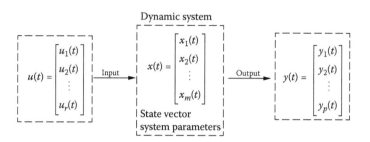

FIGURE 2.1
Functional diagram of a dynamic system with MIMO.

Let us consider a nonlinear mathematical model of a system (Figure 2.1), with multiple inputs and outputs (MIMO) as follows:

$$u(t) = \left[u_1(t)\ u_2(t)...\ u_r(t) \right]^T \tag{2.2}$$

$$x(t) = \left[x_1(t)\ x_2(t)...\ x_m(t) \right]^T \tag{2.3}$$

$$y(t) = \left[y_1(t)\ y_2(t)...\ y_p(t) \right]^T \tag{2.4}$$

where:
  $u(t)$ is the input vector
  $x(t)$ is the state vector
  $y(t)$ is the output vector

The state vector, $x(t)$, contains a set of state variables that completely characterize the state of a dynamic system at any time t. The number of state variables (in this case, m) is the *order* of the system.

The system of differential equations is nonlinear and can be brought to the following standard form:

$$\dot{x} = f\left(t,\ x,\ u\right) \tag{2.5}$$

$$y = h\left(t,\ x,\ u\right) \tag{2.6}$$

where f and h are nonlinear transformations. Equation 2.5 represents the m state differential equations, while Equation 2.6 represents the p algebraic output equations.

Considering the time interval $[t_0, t_1]$, the transformation $f(t, x, u)$ can be defined such that

$$x\left(t_1\right) = f\left(t_0,\ t_1,\ x\left(t_0\right),\ u\left(t_0,\ t_1\right)\right) \tag{2.7}$$

This means that by the causality property of a dynamic system, future states can be determined if all the inputs from the initial time up to that future moment are known (De Silva, 2009).

At any time t_1, the system output vector $y(t_1)$ is uniquely determined by the state input $x(t_1)$ and the input $u(t_1)$ at that time. That can be expressed as

$$y(t_1) = h(t_1, x(t_1), u(t_1)) \tag{2.8}$$

Equation 2.8 states that the system output at time t_1 depends on the time, the input, and the state of the system at that time.

Let us assume that the input, state, and output vectors are influenced by the following perturbations: $\delta u(t)$, input perturbation; $\delta x(t)$, state perturbation; and $\delta y(t)$, output perturbation. That can be mathematically expressed in the following form:

$$u(t) = u_n(t) + \delta u(t) \tag{2.9}$$

$$x(t) = x_n(t) + \delta x(t) \tag{2.10}$$

$$y(t) = y_n(t) + \delta y(t) \tag{2.11}$$

where $u_n(t)$, $x_n(t)$, and $y_n(t)$ are the nominal (unperturbed) quantities. Taking into account the perturbation of $\dot{x}$, Equation 2.5 becomes

$$\dot{x} = f(t, x, u) + \delta f(t, x, u) \tag{2.12}$$

where the determinant of the perturbation $\delta f(t, x, u)$ satisfies the following condition:

$$\|\delta f(t, x, u)\| \le \|f(t, x, u)\| \tag{2.13}$$

That makes the output algebraic equation become

$$y = h(t, x, u) + \delta h(t, x, u) \tag{2.14}$$

where $\delta h(t, x, u)$ is the resulting perturbation of the y output vector for a given input u and all t and x of interest.

If we consider a single-input, single-output (SISO) dynamic system, the results of first-order Taylor expansion for the functions $f(t, x, u)$ and $h(t, x, u)$ (Equations 2.5 and 2.6) lead to the following system of first-order equations of the perturbations:

$$\delta \dot{x}(t) = F(t)\delta x(t) + B(t)\delta u(t) \tag{2.15}$$

$$\delta y(t) = H(t)\delta x(t) + D(t)\delta u(t) \tag{2.16}$$

where the matrices $F(t)$, $B(t)$, $H(t)$, and $D(t)$ are partial derivatives as follows:

$$F(t) = \left.\frac{\partial f}{\partial x}\right|_{x_n, u_n}, \quad B(t) = \left.\frac{\partial f}{\partial u}\right|_{x_n, u_n} \tag{2.17}$$

$$H(t) = \left.\frac{\partial h}{\partial x}\right|_{x_n, u_n}, \quad D(t) = \left.\frac{\partial h}{\partial u}\right|_{x_n, u_n} \tag{2.18}$$

Let us review the use of a *state-space approach* to reduce nth-order linear ODEs to n first-order ODEs. We start by considering a SISO nth-order linear ODE as follows:

$$\frac{d^n y}{dt^n} + a_{n-1}\frac{d^{n-1}y}{dt^{n-1}} + \cdots + a_1\frac{dy}{dt} + a_0 y = u \tag{2.19}$$

where:
$y(t)$ is the output variable
$u(t)$ is the input variable

Converting the ODE into first-order form can be done by using the following variable change:

$$x_1 = y$$
$$x_2 = \frac{dy}{dt}$$
$$\vdots \tag{2.20}$$
$$x_n = \frac{d^{n-1}y}{dt^{n-1}}$$

Consequently, this leads to an equivalent system of n-order equations as follows:

$$\dot{x}_1 = x_2$$
$$\dot{x}_2 = x_3$$
$$\vdots \tag{2.21}$$
$$\dot{x}_n = -a_0 x_1 - a_1 x_2 - \cdots - a_{n-1}x_n + u$$

This canonical structure is known as a chained form and can be written in matrix form as

$$\dot{x}(t) = Fx(t) + Bu(t) \tag{2.22}$$

In Equation 2.22, the vector $x(t)$ is the state variable vector

$$x(t) = \begin{bmatrix} x_1 & x_2 & \cdots & x_n \end{bmatrix}^T \tag{2.23}$$

The matrix F, known as the *state matrix*, is given by

$$F = \begin{bmatrix} 0 & 1 & 0 & \cdots & 0 \\ 0 & 0 & 1 & \cdots & 0 \\ \vdots & \vdots & \vdots & \ddots & \vdots \\ 0 & 0 & 0 & \cdots & 1 \\ -a_0 & -a_1 & -a_2 & \cdots & -a_{n-1} \end{bmatrix} \tag{2.24}$$

and the matrix B is

$$B = \begin{bmatrix} 0 & 0 & \cdots & 1 \end{bmatrix}^T \tag{2.25}$$

The general SISO nth-order linear differential equation is given by

$$\frac{d^n y}{dt^n} + a_{n-1}\frac{d^{n-1}y}{dt^{n-1}} + \cdots + a_1\frac{dy}{dt} + a_0 y = b_n\frac{d^n u}{dt^n} + b_{n-1}\frac{d^{n-1}u}{dt^{n-1}} + \cdots + b_1\frac{du}{dt} + b_0 u \tag{2.26}$$

Equation 2.26 has two ODEs of an equivalent form as follows:

$$y = b_n\frac{d^n x}{dt^n} + b_{n-1}\frac{d^{n-1}x}{dt^{n-1}} + \cdots + b_1\frac{dx}{dt} + b_0 x$$

$$u = \frac{d^n x}{dt^n} + a_{n-1}\frac{d^{n-1}x}{dt^{n-1}} + \cdots + a_1\frac{dx}{dt} + a_0 x \tag{2.27}$$

where x is an intermediate variable.

To convert the above ODE into first-order form, the following variable change can be used:

$$x_1 = x$$

$$x_2 = \frac{dx}{dt}$$

$$\vdots \tag{2.28}$$

$$x_n = \frac{d^{n-1}x}{dt^{n-1}}$$

This leads to an equivalent system of n first-order equations, which in a matrix form, using the above matrices F and B, can be written as follows:

$$\dot{x}(t) = F x(t) + B u(t) \tag{2.29}$$

$$y(t) = H x(t) + D u(t) \tag{2.30}$$

The matrices H and D are given by

$$H = \begin{bmatrix} (b_0 - b_n a_0) & (b_1 - b_n a_1) & \cdots & (b_{n-1} - b_n a_{n-1}) \end{bmatrix} \tag{2.31}$$

$$D = b_n \tag{2.32}$$

The matrix representation by Equations 2.29 and 2.30 is used in system control, which is discussed in Section 2.6.

For further reading, refer to the textbooks written by Crassidis and Junkins (2004) and Nayfeh (1981).

Application 2.2: Development of a Model of Combine Vehicle Kinematics While Steering

This is an example of developing a deterministic model of combine harvester kinematics following a curve on its trajectory. The combine is a front-wheel-driven, rear-wheel-steered vehicle.

During transportation, combine speed restrictions are imposed only by traffic signs and limiting the amplitude of transport-induced pitching vibrations. These restrictions do not influence combine positioning in time on its moving path; that is, the combine harvester can be considered a nonholonomical system. Mathematically, this means the above-mentioned restrictions are not to be integrated within vehicle positioning constraints.

Let us consider that the combine vehicle moves in an absolute inertial coordinate system XYZ, as illustrated by Figure 2.2. The axis Z (not shown) points up, in a perpendicular direction to the XY plane.

The system model assumes the wheels are rigid bodies and roll on the terrain without slippage, meaning that the instantaneous velocity of a wheel is considered along the direction tangential to the wheel-rolling path. The (x, y) Cartesian coordinates give the location of the center C of the front axle. When the combine follows a curved path, no matter the drive transmission type (mechanical or hydrostatic), the driving wheels rotate in a compensatory way, similar to the effect imposed by a differential device. Consequently, the combine velocity v_c is the average of the drive wheel centers, and it is, at all times, oriented along the longitudinal axis of the machine. Similarly, the velocity v_s is the mean of the steering wheels' center velocities in the XY plane.

The exact current position and orientation of the machine is described by the following generalized coordinates that form the state variable vector $x(t) = [x \; y \; \psi_c \; \psi_s]^T$, where ψ_c is the current angle of the combine harvester orientation and ψ_s is the current angle of the steering wheels' direction.

The velocity of point C in the X direction, $\dot{x} = dx/dt$, can be determined from

$$\dot{x} = v_c \cos \psi_c$$

while its velocity in the y direction is given by

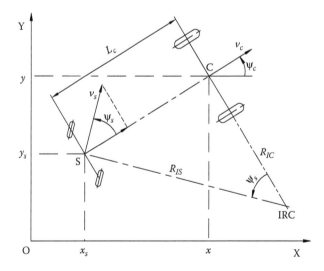

FIGURE 2.2
Variables and parameters of combine vehicle kinematics.

$$\dot{y} = v_c \sin \psi_c$$

Eliminating v_c, the above equations lead to

$$\dot{x} \sin \psi_c - \dot{y} \cos \psi_c = 0$$

In a similar way, applying the nonslippage constraint to the steering wheels, and eliminating the speed v_s, the following equation can be written:

$$\dot{x}_s \sin(\psi_c + \psi_s) - \dot{y} \cos(\psi_c + \psi_s) = 0$$

Between the points C and S, there is a constant distant L_c, which is the combine drive wheelbase.

That means

$$\cos \psi_s = \frac{v_c}{v_s}$$

At any given time t, while the combine is steering, an instantaneous rotation center (IRC) exists at the intersection of perpendiculars on the velocities' v_c and v_s directions. The angular velocity $\dot{\psi}_c$ of combine rotation around point C can be expressed as

$$\dot{\psi}_c = \frac{d\psi_c}{dt} = \frac{-v_s \sin \psi_s}{L_c}$$

In the above formula, the minus sign indicates a reverse direction with respect to the Z axis direction. Combining the last two equations yields

$$\dot{\psi}_c = -\frac{\tan \psi_s}{L_c} v_c$$

Now, the kinematic model of the combine can be written in a matrix form as follows:

$$\begin{bmatrix} \dot{x} \\ \dot{y} \\ \dot{\psi}_c \\ \dot{\psi}_s \end{bmatrix} = \begin{bmatrix} \cos \psi_c \\ \sin \psi_c \\ -\dfrac{\tan \psi_s}{L_c} \\ 0 \end{bmatrix} v_c + \begin{bmatrix} 0 \\ 0 \\ 0 \\ 1 \end{bmatrix} \omega_k$$

where $\dot{\psi}_s = \omega_k$ is the steering angular velocity of the rear wheels, that is, the rotation speed of the kingpin.

One can distinguish the two input variables that form the input vector $u(t) = [v_c \ \omega_k]^T$. Should we know or establish the ratio between the rotations of the operator's steering wheel and the rotations of the kingpin, this ratio will replace the number 1 in the last matrix, while ω_k will be replaced by the angular speed of the operator's steering wheel.

The output vector is $y(t) = \begin{bmatrix} \dot{x} & \dot{y} & \dot{\psi}_c & \dot{\psi}_s \end{bmatrix}^T$ and describes how fast the combine moves and how fast its position changes along the path.

The mathematical model is nonlinear, and theoretically, it has a singularity at $\psi_s = \pi/2 + k\pi$, $k \in N$, where the state vector has a discontinuity. This corresponds to the position when the vehicle cannot move (due to an assumed nonslippage condition) because the rear wheel direction is perpendicular to the vehicle's longitudinal axis. Practically, due to inherent limitations in the mechanical design of the steering mechanism, this situation never occurs.

A more useful model is one that describes the combine kinematics when following a defined path. The variables and parameters of vehicle kinematics are defined in Figure 2.3.

The same nonslippage conditions of the wheels are further assumed. The current normal distance between the drive axles' center C and the path is given by d_p. The angle ψ_p between the current combine direction and the tangent to the path is defined as

$$\psi_p = \psi_c - \psi_t$$

where ψ_t is the angle between the current tangent to the curve and the X axis.

The distance traveled along the path, starting at an arbitrary position, is given by the length p of the arc. Thus, the curvature $c(p)$ along the path is defined as

$$c(p) = \frac{d\psi_t}{dp} = \frac{d\psi_t}{dt}\frac{dt}{dp} = \frac{\dot{\psi_t}}{\dot{p}}$$

which implies

$$\dot{\psi_t} = c(p)\dot{p}$$

The speed of the combine at point C approaching the curve along d_p is given by

$$\dot{d_p} = v_c \sin \psi_p$$

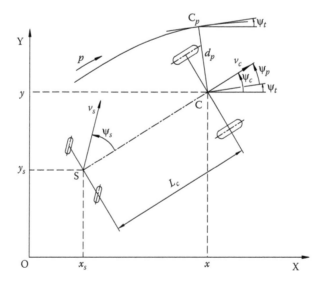

FIGURE 2.3
Variables and parameters of a path-following combine harvester.

The speed $\dot{p}$ on the curve of point C_p (corresponding to the point C movement) can be expressed as follows:

$$\dot{p} = v_c \cos \psi_p + \dot{\psi}_t d_p$$

Introducing the last two equations in the previously developed kinematics model, one can obtain the mathematical model of the path-following combine kinematics in terms of state path coordinates: $\underline{x}_p(t) = [p\ d_p\ \psi_p\ \psi_s]^T$.

$$\begin{bmatrix} \dot{p} \\ \dot{d}_p \\ \dot{\psi}_p \\ \dot{\psi}_s \end{bmatrix} = \begin{bmatrix} \dfrac{\cos \psi_p}{1 - d_p c(p)} \\ \sin \psi_p \\ -\dfrac{\tan \psi_s}{L_c} - \dfrac{c(p)\cos \psi_p}{1 - d_p c(p)} \\ 0 \end{bmatrix} v_c + \begin{bmatrix} 0 \\ 0 \\ 0 \\ 1 \end{bmatrix} \omega_k$$

One can note that such a deterministic model always generates the same output $y_p(t) = \begin{bmatrix} \dot{p}\ \dot{d}_p\ \dot{\psi}_p\ \dot{\psi}_s \end{bmatrix}^T$ for a given input $u_p(t) = \begin{bmatrix} v_c\ \omega_k \end{bmatrix}^T$ as long as the system parameters (e.g., L_c) remain unchanged.

A chained-form transformation of the above equation will be applied as an example in Section 2.6.

2.4 Stochastic Models

2.4.1 Introduction

The word *stochastic*, derived form the Greek *stohos*, means "pertaining to chance," and it is used to characterize an event, process, or system whose behavior is nondeterministic; that is, it is associated with some *randomness*. That means a stochastic system does not generate all the time the same output for a given input. According to Kac and Logan (1979), any kind of development in time that is analyzable in terms of probability can be named a *stochastic process*. By extension, a mathematical model that describes the evolution of a stochastic process or the behavior of a stochastic system is called a *stochastic model*.

A stochastic model is based on the *probability theory* for quantifying the probability distributions of potential outputs of a process or system. A probability distribution describes the indeterminacy that is associated with the process evolution or system behavior by considering which outcome is most probable.

Examples of stochastic processes performed by the units of a combine harvester are plant gathering, corn detaching from the stalk, grain threshing, grain separation, MOG fragmentation, and so forth. To further comment, even though each grain moves on a deterministic path, the exact motion of grain collection is practically and computationally unpredictable; that is, the grains' behavior under the action of bars, sieves, and so forth, exhibits stochastic characteristics. The obvious conclusion is that the variability of the above-mentioned processes can be significant and is a function of different parameters and input variables.

The following sections contain a condensed, structured review of the probability theory the reader will need to know for understanding and further applying to the development of stochastic models. When more information is needed, the reader is referred to the References and Bibliography at the end of this chapter.

2.4.2 Random Variables

A *random variable* or *stochastic variable* is a variable whose value is determined by a process/ system that is subject to chance-induced variations. A random variable is a quantity that can take on different values called *random variates*, which may or may not be predictable within the range of a variable's possible values.

A random variable associates a measurable value with the outcome of an experiment that has a sample space. The experiment might be performed or computer simulated. The latter implies computer generation of random variates.

Random variables can be classified as *discrete*, *continuous*, or *mixed*. Discrete random variables may assume any value of a countable set of exact values, while continuous random variables may assume any numerical value (an infinite number of values) within a certain interval. A random variable is mixed when its value domain is a union of a continuous interval of values and a set of discrete values.

Random variable values are usually expressed by real numbers, although random elements may be chosen, such as Boolean values, complex numbers, vectors, functions, sequences, shapes, and processes. The real-valued variables are usually considered because their *expected value* and *variance* (a measure of the extent to which the values are dispersed) can be calculated.

Stochastic models of physical phenomena are essentially mathematical relationships between random variables of the process and parameters of the system that performs that process.

2.4.3 Probability Distribution Function

The advantage of considering random variables is that they do not depend on the nature of the underlying random experiment. The modeler is interested in knowing the values a random variable can assume, and the probabilistic law that describes their occurrence. The law that defines the likelihood of a random variable X taking on values within a certain range R_X is called the *probability distribution function $P_X(x)$*, defined as follows:

$$P_X(x) = \Pr(X \le x) \tag{2.33}$$

where $\Pr(X \le x)$ is the probability that random variable X is smaller than or equal to a number $x \in (a, b)$, $a \le b$. That means the probability distribution function $P_X(x)$ allows calculating the probability interval $\Pr(X \in (a, b)$, $a \le b$.

A *discrete random variable* can take on values that can be sequentially written as a finite range (or a countable infinite range) $R_X = \{x_0, x_1, x_2, ...\}$. Let p_i be the probability that a random variable X assumes the value x_i:

$$p_i = \Pr(X = x_i), \quad i = 0, 1, 2, ... \tag{2.34}$$

That means the probability interval is given by

$$\Pr\left(X \in (a, b)\right) = \Pr\left(a < X \le b\right) = \sum_{x_i \in (a, b)} p_i \qquad (2.35)$$

Thus, the probability distribution function of a discrete random variable satisfies the normalizing condition:

$$\sum_{i=0}^{\infty} p_i = 1 \qquad (2.36)$$

As an example, the graph of the distribution function of a discrete random variable is shown in Figure 2.4.

The range R_X of a *continuous random variable* value is a finite or infinite interval. Thus, the probability distribution function of a continuous random variable is given by the (cumulative) distribution function:

$$F(x) = \Pr\left(X \le x\right), \quad x \in R_X \qquad (2.37)$$

The above distribution function, which is nondecreasing on x, has the following two properties:

$$F(-\infty) = 0$$
$$F(+\infty) = 1 \qquad (2.38)$$

For any $X \in (a, b)$, the probability interval is given by (Figure 2.5)

$$\Pr\left(X \in (a, b)\right) = \Pr\left(a < X \le b\right) = F(b) - F(a) \qquad (2.39)$$

A better understanding of the cumulative distribution function is given by a histogram (see next section). Each frequency is a probability $P_X(x)$, and its sum from left to right is the cumulative distribution function.

From lab or field experiments of combine harvesters or their processing units, researchers obtain discrete values of the variables of interest. These values can be further processed

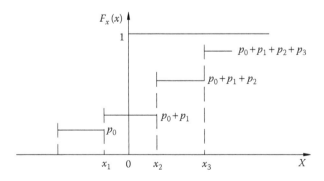

FIGURE 2.4
Graph of the distribution function of a discrete random variable.

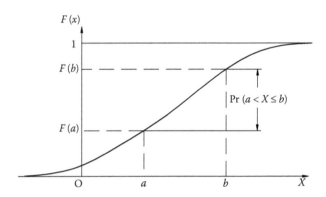

FIGURE 2.5
Graph of a continuous distribution function.

to obtain continuous functions by fitting an equation of a curve to each series of variable values through regression analysis. From this perspective, the next section deals only with significant probability density functions of continuous variables, followed by other very important functions and best practices in curve fitting.

2.4.4 Probability Density Functions

A random variable X is called *continuous* if its distribution function $F(x)$ is differentiable over the function domain. The first derivative of the distribution function,

$$f(x) = \frac{dF(x)}{dx} \tag{2.40}$$

is called the *probability density function* (pdf) of X.

Similarly, a random variable, which has the distribution function $F(x)$, is called continuous if there is a function $f(x)$ so that

$$F(x) = \Pr(X \le x) = \int_{-\infty}^{x} f(\xi)d\xi \tag{2.41}$$

The probability density function $f(x)$ corresponds to the individual probabilities $p_i(x_i)$ of a discrete random variable.

Consistent with the geometric interpretation of a definite integral (Bronshtein and Semendyayev, 1978), Equation 2.41 shows that if the function $f(\xi) > 0$ for $\xi \in (-\infty, x)$, then the integral $\int_{-\infty}^{x} f(\xi)d\xi$ represents the area of the domain bounded by $f(\xi)$ and the X axis, to the left of the vertical line corresponding to $\xi = x$. This area equals the probability distribution $F(x)$ (Figure 2.6) of a continuous random variable.

The *mean value* μ or *expected value* $E(X)$ of a continuous variable X is defined as

$$\mu = \int_{-\infty}^{+\infty} f(x)dx \tag{2.42}$$

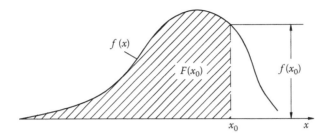

FIGURE 2.6
Probability density function and distribution of a continuous variable.

provided that

$$\int_{-\infty}^{+\infty} |x| f(x) dx < \infty \tag{2.43}$$

The *variance* of a continuous random variable σ^2, also denoted by $Var(X)$, is its mean squared deviation from its mean value (Beichelt, 2006) and can be calculated as follows:

$$\sigma^2 = \int_{-\infty}^{+\infty} (x - \mu)^2 f(x) dx \tag{2.44}$$

For a discrete random variable, the variance σ^2 measures the spread or variability of n values' (x_i) distribution and is calculated as

$$\sigma^2 = \sum_{i=1}^{n} (x_i - \mu)^2 p_i \tag{2.45}$$

where p_i is the probability of the occurrence of value x_i.
   The square root σ of the variance is called the *standard deviation* of the variable X.
   The α-*percentile* x_α of a random variable X is defined as

$$F(x_\alpha) = \alpha \tag{2.46}$$

The interpretation of the α-*percentile* is as follows: in a long series of random experiments with outcome X, about $\alpha\%$ of the observed value of X will be equal to or less than x_α.

2.4.4.1 Normal Distribution

If a continuous random variable X has a normal (or Gaussian) probability distribution, then a relative histogram of the random variable has the shape of a normal curve (bell shaped and symmetric) (Figure 2.7).

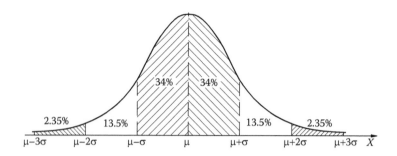

FIGURE 2.7
Normal distribution of a variable with mean μ and standard deviation σ.

The *normal density curve* of a random variable X, with the mean μ and standard deviation σ, has the following properties:

- It is symmetric about the variable mean, μ.
- Because the mean equals the median, the curve peak (the highest point) occurs at $x = \mu$.
- It has two inflection points at $\mu - \sigma$ and $\mu + \sigma$.
- The area under the curve equals 1.
- Approximately 68% of the area under the normal curve is between $x = \mu - \sigma$ and $x = \mu + \sigma$. Approximately 95% of the area under the normal curve is between $x = \mu - 2\sigma$ and $x = \mu + 2\sigma$. Approximately 99.7% of the area under the normal curve is between $x = \mu - \sigma$ and $x = \mu + \sigma$.

The probability density function of a normal curve is given by the equation

$$f(x) = \frac{1}{\sigma\sqrt{2\pi}} e^{-\frac{1}{2}\left(\frac{x-\mu}{\sigma}\right)^2} \tag{2.47}$$

The standard normal curve is used in practice based on the *standard normal curve* of variable Z with mean = 0 and standard deviation = 1, using the transform

$$Z = \frac{X - \mu}{\sigma} \tag{2.48}$$

Thus, the probability density function of the standard normal curve is given by

$$f(x) = \frac{1}{\sqrt{2\pi}} e^{-\frac{z^2}{2}} \tag{2.49}$$

The properties of the standard normal curve are as follows:

- It is symmetric about the variable mean, $\mu = 0$.
- Because the mean equals the median, the curve peak (the highest point) occurs at $z = 0$.

- It has two inflection points at $z = -1$ and $z = 1$.
- The area under the curve equals 1.
- Approximately 68% of the area under the normal curve is between $z = -1$ and $z = 1$. Approximately 95% of the area under the normal curve is between $z = -2$ and $z = 2$. Approximately 99.7% of the area under the normal curve is between $z = -3$ and $z = 3$.

For a specified Z-score, in the standard normal table one can read the area under the curve (e.g., 0.937, i.e., 93.7% probability) to the left of the Z-score ordinate ($Z = 1.53$ in this example).

2.4.4.2 Pareto Distribution

A random variable X has a *Pareto distribution* (type I) over an interval $x \in [d, \infty)$, $d > 0$, if its distribution function $F(x)$ and density function $f(x)$ are as follows:

$$F(x) = 1 - \left(\frac{d}{x}\right)^c \tag{2.50}$$

$$f(x) = cd^c x^{-(c+1)} \tag{2.51}$$

where the coefficient $c > 1$.
Then the mean and variance of the variable are

$$\mu = \frac{cd}{c-1}, \quad c > 1 \tag{2.52}$$

$$\sigma^2 = \frac{c}{(c-2)}\left(\frac{d}{c-1}\right)^2 \tag{2.53}$$

The calculation of the variance makes sense only if $c > 2$. Note that $F(x) = 0$ and $f(x) = 0$ if $x < d$. Figure 2.8 shows the graph of Pareto density functions for various coefficients c and $d = 1$.

2.4.4.3 Exponential Distribution

A random variable X has an *exponential distribution* with rate parameter λ if its distribution function $F(x)$ and density function $f(x)$ are

$$F(x) = 1 - e^{-\lambda x} \tag{2.54}$$

$$f(x) = \lambda e^{-\lambda x} \tag{2.55}$$

for $x \geq 0$ and $\lambda > 0$.

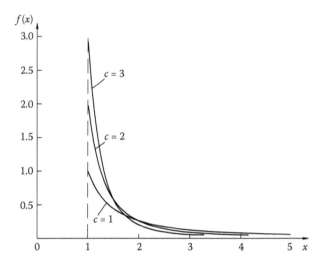

FIGURE 2.8
Pareto density functions for various coefficients c and $d = 1$.

The mean value and variance are

$$\mu = 1/\lambda \tag{2.56}$$

and

$$\sigma^2 = 1/\lambda^2 \tag{2.57}$$

Figure 2.9 shows the graph of exponential density functions for $\lambda = 1$.

2.4.4.4 Gamma Distribution

A random variable X has a *gamma distribution* with the *shape parameter* α and *scale parameter* β if its density function has the following form:

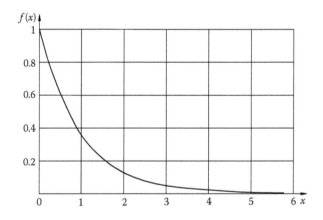

FIGURE 2.9
Exponential density function graph for $\lambda = 1$.

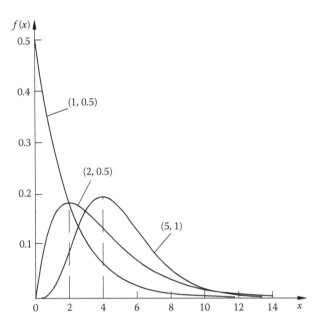

FIGURE 2.10
Gamma density function graphs for various pairs of parameters (α, β).

$$f(x) = \frac{\beta^{\alpha}}{\Gamma(\alpha)} x^{\alpha-1} e^{-\beta x} \qquad x > 0, \alpha > 0, \beta > 0 \tag{2.58}$$

In Equation 2.58, *gamma function* $\Gamma(\alpha)$ is defined as

$$\Gamma(\alpha) = \int_0^{\infty} x^{\alpha-1} e^{-x} dx, \quad \alpha > 0 \tag{2.59}$$

Figure 2.10 shows the graphs of gamma density function for a few pairs of (α, β) parameters.

The mean and variance are, respectively,

$$\mu = \alpha/\beta \tag{2.60}$$

$$\sigma^2 = \alpha / \beta^2 \tag{2.61}$$

By replacing $\alpha = 1$ and $\beta = \lambda$ in Equation 2.58, we get the exponential density function shown in Equation 2.55.

2.4.4.5 Beta Distribution

A random variable X has a *beta distribution* with parameters α and β if its density function is

$$f(x)\frac{1}{B(\alpha,\beta)}x^{\alpha-1}(1-x)^{\beta-1} \qquad (2.62)$$

where:
$0 \le x \le 1$
$\alpha > 0$
$\beta > 0$

The *beta function* $B(\alpha, \beta)$ is defined as follows:

$$B(\alpha,\beta) = \frac{\Gamma(\alpha)\Gamma(\beta)}{\Gamma(\alpha+\beta)} \qquad (2.63)$$

Because the beta distribution is limited to x values within the interval [0, 1], for larger intervals of positive values $X \in [X_m, X_M]$, one can perform a transform as defined by the next formula:

$$x = \frac{X - X_m}{X_M - X_m} \qquad (2.64)$$

The mean and variance are

$$\mu = \frac{\alpha}{\alpha+\beta} \qquad (2.65)$$

$$\sigma^2 = \frac{\alpha\beta}{(\alpha+\beta)^2(\alpha+\beta+1)} \qquad (2.66)$$

Figure 2.11 shows the graphs of beta density function for a few pairs of (α, β) parameters.

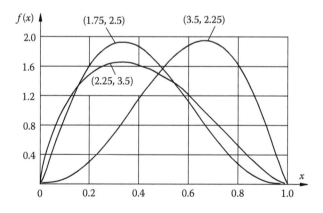

FIGURE 2.11
Beta density function graphs for various pairs of parameters (α, β).

2.4.4.6 Weibull Distribution

A random variable X has a *Weibull distribution* with scale parameter λ and form parameter c if its density function and distribution function are

$$f(x) = c\lambda^c x^{c-1} e^{-(\lambda x)^c} \tag{2.67}$$

$$F(x) = 1 - e^{-(\lambda x)^c} \tag{2.68}$$

where:
$x > 0$
$\lambda > 0$
$c > 0$

The mean and variance can be respectively calculated with the following formulae:

$$\mu = \lambda^{-1} \Gamma\left(1 + c^{-1}\right) \tag{2.69}$$

$$\sigma^2 = \lambda^{-2} \left[\Gamma\left(1 + 2c^{-1}\right) - \Gamma^2\left(1 + c^{-1}\right) \right] \tag{2.70}$$

Figure 2.12 shows the graphs of Weibull density function for a few pairs of (λ, c) parameters.

2.4.4.7 Rayleigh Distribution

The Rayleigh distribution is limited to strictly positive-valued random variables.

A random variable X has a *Rayleigh distribution* with parameter λ if its density function and distribution function are

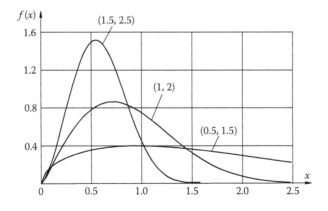

FIGURE 2.12
Weibull density function graphs for various pairs of parameters (λ, c).

$$f(x) = \frac{x}{\lambda^2} e^{-\frac{1}{2}\left(\frac{x}{\lambda}\right)^2}$$ (2.71)

$$F(x) = 1 - e^{-\frac{1}{2}\left(\frac{x}{\lambda}\right)^2}$$ (2.72)

The mean and variance can be respectively calculated with the following formulae:

$$\mu = \lambda \sqrt{\frac{\pi}{2}}$$ (2.73)

$$\sigma^2 = \lambda^2 \left(2 - \frac{\pi}{2}\right)$$ (2.74)

Figure 2.13 displays the graphs of Rayleigh density functions with various parameters λ.

2.4.4.8 Additional Functions

For process mathematical modeling, the author has derived additional functions, such as

$$f(x) = a x e^{-bx}$$ (2.75)

$$f(x) = a x^2 e^{-bx}$$ (2.76)

$$f(x) = a x^b e^{-cx}$$ (2.77)

$$f(x) = a x^b e^{cx}$$ (2.78)

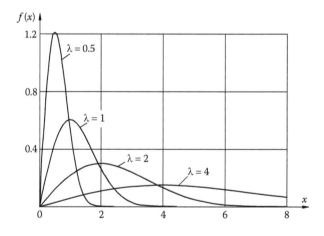

FIGURE 2.13
Rayleigh density function graphs for various parameters λ.

$$f(x) = a\, x^{-b}\, e^{c\, x} \tag{2.79}$$

where:
  $a > 0$
  $b > 0$
  $c > 0$

These functions can be used in regression analysis of experimental data.

For more information on probability density and cumulative distribution functions and their visualization, see MATLAB R2012 documentation: Statistics Toolbox Distribution Functions.

2.4.5 Distribution of Continuous Random Variables Mixture

The distribution function of any random variable depends on one or more parameters (namely, coefficients or exponents). The following theory emphasizes the dependency of such distribution on a certain parameter, which is the realization of a *corresponding random parameter* that is a variable as well.

Let's consider the distribution function $F_X(x, \lambda)$ of a continuous random variable X that depends (possibly among other parameters) on the parameter of interest λ. Let $f_L(\lambda)$ be the probability density function of corresponding continuous random parameter L, and $F_L(\lambda)$ its distribution function. One shall note that the distribution of parameter L depends on other parameters as well, not specified in the above notations.

The *mixture of probability distribution* $F_X(x, \lambda)$ with regard to the structure distribution, characterized by the density $f_L(\lambda)$, generates the *distribution* $F_Y(x, \lambda)$ *of a mixed continuous random variable* that is defined as (Beichelt, 2006)

$$F_Y(x, \lambda) = \int_0^{+\infty} F_X(x, \lambda) f_L(\lambda)\, d\lambda \tag{2.80}$$

The corresponding density function $f_Y(x, \lambda)$ can be obtained by differentiation of $F_Y(x, \lambda)$.

Example

The proportion of unthreshed grains in an axial unit decreases exponentially over the threshing unit length at a relatively constant threshing rate. This is mainly due to the fact that the active elements of the rotor and concave have a relatively uniform distribution to guarantee a continuous flow of material within the threshing space and to minimize the grain damage. However, the threshing rate may be subject to certain randomness by adopting a different design. Thus, the cumulative distribution function of the proportion of unthreshed grains may be mixed with regard to a structure distribution of the threshing rate, namely, characterized by its density function, which is to be determined. For further details on this subject, see Chapter 5.

2.4.6 Distribution of Random Variables Function

Let X denote a random variable with known density function $f_X(x)$ and distribution function $F_X(x)$. Let $Y = g(X)$ be a *real function without jump discontinuities* that transforms the random variable X into the random variable Y. If the function g is invertible, then its inverse function is denoted g^{-1}. That means

$$g^{-1}(y) = x \tag{2.81}$$

We are interested in finding the density $f_Y(y)$ and distribution $F_y(y)$ of Y that depends on X as defined by function $g(X) = Y$. That means we are looking to find the *distribution of the function of a random variable*.

Based on the definition of probability distribution function (Equation 2.37), we get

$$F_Y(y) = \Pr(Y \leq y) = \Pr\left[g(X) \leq y\right] = F_X\left[g^{-1}(y)\right] \tag{2.82}$$

The corresponding density function $f_Y(y)$ can be obtained by differentiation of $F_Y(y)$.

The mean or expected value of the variable Y can be calculated with the following formula:

$$\mu_Y = E[Y] = E\left[g(X)\right] = \int_{-\infty}^{+\infty} g(x) f_X(x) dx \tag{2.83}$$

The variance of the function Y is

$$\sigma_Y^2 = \int (y - \mu_Y)^2 f_Y(y) dy \tag{2.84}$$

The distribution of the function Y, as defined by the real function $Y = g(X)$, represents a *shifted distribution of the variable X*.

Application 2.3: Example of a Shifted Distribution Case

Grain separation in straw walkers takes place in the separating grates mounted a few steps along the straw walker channels. The proportion of grain separation along one separation grate could be considered a variable X. A function Y would transform the variable X to characterize the separation along all of the straw walkers. Additionally, one may consider another variable, Z, that takes into consideration the fact that different regions of the straw walkers are subjected to nonequal accelerations generated by a pair of spaced-apart crankshafts.

2.4.7 Laplace Transformation

The process of a system that is described by differential equations can be more easily modeled as a block diagram using Laplace transfer functions. The *Laplace transformation* converts a differentiation operation into a multiplication one by the Laplace variable s, and an integration operation into a division one by s. The reader should note the difference between *transformation* and *transform* as follows: a *transformation* is an operation, while the *transform* is the result of a transformation operation.

The Laplace transform is useful in systems control and involves mathematical transformation from the *time domain* to the *Laplace domain*, also called s-domain (Figure 2.14). The Laplace transform of a function $f(t)$ is defined as

$$L\{f(t)\}(s) \overset{def}{=} \int_0^\infty f(t) e^{-st} dt \tag{2.85}$$

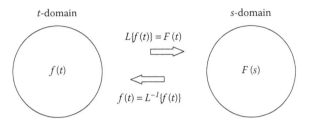

FIGURE 2.14
The Laplace transform L{f(t)} and its inverse f(t).

if the function $f(t)$ is piecewise continuous on every finite interval in the range $t \geq 0$, and it is bound by the relation

$$|f(t)| \leq Me^{-kt} \tag{2.86}$$

for some k and M parameters. Then the Laplace transform exists for all $s > k$ values (Kreyszig, 1988).

The complex variable $s = \sigma + j\omega$ has a real value ς that is chosen to be sufficiently large so that the transformation integral (Equation 2.85) is finite even when $\int f(t)dt$ is not finite. The set of s values for which the integral converges is called the *region of absolute convergence* of the Laplace transform.

In terms of systems control, we note that the *output Y(s) in the s-domain* is

$$Y(s) = L\{f(t)\} \tag{2.87}$$

The inverse Laplace transform is

$$f(t) = \frac{1}{2\pi j} \int_{\sigma - j\omega}^{\sigma + j\omega} Y(s)e^{st}ds, \quad or \quad f(t) = L^{-1}[Y(s)] \tag{2.88}$$

To obtain a time–response solution for a system/process, the following operations are necessary:

- Formulating of linear differential equations
- Performing Laplace transformations of the differential equations
- Solving the resulting algebraic equations of the variable of interest

Complex systems usually have more inputs (represented by an input vector) and more outputs (represented by an output vector). In such a case, the system representation requires several transfer functions resulting in a *transfer function matrix*.

2.4.8 Continuous Variables Convolution

The *convolution integral* is an important general property of the Laplace transformation dealing with the product of transforms. Let $f(t)$ and $g(t)$ be the density functions of two independent continuous-time variables that satisfy the existence hypothesis, as formulated in the previous section, for a Laplace integral to exist.

Assume that we know these functions $f(t)$ and $g(t)$ whose Laplace transforms are $F(s) = \mathcal{L}\{f(t)\}$ and $G(s) = \mathcal{L}\{g(t)\}$, respectively. Then, the product of their transforms $H(s) = F(s)$ $G(s)$ is the transform $H(s) = \mathcal{L}\{h(t)\}$ of the convolution $h(t)$ of $f(t)$ and $g(t)$, written $\mathcal{L}\{(f * g)(t)\}$ and defined by the formula

$$h(t) = \int_0^t f(\tau) g(y - \tau) d\tau, \, \tau \geq 0 \tag{2.89}$$

That is, the joint probability density $h(t)$ of the sum of two independent and continuous random variables with the densities $f(t)$ and $g(t)$ equals the convolution of their individual densities as expressed above.

The convolution $(f * g)$ is commutative, distributive, and associative, as is the multiplication of the numbers. The convolution is useful for calculating inverse transforms and solving differential equations.

Application 2.4: Product of Two Laplace Transforms

Let

$$H(s) = \frac{1}{(s+a)(s-b)}$$

where a and b are constant parameters. Find $h(t)$.

Solution. Applying the theory, we find that

$$f(t) = L^{-1}\left\{\frac{1}{s+a}\right\} = e^{-at}$$

$$g(t) = L^{-1}\left\{\frac{1}{s-b}\right\} = e^{bt}$$

Then we use the theorem expressed by Equation 2.89, finding that

$$h(t) = e^{-at} * e^{bt} = \int_0^t e^{-a\tau} e^{b(t-\tau)} d\tau = e^{bt} \int_0^t e^{-(a+b)\tau} d\tau = \frac{1}{a+b}\left(e^{bt} - e^{-at}\right)$$

This answers the application's question.

Using similar exponential functions, we applied the convolution theorem for stochastic modeling of threshing-separating processes in both tangential and axial threshing units (Chapter 5).

2.4.9 Multidimensional Random Variables

A group of n random variables forms an *n-dimensional random vector* $[X_1, X_2, ..., X_n]^T$ whose probability distribution is given by the *joint distribution function* of the random variables $X_1, X_2, ..., X_n$, as defined by

$$F_{X_i}\left(x_1, x_2, \ldots, x_n\right) = \Pr\left(X_1 \le x_1, X_2 \le x_2, \ldots, X_n \le x_n\right) \tag{2.90}$$

The joint probability density of the random vector $[X_1, X_2, \ldots, X_n]^T$ is given by the nth mixed partial derivative of the joint distribution function with respect to $x_1, x_2, \ldots, x_n$ as follows:

$$f_{X_i}\left(x_1, x_2, \ldots, x_n\right) = \frac{\partial^n F\left(x_1, x_2, \ldots, x_n\right)}{\partial x_1 \, \partial x_2 \ldots \partial x_n} \tag{2.91}$$

The distribution functions of variables X_i, for $i = 1, \ldots, n$, are calculated as follows:

$$F_{X_i}\left(x_i\right) = \Pr\left(X_i \le x_i\right) \tag{2.92}$$

and can be obtained from the joint distribution function

$$F_{X_i}\left(x_i\right) = F\left(\infty, \ldots, \infty, x_i, \infty, \ldots, \infty\right) \tag{2.93}$$

If these distribution functions are one-dimensional, that is, $F_{X_1}(x)$, $F_{X_2}(x)$, ..., $F_{X_n}(x)$, they are called marginal distributions of n-dimensional random vector $[X_1, X_2, \ldots, X_n]^T$. The marginal density functions are determined with the formula

$$f_{X_i}\left(x_i\right) = \int_{-\infty}^{+\infty} \int_{-\infty}^{+\infty} \cdots \int_{-\infty}^{+\infty} f\left(x_1, x_2, \ldots, x_n\right) dx_1 \cdots dx_{i-1} dx_{i+1} \cdots dx_n \tag{2.94}$$

The random variables $X_1, X_2, \ldots, X_n$ are *independent* if their joint distribution function equals the product of the distribution functions of X_i, that is,

$$F\left(x_1, x_2, \cdots, x_n\right) = F_{X_1}\left(x_1\right) F_{X_2}\left(x_2\right) \cdots F_{X_n}\left(x_n\right) \tag{2.95}$$

The mean of the vector $[X_1, X_2, \ldots, X_n]^T$ is a constant vector whose elements are the mean of the variables X_i, $i = 1, \ldots, n$, as follows:

$$\mu_{[X_i]} = \left[\mu_1, \mu_2, \ldots, \mu_n\right]^T \tag{2.96}$$

2.4.9.1 Two-Dimensional Random Vectors with Continuous Components

Let X and Y denote two continuous, random variables. Their *joint distribution function* is

$$F_{XY_i}\left(x, y\right) = \Pr\left(X \le x, Y \le y\right) \tag{2.97}$$

The joint density of X and Y is defined as the function

$$f_{XY}\left(x, y\right) = \frac{\partial F_{XY}\left(x, y\right)}{\partial x \, \partial y} \tag{2.98}$$

Equivalently, the joint density is defined as the function $f_{XY}(x, y)$ satisfying the equation

$$F_{XY}(x,y) = \int\limits_{-\infty}^{x} \int\limits_{-\infty}^{y} f_{XY}(u,v)\, du\, dv \tag{2.99}$$

The *marginal densities* $f_X(x)$ and $f_Y(y)$ can be respectively obtained by using the following formulae:

$$f_X(x) = \int\limits_{-\infty}^{+\infty} f_{XY}(x,y)\, dy \tag{2.100}$$

$$f_Y(y) = \int\limits_{-\infty}^{+\infty} f_{XY}(x,y)\, dx \tag{2.101}$$

In practical terms, Equation 2.100 describes the density of X if Y has never been measured.

2.4.9.2 Bivariate Distributions

In modeling combine harvesting processes, different *bivariate distributions* can be very helpful. The *Farlie–Gumbel–Morgenstern* families of bivariate distributions have the distribution function given by

$$F_{XY}(x,y) = F_X(x) F_Y(y) \left\{ 1 + \delta \left[1 - F_X(x) \right] \left[1 - F_Y(y) \right] \right\} \tag{2.102}$$

where $|\delta| \le 1$ for a bivariate distribution to exist, and the marginal distributions are $F_X(x)$ and $F_Y(y)$. In the case of independence, $\delta = 0$, although included, *correlation* of X and Y variables does not exceed 1/3 in absolute value (Schucany et al., 1978).

Using Equation 2.102, we can derive the *bivariate exponential distribution* with the following form:

$$F_{XY}(x,y) = \left(1 - e^{-\lambda x}\right)\left(1 - e^{-\beta y}\right)\left(1 + \delta e^{-\lambda x - \beta y}\right) \tag{2.103}$$

Based on Weibull distribution (Equation 2.68) and Equation 2.102, the *bivariate Weibull distribution* has the following distribution function:

$$F_{XY}(x,y) = \left(1 - e^{-(\lambda x)^c}\right)\left(1 - e^{-(\beta y)^d}\right)\left(1 + \delta e^{-(\lambda x)^c - (\beta y)^d}\right) \tag{2.104}$$

One can simply note that when the parameters $c = 1$ and $d = 1$, we get Equation 2.103.

2.4.10 Data Regression Analysis

While deterministic models greatly depend on experimental data for ultimate validation, stochastic models attempt to explicitly deal with observed scatter in experimental data. The experiments suggest causes and effects among process variables that help us to understand them and apply our educated intuition to explain the process evolution. Thus, this leads to the development of process-governing equations whose predictions will later be compared to new data that is not part of the original data set. Thus, experimental data is a prerequisite need for formulation, derivation, and validation of stochastic models.

The stochastic model equations can be further extended and interpreted in a deterministic way to generate deterministic mathematical models. The advantages of fitting a stochastic model to experimental data include quantification of uncertainty in the deterministic model's parameter estimates, estimation of the model error rate, and checking the fit of the model nontriviality. However, getting the necessary and valid amount of data is a matter of experience in design of experiments (DOE), modeling, simulation, and optimization of processes.

This section focuses on mathematical data processing techniques as a very important step in developing stochastic models, followed by deterministic models, of processes and systems. Such techniques may be successfully used, beyond the purpose of this book, by scientists and engineers in other engineering specialties.

2.4.10.1 Least Squares Regression

Any variable or parameter in estimation has three quantities associated with it: the *true value*, the *measured value*, and the *estimated value*. The true value, denoted by x, is usually not known, but we aim to find it. The measured value, $\tilde{x}$, is determined with different instrumentation, including sensors, transducers, and data acquisition systems. The difference between the measured value and the true value is called the *observational* or *measurement error* ε. That means

$$\tilde{x} = x + \varepsilon \tag{2.105}$$

Statistically, the measurement error is not a mistake due to the variability (inherent in a statistical process) of the quantity that is being measured or the measurement process and instrumentation.

Let us consider a series of measurements $\tilde{x}$ of a random variable x with a univariate distribution. We need to estimate its values—in other words, find the closest values to the true ones. The *estimated values* are denoted by $\hat{x}$. By similitude with Equation 2.105, we can write

$$\tilde{x} = \hat{x} + \hat{\varepsilon} \tag{2.106}$$

where $\hat{\varepsilon}$ is the *residual* or *fitting error*, which is an observable (i.e., explicitly known) estimate of the statistical error. Based on the above equation, the residual error can be easily calculated once an estimated value has been found.

To determine a mathematical model $\hat{y}(x)$ (i.e., the estimated model), using a series of measurements, one would apply a *regression technique* for finding the *best-fitting curve* to the given set of points with coordinates $(\tilde{x}_i, \tilde{y}_i)$, where $i = 1, ..., n$. If n is big enough, the residual errors (i.e., between the measured values and estimated values) will have a normal distribution with an absolute mean μ_ε and standard deviation ς_ε that can be calculated as follows:

$$\mu_\varepsilon = \frac{1}{n} \sum_{i=1}^{n} \left[\tilde{y}(x_i) - \hat{y}(x_i) \right] \tag{2.107}$$

$$\sigma_\varepsilon = \frac{1}{n-1} \sum_{i=1}^{n} \left\{ \left[\tilde{y}(x_i) - \hat{y}(x_i) \right] - \mu_\varepsilon \right\}^2 \tag{2.108}$$

The *least squares fitting method* involves the application of a mathematical procedure for finding the equation (mathematical model) of the best-fitted curve to the given set of points by minimizing the *sum of the squares of the residuals* (vertical offsets) of the points from the curve. In such a case, the minimum residual error $\varepsilon_{y\min}$ of the estimated values $\hat{y}(x)$ will be

$$\varepsilon_{y\min} = \min \left\{ \sum_{i=1}^{n} \left[\tilde{y}(x_i) - \hat{y}(x_i) \right]^2 \right\} \tag{2.109}$$

The sum of the squares of the residuals is used instead of the absolute values of residuals because this allows the residuals to be considered a continuous differentiable quantity.

Let suppose that the estimated model $\hat{y}(x)$ depends on some parameters (coefficients or exponents) a_j, where $j = 1, 2, \dots$.

Based on mathematical analysis of the functions, the minimum of residual errors ε_y function can be found by solving an equation system of partial derivatives of the error ε_y with respect to each unknown parameter a_j that in analytic form can be written

$$\frac{\partial \varepsilon_y}{\partial a_j} = 0, \, j = 1, 2, \dots \tag{2.110}$$

For linear equations, the above written equation system can be solved explicitly to find the parameters a_j that will be used to generate the curve of the regression model.

For nonlinear equations, one usually cannot solve such a system; therefore, various iterative techniques are used instead, such as *Levenberg–Marquardt*, which is the most widely used iterative procedure, then *Gauss–Newton* and *steepest descent* algorithms. A review of these very interesting techniques is beyond the purpose and size of this textbook. Besides, these methods are used by the best software packages in statistics and mathematics. It is very important to know that nonlinear equation fitting starts with *initial values of parameters* a_{j0} that are established on an educated-guess basis.

Curve fitting to experimental data provides the following:

1. A completely defined equation whose shape and structure are already known for describing the current data, by similitude with a previously studied process
2. An equation that has been proven to describe the data in a better way than other investigated mathematical models

For a linear model that is fitted to the data, the quality of fit is assessed in terms of the *coefficient of determination* value R^2. In nonlinear regression, such a measure is not readily defined due to the need for an *intercept* of the curve (which most nonlinear curves do not have), but one can use a *pseudo-R^2*. For both cases we may write

$$R^2 = 1 - \frac{SS_\varepsilon}{SS_{\tilde{y}}} \qquad (2.111)$$

where the sum of the squares of residuals SS_ε is

$$SS_\varepsilon = \sum_{i=1}^{n} \left[\tilde{y}(x_i) - \hat{y}(x_i) \right]^2 \qquad (2.112)$$

and the total sum of the squares is

$$SS_{\tilde{y}} = \sum_{i=1}^{n} \left[\tilde{y}(x_i) - \mu_{\tilde{y}} \right]^2 \qquad (2.113)$$

The mean of the experimental data of the dependent (response) variable is

$$\mu_{\tilde{y}} = \frac{1}{n} \sum_{i=1}^{n} \tilde{y}(x_i) \qquad (2.114)$$

The coefficient of determination takes values as expressed by the following inequality: $0 < R^2 < 1$, with the value 1 for a perfect fit.

Properly formulation and application of a mathematical model for fitting to experimentally acquired data require an intimate knowledge of the process and basic principles underlying the phenomena that compose the process, coupled with a very good knowledge of mathematical functions and understanding of estimation theory.

Conversely, regression analysis of experimental data requires *a priori* decisions regarding which are the dependent and independent variables of the process, the range of the variables of interest, the size of the set of points, the frequency of collecting data, and the necessary measurement accuracy. This activity requires knowledge of the DOE.

A set of important rules and practices for curve fitting, determined as a result of the author's modeling activity, are discussed in Section 2.4.10.3.

2.4.10.2 Weighted Least Squares Regression

The method of weighted least squares regression of experimental data takes into consideration that different data measurements or groups of measurements have been done with unequal precision; in such cases, weighted least squares regression implies selecting and assigning different weights to data so that the weights shall be inversely proportional to the estimated data accuracy. Practically, there are no measurements with zero error or infinite error; that means the weights are positive real numbers, selected from a finite range, and associated with each data point in the fitting criterion. The weight for each measurement data is given relative to the weights of the other observations.

Supposing that the best-fitting equation (model) has been identified, the question arises as to what is the proportion of intuitively selected weights to the weights selected based on prior statistical analysis of measurement errors.

A statistically optimal (maximum likelihood) choice for the weights' values is the reciprocal of the measurement error variance (Beichelt, 2006). That is, the weights w_i for the independent variables x_i are inversely proportional to the variance at each level of the independent variable:

$$w_i = \frac{1}{\sigma_{x_i}^2} \qquad (2.115)$$

This higher level of data processing offers an ability to perform regression analysis of data of a varying quality. This method is also called the *robust fitting* method because it offers the possibility of knocking out outlying data points.

For a deeper review of this subject, refer to other statistics books as indicated at the end of this chapter.

2.4.10.3 Rules for Regression Analysis

In curve-fitting applications, some requirements are to be met, as follows:

- The practitioner has an intimate knowledge of the field in which the estimation problem is embedded.
- Experimental data acquired for each couple of dependent–independent (response–explanatory) variables exists for a *minimum set of points* $(x_i, \tilde{y}_i)$ to support the proper fitting of a known curve equation or to suggest the proper shape of a curve to be fitted. We recommend at least six data points because:
  - Two points suggest a straight line.
  - Three points suggest a curve, but due to possible measurement errors, there will be no safe indication about a minimum or maximum.
  - Four or five points may suggest a spline, a straight line, or a curve.
  - Six points is the minimum number to ensure a good value of the coefficient of determination, if the fitting model was properly chosen. Discussion: If a dependent variable is a function of three independent variables, a number of $6^3 = 216$ experimental tests must be done. That will help in developing a very good mathematical function of three unknowns.
- Experimental data points cover the entire region of the fit. This is required to help find *maximum*, *minimum*, and *inflection points* of the fitting curve. Ideally, the data points should be isotropically distributed over the entire range; that is, there is no directional bias in regard to the placement of the points.
- Accuracy of the data is at least satisfactory relative to the variable variation range as well as to the variable weight relative to the weights of the other variables.

Depending on the process type being modeled and the data points' compliance with the above-mentioned requirements, the mathematical model to be fitted to experimental data must satisfy certain conditions as follows:

- Acceptance of zero values for curves that start in or pass through the (0, 0) point.
- Exclusion of "division by zero" errors.
- Asymptotic tendency of the dependent variable to a value that makes sense and can be determined by calculation or in an intuitive way.
- Flexible shape in regard to positioning of maximum, minimum, and inflection points.
- Minimum number of parameters. The more parameters, the more difficult is the physical interpretation of the equation structure.
- Correct physical meaning of the parameters, especially at the extreme points of the independent variable. For example, in many papers, the authors use the second-order polynomial, $y(x) = ax^2 + bx + c$, for data fitting. That means, when $x = 0$, the value $y = c$ shall allow a correct meaning in regard to the physical process. We strongly discourage the use of this model; instead, use combinations of exponential functions as shown in numerous examples throughout this book.
- Reasonable estimates of initial parameter values; otherwise, there will be no convergence in the fitting process.
- Reasonable values of residuals (vertical distance between the measured data point and its corresponding point on the fitting curve). This is when a limited number of outliers can be identified based on their relatively high residual values. Particular attention should be paid to eliminating possible outliers before repeating the regression process.
- Reasonable values of estimated parameters, for example, within the [–5, 5] interval; unless some physical interpretation can be proven, getting estimated parameter values comparable with $10^{\pm3}$ shall be avoided by changing the fitting function structure or definition.
- Good fitting result ($R^2 > 0.9$). High values of the coefficient of determination further help in getting a good fit of a function of multiple variables.
- Very small standard error value as well as a 95% confidence interval relative to each parameter's estimated values.

Good curve-fitting software computes and displays graphically all data points, fitted curves, residuals, R^2, estimated standard values, errors, and 95% confidence intervals, along with comments to help the researcher make further decisions in system/process mathematical modeling.

2.5 System Simulation and Optimization

2.5.1 Introduction

In its broadest sense, *system simulation* is a convenient development tool for testing, evaluating, and suggesting improvements of an existing or envisioned system/process, under various physical configurations and functional settings of interest, over a compressed period of time.

System simulation can be classified in simulation based on downscaled systems (*traditional simulation*), *computer simulation*, and simulation based on *training simulators*. All three materialization forms of simulation aim to reproduce similar systems (physical, virtual, and combined, respectively) for the purpose of experimenting and investigating phenomena and processes associated with the system of interest. In regard to *simulation of combine harvester* processes, the most important is a computer simulation, accompanied by training simulation, with simulators that emulate the automatic control functions of the combine subsystems and help combine operators become accustomed to the machine. The related subject of the book is centered mostly on computer simulation of combine harvesters.

Computer simulation is the process of emulating and experimenting with a physical system through development of a computer application using a mathematical model, data input/output interfaces, visual display, and animation features, followed by running the virtual system according to real and diverse functioning and environment conditions of the system of interest.

The *purposes of system/process simulation* are

- Gaining insight into the system operation/process performance and capability
- Testing system operation/process performance for finding errors/functional limitations
- Developing and implementing new concepts, design, and operating procedures of the simulation model
- Acquiring knowledge to be implemented in a (new) real system/process
- Saving product development time and costs (in most cases)

Dynamic, complex systems, such as combine harvesters, and their processes are difficult to understand, develop, and integrate without extensive testing, analysis, and results implementation. Once a simulation model becomes operational, it may be tested for a variety of extreme working and environmental conditions, while the performance is continuously monitored. Testing of a simulation model helps in the detection of deviations in design, improper functioning of the system, and process performance limitations. Testing also adds value to the product by compliance with customer requirements or expectations (requirement-based testing). Based on the knowledge acquired by simulation, changes in design, operation procedures, and resource policies may be applied to the real system to be developed.

Running the simulation model on a computer offers the advantage of compressed-time experimentation, repetitive runs, and real cost savings in comparison with extended time and costs of material and labor for real system testing. Then, simulation helps eliminate unforeseen bottlenecks and constitutes a research tool for further optimization of the system and associated process. On top of these advantages, using an animation with continuous display of system operating results helps strengthen model credibility and product development feasibility.

System/process simulation also has disadvantages that are related to the expectations associated with the simulation process. Such *possible disadvantages* refer to

- Questionable veracity and accuracy of results.
- System/process analysis limitations due to simplifying assumptions or hypotheses formulated for simulation model development.
- Inefficiency of simulation process if its results are not implemented on the real system. That means simulation alone does not solve the problems.

The veracity and accuracy of system/process simulation results depend on the accuracy of modeling of the real system, input data, and output interpretation. Many times, developing a mathematical model requires data from experimenting on existing real models. Extrapolation of the results in simulation package development may also have a major impact on simulation results.

Let us consider a simulation mathematical model of a real, continuous-time system, as represented in Figure 2.1 and described by the nonlinear system of Equations 2.5 and 2.6, with n parameters p_i, where $i = 1, \ldots, n$ are used. Assumptions about the structure of derivative vector $f(t, x, u; p)$ and parameters p_i are tested and refined using experimental data acquired from the system testing. Tuning this simulation model is an iterative process, as shown in Figure 2.15. An essential component of *model validation* and the *parameter identification* process is minimization of an *error function* $\varepsilon(f, p)$, which quantifies the difference between outputs of the actual system and those of the simulation model (Klee and Allen, 2011).

System/process optimization is the process coupled with a method of identifying certain design and functional parameters of a system and its associated process, with the purpose of *simultaneous minimization and maximization* (i.e., optimization) of certain performance indices, without violating the other constraints. The most common performance indices of a combine harvester are discussed in the Chapter 1 (Section 1.8); here are a few: throughput of material (grains and MOG), proportion of separated grains, proportion of grain losses, grain damage, and specific processing power requirements. System/process optimization takes into consideration the following:

- Specific properties of material that is to be processed.
- *Specific design parameters* of the equipment, whose values influence substantially the system function (e.g., distance between every two consecutive bars or rods of a threshing concave, crankshaft throw radius, etc.).
- *Functional (process) parameters* (e.g., angular speed of a shaft, combine speed, distance between rotor and concave, cleaning sieve angle, etc.).
- *Operating procedures* and *process control algorithm.* The operating procedures specify the succession of process transition phases and associated settings, as well as the correlation of steady-state process parameters. The control algorithm facilitates the application of certain operating procedures and monitors the output (performance) parameters versus the input and state parameters of the system.

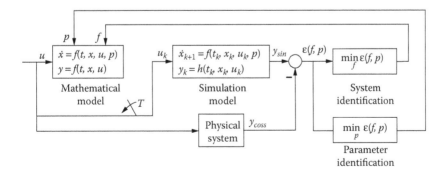

FIGURE 2.15
Iterative process of simulation model validation.

2.5.2 Optimization Criteria and Algorithms

The task of optimization identifies the optimal choice that ensures the minimum or maximum value of at least one *optimization criterion* (single-objective problem) or more *criteria* (multiobjective problem). An optimization criterion is a system/process performance index whose value is established by a mathematical function or inequality. System/process optimization implies the existence of a suitable predictive mathematical model that can simulate the behavior of the real system. This is materialized into a set of equations and inequalities that are called *constraints*.

An *optimization model* implies an intrinsic part of a simulation model coupled with a *mathematical optimization method* transposed into a *computer algorithm*, as shown in Figure 2.16. Understanding fundamental properties of optimization problems and algorithms includes key elements referring to the existence of mathematical solutions, a convergence rate, and the stability of the algorithm. The solution of a typical optimization problem in the field of combine harvesters must simultaneously satisfy *multiobjective requirements*, that is, optimize multiple performance criteria.

Before optimization is attempted, the problem at hand must be properly formulated.

Suppose we already have experimental data from the testing of an existing working unit (the system), which performs a certain process (e.g., cleaning unit/grain cleaning process). An *optimal cleaning unit* produces a perfect cleaning process of a relatively high output, with maximum cleaned separated grain and minimum losses (including minimum not-rethreshed ears), while preserving crop quality (minimum damaged grain) and minimizing the specific power requirement. Thus, the system/process performance is assessed by a series of performance criteria denoted by functions that have to be expressed mathematically and optimized using a certain optimization method.

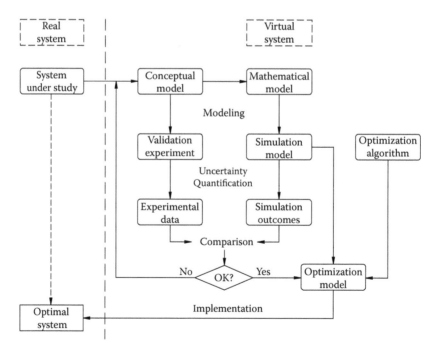

FIGURE 2.16
Integration of system modeling and simulation, and implementation of optimization results.

Using the process theory, findings from literature, and available experimental data, complemented by engineering innovation and skills in mathematics, one can finally build a deterministic mathematical model whose functions are continuously differentiable (first and second derivatives).

The system receives a series of inputs, its process is characterized by state variables, and it generates a series of outputs. Regarding the cleaning unit, the inputs are represented by specific crop properties, design parameters (e.g., crankshaft throw radius, sieve length, sieve opening shape, size, distribution, proportion of total opening area, etc.), and a series of functional parameters (e.g., crankshaft revolutions per minutes, sieve angle, airflow direction, velocity, distribution, etc.).

The state parameters in this case would be the distribution of segregated grains within the MOG layer, grain segregation speed, material speed over the sieve, percentage of recovered unthreshed grains, and so forth.

Each objective function (performance criterion function) may depend on a large number ρ of variables u_i, $i = 1, ..., \rho$, that can be expressed as a row vector:

$$u = \left[u_1 \; u_2 \; ... \; u_\rho \right]^T \qquad (2.116)$$

A performance criterion F_k, $k = 1, ..., \psi$, must be derived in terms of ρ parameters as follows:

$$F_k = f_k \left(u_1, u_2, ..., u_\rho \right) \qquad (2.117)$$

Since we have a number ψ of performance criteria, that is, a number of distinct functions of u that need to be optimized, we will construct a vector

$$F(u) = \left[f_1(u), f_2(u), ..., F_\psi(u) \right]^T \qquad (2.118)$$

The problem has to be solved by finding a point u^* such that

$$F(u^*) = 0 \qquad (2.119)$$

Sometimes, the point u^* may not exist, but an approximate solution that simultaneously reduces all functions or inequalities $f_k(u^*)$, $k = 1, ..., \psi$, close to zero.

Multiobjective optimization often means a compromise among contradictory criteria (e.g., maximization of the material throughput and minimization of the proportion of unthreshed grains). In such a case, additional criteria may need to be introduced, such as a weight-based quantification of the importance of each optimization criterion. However, solving the problem may result in not just one distinct solution (a maximum or a minimum), but a few global and local optima (Figure 2.17). This happens because the mutual influences between optimization objectives can be complicated and are not always obvious.

Generally, optimization algorithms can be divided into two main categories: deterministic and probabilistic algorithms.

A *deterministic optimization algorithm* is based on the state-space search of the solution based on deterministic model equations if the dimensionality of the search space is not very high.

The taxonomy of *probabilistic optimization algorithms* identifies two basic groups of algorithms, Monte Carlo algorithms and evolutionary algorithms (Weise, 2009), as shown in Figure 2.18. Since an exhaustive treatment of these algorithms is beyond the size of this

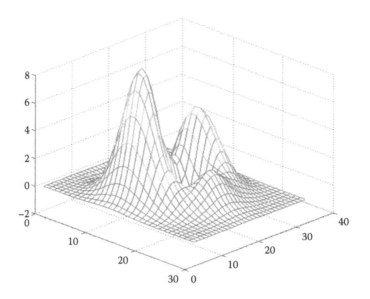

FIGURE 2.17
(See color insert.) Global and local optima of a two-dimensional function.

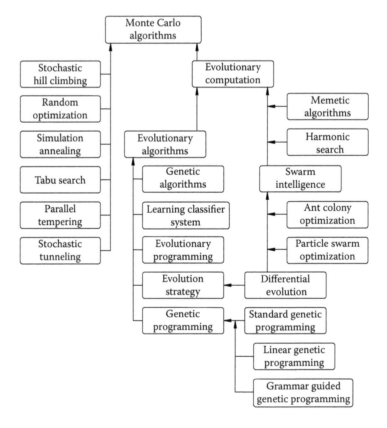

FIGURE 2.18
Classification of probabilistic optimization algorithms. (Adapted from Weise, T., *Global Optimization Algorithms: Theory and Applications*, 2nd ed., 2009 http://www.it-weise.de/projects/book.pdf.)

book, we will briefly describe a few of them that, in our opinion, can be very well used in process optimization of combine harvester processes, as combines have to work with many types of crops in different environment conditions and work quality standards. Out of all these optimization techniques, the genetic algorithms' development and use are particularly related to the author's research experience.

2.5.2.1 Evolutionary Algorithms

An *evolutionary algorithm* is a subset of evolutionary computation algorithms. An evolutionary algorithm is a population-based optimization algorithm that emulates the natural evolution of species through reproduction, mutation, migration, recombination, and selection, that is, evolution, to iteratively refine a set of *candidate solutions* for the optimization problem at hand. The *fitness function* ascertains the environment to which the candidate solutions (certain individuals of the population) are fitted.

The evolutionary algorithm method has the advantage of a "black box" character that implies only a reduced number of assumptions about the underlying objective functions. Usually, an evolutionary algorithm deals with multiobjective optimization criteria by following a cyclic optimization process, as represented in Figure 2.19.

The initial population of individuals (solution candidates) is randomly generated from a certain range for each of them, for example, the revolutions per minute between 20 and 38 m/s, the total throughput from 1 to 15 kg/s, and so forth. The objective functions are then evaluated for these individuals. Individuals are selected according to their fitness for offspring generation. Then, the parents are recombined and some offspring undergo mutations with a certain probability. The fitness of the offspring is then computed and the cycle can be repeated. The optimization cycle is followed until the termination criteria are met; thus, the best individuals represent the solution of the optimization problem.

There are some significant differences between the evolutionary algorithms and other search and optimization methods. The evolutionary algorithms distinguish themselves from other search and optimization methods by

- Parallel searching a population of potential solutions, not just a single parameter
- Not requiring derivative knowledge, but only the objective functions and their satisfactory fitness levels

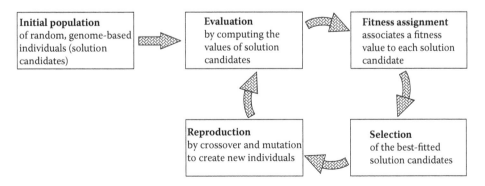

FIGURE 2.19
Optimization cycle of an evolutionary algorithm.

- Using probabilistic transition rules, not deterministic ones
- Being straightforward methods to apply because no restrictions for the objective functions need to be imposed
- Delivering very good results when the objective function definitions are weight balanced well
- Potentially providing more solutions to a given problem and providing a convenient choice to the user

The step sequence of an evolutionary algorithm is given in a pseudocode fashion as follows:

```
BEGIN
      F_i ↓ INPUT n objective functions (i = 1, ..., n);
      S_p ↓ INPUT initial population size;
      t ↓ 0;
      Pop ↓INITIALIZE population of randomly chosen solution candidates;
      EVALUATE each solution candidate;
      WHILE termination criterion () DO
            Mate ↓ SELECT mating pool (Pop, S_p, F_i);
            Offs ↓ CROSSOVER pair of parents for offspring generation;
            MOff ↓ MUTATE randomly a certain proportion of offspring (Offs,
            Pr);
            t ↓ t + 1;
            Pop ↓ FORM new population (Offs-MOffs, MOffs);
      RETURN Extract Phenotypes (Extract Optimal Set (Pop));
END
```

During selection, the close-fitted individuals have a better chance of being selected for the new population. The evolution of individuals (solution candidates) is possible as an adaptation process to the environmental requirements represented here by the objective functions. Matching these objectives increases the viability, reflected by the best-fitted offspring (and possibly some individuals from the initial population).

2.5.2.2 Genetic Algorithms

Genetic algorithms are a subclass of the evolutionary algorithm group. Genetic algorithms are parameter iterative search techniques that rely on analogies to the natural evolution of species and individual selection based on Darwin's principle of the *survival of the fittest*. Genetic algorithms work simultaneously on a set (population) of potential solutions (individuals). The degree to which solutions meet certain performance requirements is evaluated and used to select *"surviving"* individuals that will *"reproduce"* and generate a new population. Then individuals will undergo alterations similar to the natural genetic mutation and crossover.

From a programming perspective, in genetic algorithms, the elements of search space are binary strings or arrays of other elementary types. As represented in Figure 2.20, the genotypes are used in the reproduction operations, whereas the values of the objective functions are computed on the basis of phenotypes that are obtained via the genotype–phenotype mapping (Weise, 2009).

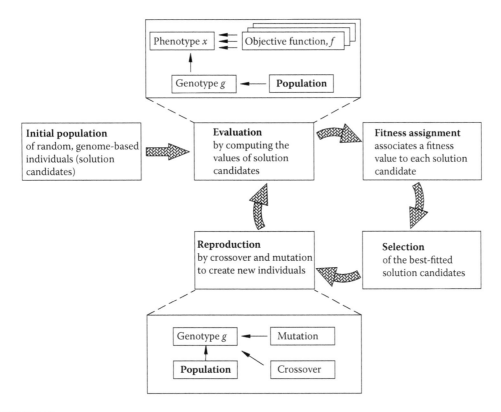

FIGURE 2.20
Optimization cycle of a genetic algorithm. (Adapted from Weise, T., *Global Optimization Algorithms: Theory and Applications*, 2nd ed., 2009 http://www.it-weise.de/projects/book.pdf.)

Genetic algorithms start with an initial randomly chosen population of potential solutions (solution candidates). Each individual is characterized by a genetic representation (chromosome). That means every set of parameters is coded to allow easy application of genetic operators (crossover and mutation). Evaluation of each chromosome in each generation gives the measure of its fitness. The fittest individuals are selected to reproduce, that is, to form a new generation (for the next iteration). Genetic operators are further applied to alter the genetic information, and hence introduce new potential solutions into the search scheme.

The selection scheme is biased toward high-performance solutions. A careful selection of genetic algorithm structure and parameters guarantees a good chance of reaching the global optimum solution after a reasonable number of iterations.

In Chapter 5, we describe in detail an example of a *genetic algorithm for axial threshing unit/process* optimization, including the source code developed by the author.

2.5.2.3 Genetic Programming

Genetic programming can be defined as a set of evolutionary programming algorithms that breed programs, algorithms, or similar constructs (Weise, 2009). Genetic programming is an optimization technique because it automatically generates computer programs to perform specific tasks (Koza, 1992; Montana and Beranek, 2002). The genetic programming method is based on using a genetic algorithm to search through a space of possible

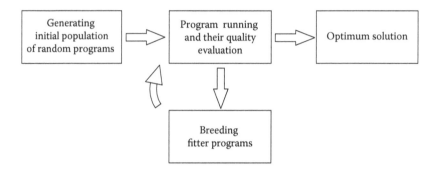

FIGURE 2.21
Optimization cycle of a genetic programming algorithm.

computer programs for the optimal one to perform a certain task. Basically, genetic programming is an evolutionary computation technique that automatically solves problems without requiring the user to *a priori* know or specify the form or structure of the solution (Poli et al., 2008). In comparison with genetic algorithms where the population is composed of individuals (with assigned chromosomes), in genetic programming algorithms, a population of computer programs evolve toward hopefully better populations of programs (cf. Figure 2.21). The genetic programming algorithm estimates how well a program makes predictions by running it, and then compares its behavior with some ideal.

As an example, genetic programming can be used for solving a multiple regression problem that requires finding a multiple-valued function, in symbolic form, that fits an available set of experimental values of dependent variables associated with particular given values of the independent variables.

As another example, in regard to this book's objective, *developing a combine process control* may require identifying a scheme based on receiving information from sensors about the state of various combine units, and then, through genetic programming, using this information for selecting an optimum sequence of actions that change the state of combine working units and their associated processes.

For further information on the subject of genetic programming, see Koza (1992) and Poli et al. (2008).

2.5.2.4 Memetic Algorithms

To solve an optimization problem, the principle of *memetic algorithms* is to simulate the social evolution (where *behavioral patterns* are passed on in *memes*), unlike the genetic algorithms that simulate the natural evolution by encoding the phenotypic features in genes (Weise, 2009).

Thus, the memetic algorithms are a special kind of genetic algorithms with a local search made by each of the individuals to improve their fitness to the objective functions' value. Thus, higher-quality individuals are selected for a new generation of solution candidates. Then, like in genetic algorithms, once the parents have been selected, their chromosomes are combined and the crossover operators are applied to generate new individuals. These individuals will be enhanced based on a *hill-climbing* local search technique whose corresponding algorithm is based on a loop that continuously moves toward increasing quality value. The definition of local search operators applied within a memetic algorithm may also be changed during the course of optimization. This type of memetic algorithm was named *coevolving memetic algorithm*. It maintains two populations: one for genes encoding for solution candidates and one for memes encoding for local search operators to be used within the memetic algorithm. It follows that even if a population contains only local optima, changing the local search operators may provide a means of progression in addition to recombination and mutation results (Smith, 2007).

2.6 System/Process Control

Combine harvester modeling and simulation efforts are directed toward supporting specific applications from a system of systems perspective aimed at optimization, design, testing, evaluation, and *control* of combine working processes during the harvesting of a multitude of crop types and sorts. The *control theory* is very well elaborated, and the diversity of control systems seems to make understanding them difficult, as well as our choice for development of a common analysis and design strategy. While a thorough review of control systems engineering is far beyond the scope and extent of this book, in the following sections we attempt to review the basic concepts and theory of control systems engineering to facilitate their integration with modeling, simulation, and design, and to ease their application by the intended reader. Then, in each following chapter, we will highlight the connection between the emerging theory, modeling, and design of combine harvesters with various aspects of the control design.

2.6.1 Introduction in System/Process Control

A *control system* is a structured assembly consisting of *subsystems* and *processes* (*plant*) that was built with the purpose to achieve and maintain, through deliberate guidance, a desired *output* of the plant, at a certain *performance* level, given an *input*, which must be specified in a convenient and proper form.

There are a few important advantages offered by control systems:

- Accuracy/precision enhancement
- Power amplification
- Convenience of input form
- Compensation for disturbances
- Remote control

Based on the linear system theory, the analysis of a system of interest assumes a *cause–effect* relationship for the components of the system deriving from the processing of the input variables by the system components to generate corresponding output variables. Thus, each component (subsystem) is represented by a block with inputs and outputs. A *block diagram* representation of component relationships consists of unidirectional, operational blocks, so that each block illustrates the transfer function of the variable of interest. The output from one component must be a compatible input for the adjacent component. When multiple system components are interconnected, additional schematic elements, such as *summing junctions* and *pickoff points*, are required. For generating and simplifying block diagrams, the basic relationship displays in Table 2.1 are very useful. Since the transfer functions represent linear systems, the multiplication operation is commutative.

TABLE 2.1

Equivalence of Block Diagrams

Description	Actual Diagram	Equivalent Diagram
1. Summing junction	x_1 — x_2, x_3 (summing junction)	$x_3 = x_1 + x_2$
2. Cascade (series) connection	$x_1 \to \boxed{G_1} \to \boxed{G_2} \to x_2$	$x_1 \to \boxed{G_1 G_2} \to x_2$
3. Parallel connection	$x_1 \to \boxed{G_2}$, $x_1 \to \boxed{G_1} \to$ (summing) $\to x_2$	$x_1 \to \boxed{G_1 + G_2} \to x_2$
4. Shifting signal: pickoff point	$x_1 \to \boxed{G} \to x_2$, $\downarrow x_2$	$x_1 \to \boxed{G} \to x_2$, $x_1 = x_2/G$
5. Shifting signal: application point	$x_1 \to \boxed{G} \to$ (summing, x_2) $\to x_3$	x_2/G, $x_1 \to$ (summing) $\to \boxed{G} \to x_3$
6. Reduction of feedback loop	$x_1 \to$ (summing) $\to \boxed{G} \to x_2$, $\boxed{H}$	$x_1 \to \boxed{\dfrac{G}{1+GH}} \to x_2$

A negative sign of incoming signal at a summing junction means that the signal is subtracted. Let us briefly discuss the current block diagram shown at item 6 in Table 2.1. The incoming feedback signal at the junction function equals $-Hx_2$. Thus, the resulting input signal of block G is $x_1 - Hx_2$. Because $x_2 = G(x_1 - Hx_2)$, we get

$$x_2 = \frac{G}{1+GH} x_1 \qquad (2.120)$$

In the above example, the function H is a *feedback transfer function* of the output x_2.

In terms of their internal architecture, the control systems are categorized as being either open-loop or closed-loop systems. An *open-loop control system* relies only on *preset processing operations* of the input (*reference*) command(s) to deliver the corresponding output (*controlled*) *variable*(s), as represented in Figure 2.22. The *controller* of the process receives the input commands through an *input module* (*transducer*) and delivers its commands to the process (*plant*). Because an open-loop control system lacks the *feedback*, any possible *disturbances* may interfere with the controller and process outputs via corresponding *summing*

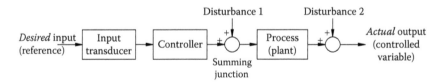

FIGURE 2.22
Block diagram of a generic open-loop control system.

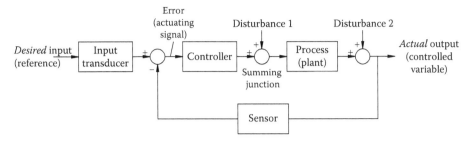

FIGURE 2.23
Block diagram of a generic closed-loop control system.

junctions. The open-loop system does not take corrective actions to alleviate undesirable influences induced by disturbances.

In a *closed-loop control system* the output is fed back through a functional block with a feedback transfer function (Figure 2.23). An *output transducer,* or *sensor,* measures the output and converts it into a *feedback signal* whose form is accepted by the controller. The signal resulting from the difference between the reference signal and the feedback signal is called the *error* or *actuating signal.* If the error is zero, the process response is the desired response; otherwise, any change in the output causes a change of the actuating signal toward correcting the output value.

Closed-loop systems also compensate for disturbances by feeding back the output response through a feedback path, and comparing that response to the input at the summing junction.

In comparison with open-loop systems, closed-loop control systems have the advantages of greater control accuracy, lower sensitivity to disturbances and noise, and greater flexibility by adjusting the *gain (amplification)* in the loop. However, they are more complex, that is, expensive, and require attention to possible issues of *response stability.*

In *time domain,* if we want to represent the *state-space model* expressed by Equations 2.29 and 2.30, we apply the Laplace transformation to each vector component, which, assuming zero initial conditions, will lead to the following equation system:

$$s X(s) = F X(s) + B U(s) \tag{2.121}$$

$$Y(s) = H X(s) + D U(s) \tag{2.122}$$

Solving Equation 2.121 for *X(s)*, we get

$$X(s) = (s I - F)^{-1} B U(s) \tag{2.123}$$

where *I* is the identity matrix.

Thus, solving for *Y(s)* yields

$$Y(s) = \left[H (s I - F)^{-1} B + D U(s) \right] \tag{2.124}$$

Now, we can develop the control system block diagram shown in Figure 2.24, where the feedback path corresponding to matrix *F* represents a *natural feedback control.* The path corresponding

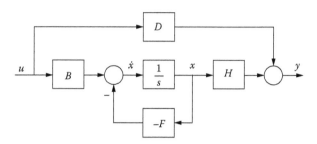

FIGURE 2.24
Block diagram of a system state-space model.

FIGURE 2.25
Actual and desired path of a combine harvester in the field.

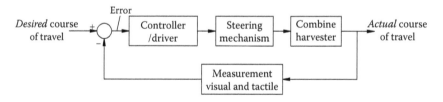

FIGURE 2.26
Block diagram of combine steering control system.

to D is called a *feedforward path*. In *feedforward control*, unknown, measured inputs are used to generate control signals that can reduce errors due to possible variations in these inputs.

Let us consider the *steering control system* of a combine harvester moving in the field, as shown in Figure 2.25. The desired course is compared with a measurement of the actual course to find the error, as displayed by the control system block diagram of Figure 2.26. The measurement is obtained by a visual or tactile feedback.

2.6.2 Control System Performance

The primary goal in designing a control system is to build a system that is highly sensitive to the control input (*sensitivity* feature), achieves the desired output level as fast as possible (*speed of response* feature), and maintains it without, or with a relatively small, variation (*stability* and *steady-state error* features). That is, the system has to have a low sensitivity to noise, external disturbances, modeling errors, and parameter variations (*robustness* feature), as well as a reduced coupling among system variables (*cross-sensitivity* or *dynamic coupling* feature). For system design and performance assessment, the above specifications have to be indicated in a quantitative manner.

Two commonly used specifications in the time-domain design of control systems are the *speed of response* and *degree of stability*. If the system is stable, the response to a specific input signal indicates several measures of the system performance. In terms of response, conventional performance specifications are given by an oscillatory (underdamped) system to a unit step input, as shown in Figure 2.27. From the step response, the following performance specifications can be derived:

- Rise time: Time required for the response to pass through the steady-state value.
- Peak time: Time at which the response reaches its first peak value.
- Peak magnitude: Response value at the peak time.
- Settling time, 2%: Time required for the response to settle within 2% of the steady-state value.

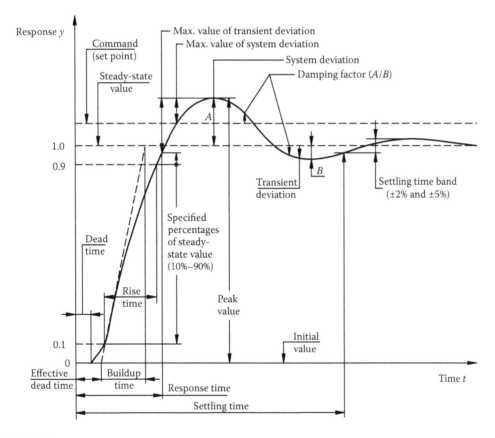

FIGURE 2.27
Conventional performance specifications used in the time-domain design of control systems.

If the system does not settle to its steady-state value, it is considered unstable; that is, the settling time can be considered a relative measure of stability.

Besides the step input, two other standard test signals are commonly used: the ramp input and the parabolic input. The ramp signal is given by the integral of the step input, and the parabolic signal is represented by the integral of the ramp input.

The output response of a control system is the sum of two components: the *forced response* (*steady-state response*) and the *natural response*. The amplitude of the forced response is determined by a *pole of the input function*. The amplitude of the natural response is determined by a *pole of the transfer function*. The poles of a function are the roots of the characteristic equation, which is the function denominator polynomial set equal to zero. The roots of the numerator polynomial are called *zeros*. Both poles and zeros are critical frequencies, and they are represented in an *s plane* (σ, $j\omega$).

In the following, we briefly discuss the performance of a single-loop-feedback, *second-order system* whose output, for a unit step input, is

$$Y(s) = \frac{\omega_n^2}{s\left(s^2 + 2\zeta\omega_n s + \omega_n^2\right)} \tag{2.125}$$

where ω_n, the natural frequency, is the frequency of oscillation of the system, and ζ is the damping ratio.

The transient output of the Laplace transform (2.125) is

$$y(t) = 1 - \frac{1}{\sqrt{1-\zeta^2}} e^{-\zeta\omega_n t} \sin\left(\omega_n t \sqrt{1-\zeta^2} + \cos^{-1}\zeta\right) \tag{2.126}$$

The responses of this second-order underdamped system for various values of the damping ratio ζ are shown in Figure 2.28. The code for the MATLAB m-file that generates

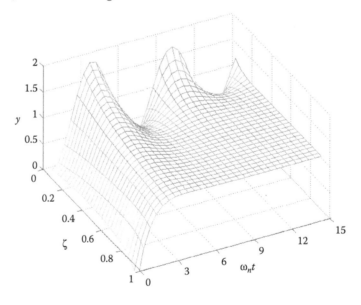

FIGURE 2.28
(See color insert.) Transient responses of the second-order underdamped system for different damping ratio values.

```
% Figure 2.29 code
% Transient responses of a second-order, under-damped system
clear all;
x = (0.03:0.03:0.9);
y = (0.5:0.5:15);
% xx is the damping ratio matrix
% yy is the matrix of natural frequency and time product
[xx,yy] = meshgrid(x,y);

b = sqrt(1-xx.^2);
teta = 1./cos(xx);
% Z is the system output
Z = 1-(1./b).*(exp(-xx.*yy)).*sin(yy.*b+teta);

% Graph settings
set(gcf,'Color',[1,1,1]);
set(gcf,'InvertHardCopy','off');
mesh(xx,yy,Z);
axis([0 1 0 15 0 2]);
colormap(hsv);
set(gca,'FontSize',[14]);
set(gca,'ZTick',[0 0.5 1 1.5 2]);
set(gca,'XTick',[0 0.2 0.4 0.6 0.8 1]);
set(gca,'YTick',[0 3 6 9 12 15]);
```

FIGURE 2.29
MATLAB m-file code for the response of a second-order system to a unit step input.

the surface plot from Figure 2.28 is given in Figure 2.29. As the damping ratio decreases, the response is more oscillatory. The control system designer may select several alternative performance measures from the transient response of the system. The reader should reference the numerous books on control systems engineering, a few of which are listed at the end of this chapter.

2.7 General Simulation and Control Model of a Combine Harvester

A combine harvester working in the field is a variable-mass terrain vehicle, with moving parts and variable positions of the mass center, that must follow a certain path and perform various, interrelated processes of the plants and their components, while achieving specified quality performance indices. There is an increasing need for the automatic control of the machine while working in the field due to higher productivity and quality performance requirements, along with eliminating the differences in skills and performance of combine operating individuals. On the other hand, autonomous vehicle systems form a new, growing area that promises to allow humans remote control of them in various locations while performing different tasks.

This section has two main purposes: (1) to describe the general development concept, structure, and block diagrams of a simulation package of combine harvester motion and task in the field, and (2) to establish a guiding simulation frame to be followed throughout the book and customized by the practitioner according to the specific needs of the project of interest.

The conceptual framework of the simulation model is formulated to meet key objectives, such as modularity, flexibility, and portability. A hierarchical modular structure has been defined that is highly flexible such that it can be adapted and organized to meet specific simulation modes and combinations of modes, such as system behavior analysis and prediction, testing and evaluation, personnel training, and software and hardware simulation in the loop.

For simulation purposes and advanced control, no matter whether the combine is operated by a simulation operator or controlled by a (GPS-like) based control system, a system manager shall provide *a priori* all necessary data to fully characterize the vehicle, its path in the field, its processing systems, the crop to be harvested, and field location coordinates.

The *general simulation package* (Figure 2.30), adapted from Perhinschi et al. (2010), is composed of an *input* module, a *nucleus* module, and an *output* module and can be managed by two categories of personnel: the human manager and the operator (user). The simulation manager should not be confused with the other virtual managers or the human manager. For a given combine harvester, the simulation manager provides the complete menu of compatible simulation options based on the operator's options and data availability, processes the data, and distributes it to the nucleus module.

The *input module* includes the following submodules: *operator's interface, system data files, and simulation manager's input* (Figure 2.31). The operator uses the operator interface to interact with the system for actual data input, simulation definition, and initialization. The human manager provides all general vehicle and working units' data files, which are necessary for setting the simulation configuration for a particular combine harvester. The simulation manager is a software piece that performs data identification, processing, and distribution, followed by the setting of the simulation configuration.

The *auxiliary module* is composed of submodules representing systems whose action may be considered passive in nature, replaceable, and somehow redundant. The auxiliary module includes the following submodules: the *human operator model*, the *precision harvesting module*, and the *terrain model* (Figure 2.32). The human operator controls certain functions of the combine harvester by commands that can be recorded and further used in simulations with a mathematical model of a real operator. The precision harvesting module takes into consideration the variability of crop distribution and crop yield in space and time, as well as the vehicle motion models and tracking methods. The submodule of *path planning* in the field computes the desired trajectory subject to an optimization process

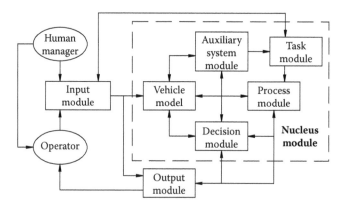

FIGURE 2.30
Block diagram of general simulation package. (Perhinschi, M.G., M.R. Napolitano, and S. Tamayo. Integrated simulation environment for unmanned autonomous systems: Towards a conceptual framework. *Modeling and Simulation in Engineering*. Article 736201, 2010.)

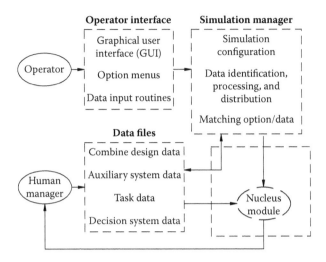

FIGURE 2.31
Block diagram of the input module.

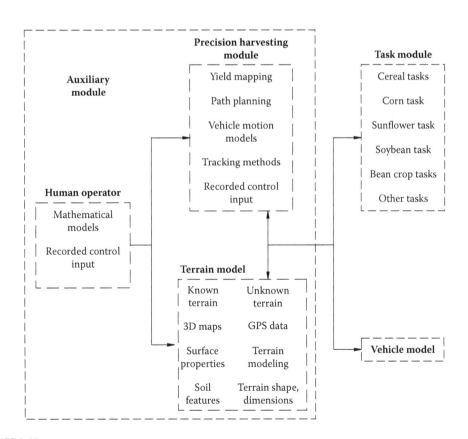

FIGURE 2.32
Block diagrams of the auxiliary module and the task module.

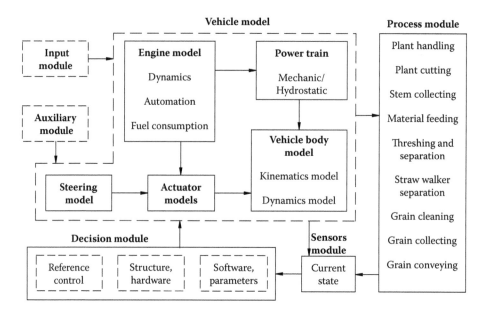

FIGURE 2.33
Block diagrams of the vehicle model, process module, and decision module.

(e.g., minimization of the number of turnings) unless the path is not imposed by the operator. Data from the terrain model, such as obstacles and interdiction areas, modeled as path constraints, is used for the planning of the combine path in the field.

The block diagrams of the vehicle model, process module, and decision module (Figure 2.33) are self-explanatory.

Designing such a system in detail is a tremendously challenging task. For simulation and control purposes, most of the modules of this simulation package need a lot of innovation for their integration, accompanied by information yet to be defined and described, gathered, processed, further developed, and assembled in a proper format. This book aims to represent a substantial contribution toward achieving this goal.

References

Ayyub, B.M. and G.J. Klir. 2006. *Uncertainty Modeling and Analysis in Engineering and the Sciences.* Boca Raton, FL: Chapman & Hall/CRC.

Beichelt, F. 2006. *Stochastic Processes in Science, Engineering, and Finance.* Boca Raton, FL: Chapman & Hall/CRC/Taylor & Francis.

Blanchard, B.S. and W.J. Fabrycky. 2011. *Systems Engineering and Analysis.* 5th ed. Upper Saddle River, NJ: Prentice Hall.

Bronshtein, I.N. and K.A. Semendyayev. 1978. *Handbook of Mathematics.* Berlin: Springer.

Crassidis, J.L. and J.L. Junkins. 2004. *Optimal Estimation of Dynamic Systems.* Boca Raton, FL: Chapman & Hall/CRC.

De Silva, C.W. 2009. *Modeling and Control of Engineering Systems.* Boca Raton, FL: CRC Press/Taylor & Francis Group.

Einstein, A. and L. Infield. 1938. *The Evolution of Physics.* New York: Simon & Schuster.

Emery, A.F. and A.V. Nenarokomov. 1998. Optimal experiment design. *Measurement Science and Technology* 9:864–876.

Grimvall, G. 2008a. Accuracy of models. In *Scientific Modeling and Simulations*, ed. S. Yip and T.D. de la Rubia, 41–57. Berlin: Springer.

Grimvall, G. 2008b. Extrapolative procedures in modeling and simulations: The role of instabilities. In *Scientific Modeling and Simulations*, ed. S. Yip and T.D. de la Rubia, 5–20. Berlin: Springer.

Heisenberg, W.K. 1958. *Physics and Philosophy: The Revolution in Modern Science*. London: George Allen and Unwin.

Kac, M. and J. Logan. 1979. Fluctuations. In *Studies in Statistical Mechanics*. Vol. VII. *Fluctuation Phenomena*, ed. E.W. Montroll and J.L. Lebowitz. New York: Elsevier.

Klee, H. and R. Allen. 2011. *Simulation of Dynamic Systems with MATLAB and SIMULINK*. Boca Raton, FL: CRC Press/Taylor & Francis.

Koza, J.R. 1992. *Genetic Programming*. Cambridge, MA: MIT Press/Bradford Books.

Kreyszig, E. 1988. *Advanced Engineering Mathematics*. 6th ed. New York: John Wiley & Sons.

Montana, D.J. and B. Beranek. 2002. *Strongly Typed Genetic Programming*. Cambridge, MA: Newman.

Nayfeh, A.H. 1981. *Introduction to Perturbation Techniques*. New York: John Wiley.

Perhinschi, M.G., M.R. Napolitano, and S. Tamayo. 2010. Integrated simulation environment for unmanned autonomous systems: Towards a conceptual framework. *Modeling and Simulation in Engineering*. Article 736201, doi: 10.1155/2010/736201.

Poli, R., W.B. Langdon, and N.F. McPhee. 2008. *A Field Guide to Genetic Programming*. San Francisco: Creative Commons.

Schucany, W., W. Parr, and J. Boyer. 1978. Correlation structure in Farlie–Gumbel–Morgenstern distributions. *Biometrika* 65:650–653.

Smith, J.E. 2007. Coevolving memetic algorithms: A review and progress report. *IEEE Transactions on Systems, Man, and Cybernetics—Part B: Cybernetics* 37(1):6–17.

Weise, T. 2009. *Global Optimization Algorithms: Theory and Applications*. 2nd ed. http://www.it-weise.de/projects/book.pdf (accessed April 3, 2012).

Bibliography

Agilent Technologies. 1998. Control systems development using dynamic signal analyzers. Application Note 243-2. Santa Clara, CA: Agilent Technologies.

Antoniou, A. and W.S. Lu. 2007. *Practical Optimization, Algorithms, and Engineering Applications*. New York: Springer.

Benaroya, H. and S.M. Han. 2005. *Probability Models in Engineering and Science*. Boca Raton, FL: Taylor & Francis.

Biegler, L. 2010. *Nonlinear Programming: Concepts, Algorithms, and Applications*. MOS SIAM Series of Optimization. Philadelphia, PA: Society for Industrial and Applied Mathematics.

Chung, C.A. 2004. *Simulation Modeling Handbook: A Practical Approach*. Boca Raton, FL: CRC Press.

Conway, D. 1983. Fairlie-Gumbel-Morgenstern distributions. In *Encyclopedia of Statistical Sciences*, 28–31. Vol. 3. New York: John Wiley & Sons.

Dorf, R.C. and R.H. Bishop. 2011. *Modern Control Systems*. 12th ed. Upper Saddle River, NJ: Prentice Hall.

Henderson, D. and P. Plaschko. 2006. *Stochastic Differential Equations in Science and Engineering*. Hackensack, NJ: World Scientific.

Jarne, G., J. Sanchez-Choliz, and F. Fatas-Villafranca. 2007. S-shaped economic dynamics. The logistics and Gompertz curves generalized. *Evolutionary and Institutional Economic Review* 3(2):239–259.

Koza, J.R. 1990. Genetic programming: A paradigm for genetically breeding populations of computer programs to solve problems. Stanford, CA: Stanford University.

Maria, A. 1997. Introduction to modeling and simulation. In *Proceedings of the 1997 Winter Simulation Conference*, ed. S. Andratomir, K.J. Healy, D.H. Withers, and B.L. Nelson, 287–295. Piscataway, NJ: Institute of Electrical and Electronics Engineers.

MathWorks. 2012. MATLAB R2012a: Documentation—Statistics toolbox. Natick, MA: MathWorks.

Michalewicz, Z. 1996. *Genetic Algorithms + Data Structure = Evolution Programs.* Berlin: Springer.

Miu, P. 2001a. Optimal design and process of threshing units based on a genetic algorithm I. Algorithm. Paper presented at the annual meeting of the American Society of Agricultural Engineers, ASAE Paper 01-3124. St. Joseph, MI: American Society of Agricultural Engineers.

Miu, P. 2001b. Optimal design and process of threshing units based on a genetic algorithm II. Application. Paper presented at the annual meeting of the American Society of Agricultural Engineers, ASAE Paper 01-3125. St. Joseph, MI: American Society of Agricultural Engineers.

Miu, P., M.G. Perhinschi, and H.D. Kutzbach. 2000. Evolutionary optimization of threshing units design and operation: Implementation technique. Presented at the annual meeting of the European Society of Agricultural Engineers, Warwick, Paper 00-PM-055, CD-ROM Abstract Proceedings, Part 2: 211–212.

Nise, N.S. 2011. *Control Systems Engineering.* 6th ed. John Wiley & Sons.

Perhinschi, M.G. 1997. Controller design for autonomous helicopter model using a genetic algorithm. Paper presented at the 53rd annual forum of the American Helicopter Society, Virginia Beach, VA.

Perhinschi, M.G. 2000. A simulation model of an uninhabited aircraft. AIAA 2000-4188. Reston, VA: American Institute of Aeronautics and Astronautics.

Samson, C. 1995. Control of chained systems. Application to path following and time-varying point-stabilization of mobile robots. *IEEE Transactions on Automatic Control* 40(1):64–77.

Schaller, C. 2004. Concepts of model verification and validation. LA-14167-MS. Los Alamos, CA: Los Alamos National Laboratory.

Sieniutycz, S. and J. Jezowski. 2009. *Energy Optimization in Process Systems.* Oxford: Elsevier.

3

Crop Harvesting Data and Plant Properties

3.1 Introduction

Plant processing performed by combine harvesters derives essentially from considering crop data, physical and mechanical properties of the plants (stems and seeds), requirements for final product delivery in terms of form and quality, and combine work productivity. During combine harvesting, the plants are exposed to various mechanical stresses (tensile, compression, bending, shearing, etc.), thermal stress, and acoustical and possibly electrostatic effects.

The behavior of the plants during combine processing is the result of a combination of their elastic–plastic material properties, granular material properties, and non-Newtonian liquid-flow characteristics. Such behavior, involving complicated interactions and nonlinear relationships, makes applying or adapting well-known theories of homogenous materials, as well as the elaboration of reliable theories, difficult. Thus, it becomes more difficult to mathematically model these properties for use in combine process modeling, simulation, optimization, and control.

On the one hand, this chapter summarizes the crop data, and mechanical and physical properties of the plants that are combine harvested. Some properties and data may not be available from the published sources. On the other hand, in this chapter we attempt to develop, where possible, several mathematical models that will help the readers beyond the purpose of modeling, simulation, and design of combine harvesters.

3.2 Crop Data

For a good reference in modeling and design of combine harvesters, the crop data presented in Tables 3.1 and 3.2 is the result of comparison and further statistical processing of information acquired with predilection from countries with developed agricultural practices. Since the amount and quality of information related to such a subject are continuously evolving, in time a slight update may be welcome. In imperial units, the grain yield (bu/acre) for each crop type is corrected to the moisture and specific weight per the ANSI/ASAE D241.4 NOV98 standard.

TABLE 3.1

Cereal Crop Data (above average)

Crop Type	Wheat	Barley	Rye	Triticale	Oats	Rice	Flax (Linseed)
Grain yield,[a] kg/ha (bu/acre)	4705–5915 (70–88)	3335–4415 (62–82)	3140–8500 (50–135)	2900–3760 (52–67)	2690–5100 (75–142)	5500–7500 (109–149)	1350–1570 (21.5–25)
Grain moisture[b] (wet basis), %	14–18	11–15	18–22	15–19	14–18	17–21	10–12
MOG moisture[b] (wet basis), %	7–11	8–13	11–16	8–12	8–12	12–16	
Plants population, plants/m² (plants/ft²)	180–290 (16.7–27)	180–290 (16.7–27)	194–237 (18–22)	215–300 (20–28)	130–240 (12–22.3)	15–23[c] (1.4–2.14)	540–800 (50–75)

[a] Grain yields (bu/acre) are corrected according to the standard ANSI/ASAE D241.4 NOV98.
[b] During harvesting.
[c] Seedlings/m² (seedlings/ft²).

TABLE 3.2

Row Crop Data (above average)

Crop Type	Soybean	Sunflower	Corn (Maize)	Sorghum	Canola (Rape)	Beans
Grain yield,[a] kg/ha (bu/acre)	4538–4840 (67.5–72)	1680–2240 (46–62)	7200–9500 (115–150)	4080–6280 (65–100)	1500–2000 (25.7–34)	1500–3500 24–56
Grain moisture[b] (wet basis), %	12–16	9.5–12	18–23	16–20	10–14	15–18
MOG moisture[c] (wet basis), %	12–16		20–38		12–22	
Distance between rows, m (in.)	0.2–0.75 (7.5–30)	0.4–0.91 (16–36)	0.38–0.76 (15–30)	0.51–0.91 (20–36)	0.18–0.35 (7–14)	0.3–0.76 (12–30)
Plant population, plants/m² (plants/ft²)	25–41 (2.3–4)	4–5.5 (0.37–0.51)	7.4–8 (0.7–0.75)	8.7–17.5 (0.8–1.6)	44–68 (4.1–6.3)	30–43 (2.75–4)

[a] Grain yields (bu/acre) are corrected according to the standard ANSI/ASAE D241.4 NOV98.
[b] During harvesting.

3.3 Plant Properties and Behavior Modeling

3.3.1 Physical and Mechanical Properties of Plant Stems

Physical and mechanical properties of plant stems govern their behavior under the effect of mechanical forces, applied during combine harvesting, in terms of plant bending (toward the cutter bar), cutting (shearing stress), ear removal (stem elongation), and further deformation by compression, friction, and stem, ear, or panicle drawing (e.g., during the threshing process).

Plants are rheological materials whose properties follow *non-Newtonian laws* as derived from their behavior in terms of plasticity and elasticity.

TABLE 3.3

Properties of Cereal Stem at Harvesting Moisture

Crop Type		Wheat	Barley	Rye	Oats	Rice
Plant height,[a] h_{pt}, m (in.)		0.76–1.05 (30–41.3)	0.68–1.1 (26.8–47)	1.0–1.7 (39.4–66.9)	0.6–1.74 (23.6–68.5)	1–1.8 (39–70)
Stem height,[a] h_{be}, m (in.)		0.67–0.93 (26.4–36.6)	0.56–0.96 (22–37.8)	0.9–1.55 (35.4–61)	0.52–1.5 (20.5–59.06)	0.82–1.52 (32.3–59.8)
Height to plant center of gravity,[a] h_{cw}, m (in.)		0.38–0.52 (15–20.5)	0.3–0.55 (11.8–21.7)	0.51–0.92 (20–36.2)	0.31–0.9 (12.2–35.4)	0.32–0.60 (12.6–27.6)
Stem diameter, mm (in.)	Min	2.95 (0.116)	2.9 (0.114)	2.9 (0.114)	2.1 (0.083)	3.2 (0.247)
	Mean	4.2 (0.165)	4.5 (0.177)	3.6 (0.142)	3.8 (0.150)	6.2 (0.247)
	Max	4.75 (0.187)	5.1 (0.201)	4.5 (0.177)	5.7 (0.244)	8.4 (0.33)
Stem wall thickness, mm (in.)		0.25–0.45 (0.01–0.018)	0.26–0.38 (0.01–0.015)			0.25–0.35 (0.01–0.014)
Stem tensile stress, N/mm² (lb/in.²)		6.8–12 (986.2–1,740.4)	11.1 (1610)	9.5–10.5 (1,378–1,523)		
Stem bending stress, N/mm² (lb/in.²)		17.5–26.8 (2,538–3,887)	6.3–12.4 (913.7–1,798)			4.7–8.3 (681.6–1,203.7)
Stem shearing stress, N/mm² (lb/in.²)		3.8–6.8 (551.1–986.2)	7.2–9.2 (1,044–1,334)			5.4–10.2 (783.2–1,479.3)
Stem modulus of elasticity, N/mm²		2,000–6,799	11,064	10,500		390–1,210

Note: 1 N/mm² = 1 Mpa.

[a] Symbols shown in Figure 3.1.

To facilitate modeling, simulation, and optimization of combine processes, it is absolutely necessary to consider reliable experimental data, as shown in Tables 3.3 and 3.4. The symbols are shown in Figure 3.1. Presented data characterizes the stem material behavior corresponding to plant maturity and environment conditions at harvesting time. That means data from Table 3.3 should be simultaneously considered with data from Table 3.1, and data from Table 3.4 should be paired with data from Table 3.2.

Although data presented in Tables 3.3 and 3.4 are self-explanatory, the reader should review the mechanics theory if needed. This will facilitate a better understanding of the plant bending model (Section 3.3.2), which is very important in the process modeling and engineering design of the reel and header of a combine.

TABLE 3.4

Properties of Row Crop Stem at Harvesting Moisture

Crop Type		Soybean	Sunflower	Corn (Maize)	Sorghum
Plant height,[a] h_{pt}, m (in.)		0.35–1.15	0.75–1.15	0.98–2.50	0.79–2.76
Stem height,[a] h_{be}, m (in.)		0.29–1.05	0.7–1.08	0.89–2.18	0.8–2.47
Stem diameter, mm (in.)	Min	6.38	21	14.0	5.1
	Mean	8.05	25	21.8	11.5
	Max	9.23	28	32.0	16.7

Note: 1 N/mm² = 1 Mpa.

[a] Symbols shown in Figure 3.1.

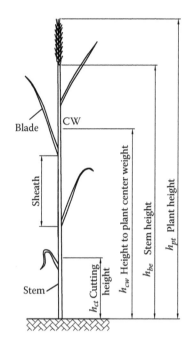

FIGURE 3.1
Generic sketch of a cereal plant with symbols.

3.3.2 Plant Bending Modeling

Modeling of plant stem bending aims to mathematically describe the curved shape and deflection of a stem when it is pushed laterally by a certain force (e.g., as exercised by the reel bar). Mechanically, a simplified stem (a continuous medium, without leaves) can be associated with a *cantilevered, tapered beam* with two possible types of cross sections:

1. *Solid* cross section (corn, sorghum)
2. *Hollow* cross section, with a variable wall thickness (wheat, barley, rye, oats, triticale, etc.)

The shape of both types of cross sections could be either *circular* or oval (*elliptical*).

Mechanics theory already contains mathematical formulae for deflection of such cantilevered, tapered beams. We are interested in preparing the stem bending model as the stem interacts with the reel bar.

In the following, the case of a cantilevered, tapered stem, with a circular hollow cross section and linearly variable wall thickness (Figure 3.2) will be considered. The stem, with the height h_{be}, is naturally supported in a vertical position. A force P is considered to act perpendicularly on the longitudinal axis of the stem and maintain the contact point, initially established at the height s_p, on the stem. The length of the stem is h_{be}, and its outside diameters are d_1 at the lower end and d_2 at the upper end (below the ear).

Let us consider the current cross section with the outside diameter d and the wall thickness t. Then, y is the current vertical position, x is the corresponding deflection (horizontal displacement), and θ is the deflection angle.

The moment of inertia of current cross section, with thin wall ($\tan(\theta) = \theta$), is

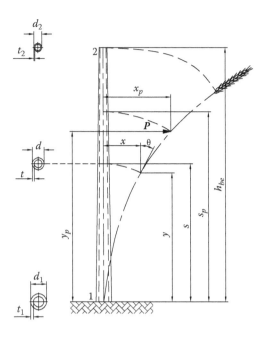

FIGURE 3.2
Geometry of plant bending.

$$I = \frac{\pi t (d - t)^3}{8} \tag{3.1}$$

where d is the outside diameter of the cross section.

According to Thales' theorem, we can write

$$d = d_1 - (d_1 - d_2)\frac{s}{h_{be}} \tag{3.2}$$

$$t = t_1 - (t_1 - t_2)\frac{s}{h_{be}} \tag{3.3}$$

We denote by s the arc length that corresponds to the position of the current cross section, by E the modulus of elasticity of the stem (Tables 3.3 and 3.4), and by x the stem deflection corresponding to the ordinate s.

The following formula (adapted from Hopkins (1970)) helps in estimating the *maximum deflection* x_p of the stem:

$$x_p = \frac{Ps_p^3}{2\pi E t_p (r_1 - r_p)^3}\left[2\ln\left(\frac{r_1}{r_p}\right) - \left(\frac{r_1 - r_p}{r_1}\right)\left(3 - \frac{r_p}{r_1}\right)\right] \tag{3.4}$$

with $r_1 > r_p$ and t_p is the wall thickness of the cross section, where the force P is applied.

The radii r_1 and r_p are calculated to midthickness of the wall as follows:

$$r_1 = \frac{d_1 - t_1}{2} \tag{3.5}$$

$$r_p = \frac{d_p - t_p}{2} \tag{3.6}$$

and d_p is the outside diameter of the stem corresponding to the position of applied force P.

According to the theory of cantilevered beam deflection, we can develop the following formula to describe the deflection of the stem:

$$x = \frac{Ps^2}{2\pi Et \left(r_1 - r\right)^3} \left[2\ln\left(\frac{r_1}{r}\right) - \left(\frac{r_1 - r}{r_1}\right)\left(3 - \frac{r}{r_1}\right) \right]\left(2s_p - s\right) \tag{3.7}$$

where the radius r is calculated to midthickness of the wall as follows:

$$r = \frac{d - t}{2} \tag{3.8}$$

To find the center coordinates of a stem cross section where the force P is applied, we will adapt the formula developed by Parkinson et al. (1996) as follows:

$$y_p = s_p\sqrt{1 - \left(\frac{x_p}{c_{bp}s_p}\right)^2} \tag{3.9}$$

In the above equation, c_{bp} is a *bending coefficient* that depends on the *moments of inertia ratio*, as follows:

$$c_{bp} = 0.030432\ln\left(\frac{I_p}{I_1}\right) + 0.91897 \tag{3.10}$$

where:
  I_p is the moment of inertia of the cross section
  I_1 is the moment of inertia at the bottom end of the stem

By extension, the stem curve equation is

$$y = s\sqrt{1 - \left(\frac{x}{c_b s}\right)^2} \tag{3.11}$$

where

$$c_b = 0.030432\ln\left(\frac{I}{I_1}\right) + 0.91897 \tag{3.12}$$

The moment I of inertia corresponds to the current cross section.

For $s_p < s < h_{be}$ we shall use the following equation to describe the deflection of the stem:

$$x = \frac{Ps^2}{2\pi Et(r_1 - r)^3}\left[2\ln\left(\frac{r_1}{r}\right) - \left(\frac{r_1 - r}{r_1}\right)\left(3 - \frac{r}{r_1}\right)\right](s - s_p) \tag{3.13}$$

If the hollow cross section of the stem has an elliptical shape, the moment I of inertia is calculated with the following formula (Gere and Timoshenko, 1997):

$$I = \frac{\pi}{4}\left[r_A r_B^3 - (r_A - t)(r_B - t)^3\right] \tag{3.14}$$

where:
   r_A is the major radius of the elliptical cross section of the stem
   r_B is the minor radius of the elliptical cross section of the stem

   Deger et al. (2010) have published graphs of stem diameter and wall thickness variation in wheat and barley. Before measurement, the leaves wrapped around the stem were removed; therefore, their additional material influence is not implicitly considered. Nonetheless, based on such graphs and through regression analysis of their data, we have developed mathematical models that can be used in a more rigorous approach for plant stem modeling. It is understood that any misrepresentation in the graphical data may have influenced our results.

   The mathematical model that describes the variation of the *average outside diameter of plants* with the stem height is

$$d = a(1 + h_{be})e^{-bh_{be}^2} \tag{3.15}$$

The coefficients a and b have the following values:

- In wheat: $a = 3.6481$ mm and $b = 1.7637$ mm^{-1} ($R^2 = 0.995$)
- In barley: $a = 3.8077$ mm and $b = 1.962$ mm^{-1} ($R^2 = 0.96$)

   One of the difficulties is choosing the right value of modulus of elasticity that, for different crop varieties and field locations, may vary significantly due to individual structure and composition of the stems, whose anisotropic behavior is consistent with composite structures. That is why, in modeling and simulation of the plant bending process, the practitioner should consider different practical situations with their corresponding parameters and variables.

Application 3.1: Required Reel Force and Stem Curve Representation

Let us consider a wheat crop whose physical characteristics have the following mean values:

$h_{be} = 0.78$ m
$d_1 = 4.75$ mm
$d_2 = 3$ mm
$t_1 = 0.2$ mm
$t_2 = 0.4$ mm
$E = 4200$ N/mm^2

The height of the initial contact point with the header reel is

$$s_p = 0.716 \text{ m}$$

and the last contact point of the stem with the reel bar has the ordinate

$$y_p = 0.63 \text{ m}$$

The stem diameter and wall thickness at the initial contact point with the reel bar are

$$d_p = d_1 - (d_1 - d_2)\frac{s_p}{h_{be}} = 4.75 - (4.75 - 3)\frac{0.716}{0.78} = 3.144 \text{ mm}$$

$$t_p = t_1 - (t_1 - t_p)\frac{s_p}{h_{be}} = 0.4 - (0.4 - 0.2)\frac{0.716}{0.78} = 0.216 \text{ mm}$$

The stem moment of inertia at the initial point of force application is

$$I_p = \frac{\pi t_p (d_p - t_p)^3}{8} = \frac{\pi 0.216 (3.337 - 0.216)^3}{8} = 2.579 \text{ mm}^4 = 2.579 * 10^{-12} \text{ m}^4$$

The moment of inertia at the stem base has the value

$$I_1 = \frac{\pi t_1 (d_1 - t_1)^3}{8} = \frac{\pi 0.4 (4.75 - 0.4)^3}{8} = 12.93 \text{ mm}^4 = 12.93 * 10^{-12} \text{ m}^4$$

The ratio of the moments of inertia is $I_{rp} = 2.579/12.93 = 0.1995$.
The bending coefficient corresponding to the initial point of force application is

$$c_{bp} = 0.030432\ln\left(\frac{I_p}{I_1}\right) + 0.91897 = 0.030432\ln 0.1995 + 0.91897 = 0.87$$

That means the reel bar will produce a deflection x_p of the stem:

$$x_p = c_{bp} s_p \sqrt{1 - \left(\frac{y_p}{s_p}\right)^2} = 0.87 * 0.716\sqrt{1 - \left(\frac{0.63}{0.716}\right)^2} = 0.296 \text{ m}$$

We will now calculate the radii r_1 and r_p as follows:

$$r_1 = \frac{d_1 - t_1}{2} = \frac{4.75 - 0.4}{2} = 2.175 \text{ mm}$$

$$r_p = \frac{d_p - t_p}{2} = \frac{3.144 - 0.216}{2} = 1.464 \text{ mm}$$

From the equation

$$x_p = \frac{Ps_p^3}{2\pi Et_p(r_1 - r_p)^3}\left[2\ln\left(\frac{r_1}{r_p}\right) - \left(\frac{r_1 - r_p}{r_1}\right)\left(3 - \frac{r_p}{r_1}\right)\right]$$

we get

$$296 = \frac{P * 716^3}{2\pi * 4200 * 0.216(2.175 - 1.464)^3}\left[2\ln\left(\frac{2.175}{1.464}\right) - \left(\frac{2.175 - 1.464}{2.175}\right)\left(3 - \frac{1.464}{2.175}\right)\right]$$

That means the force from the reel has the magnitude

$$P = \frac{0.296}{5.564} = 0.053\ \text{N/plant}$$

This force shall be just a little big bigger due to friction of the reel bar with the plant.
Using Equation 3.7, for $0 < s < s_p$ we get the equation that describes the deflection of the stem as follows:

$$x = \frac{Ps^2}{2\pi Et(r_1 - r)^3}\left[2\ln\left(\frac{r_1}{r}\right) - \left(\frac{r_1 - r}{r_1}\right)\left(3 - \frac{r}{r_1}\right)\right](2s_p - s)$$

$$x = \frac{0.053s^2}{2\pi 4200t(2.175 - r)^3}\left[2\ln\left(\frac{2.175}{r}\right) - \left(\frac{2.175 - r}{2.175}\right)\left(3 - \frac{r}{2.175}\right)\right](2 * 0.716 - s)$$

where

$$r = \frac{d - t}{2}$$

For $s_p < s < h_{be}$ we will use Equation 3.13. That is,

$$x = \frac{0.053s^2}{2\pi 4200t(2.175 - r)^3}\left[2\ln\left(\frac{2.175}{r}\right) - \left(\frac{2.175 - r}{2.175}\right)\left(3 - \frac{r}{2.175}\right)\right](s - 0.716)$$

Then we use Equation 3.11 to describe the curve of the bent stem. The graph of the stem curve is shown in Figure 3.3.

3.4 Plant Grain Properties

3.4.1 Physical and Mechanical Properties of Plant Grains

Most of the processes performed by a combine harvester are influenced decisively by grain (seeds) physical properties such as shape, size, and moisture, as well as by their mechanical properties. The dimensions of the grains are not uniform, but for a certain crop, they

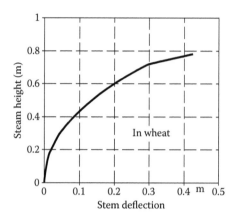

FIGURE 3.3
Wheat plant stem bending curve.

scatter around a mean value. Thus, it is absolutely necessary to determine the *distribution of physical properties* to help in the decision process of combine harvester components, for example, the opening size of the cleaning sieves. Chapter 2 explains in detail the analysis of the values of a variable; therefore, the reader is referred there to apply the appropriate knowledge in finding the mean and deviation of each variable of interest.

Then, different *measuring indices* are also useful, such as *roundness, sphericity,* and equivalent diameter. Sitkei (1986) has described such indices quite extensively. Here we will give two definitions of the *equivalent geometric mean diameter d_{eg}* (mm) of the grains. One formula implies the use of the mean values of seed dimensions and weight (Mohsenin, 1980; Song and Litchfield, 1991):

$$d_{eg} = (abc)^{1/3} \qquad (3.16)$$

where:
   a is the mean length of the grains, mm
   b is the mean width of the grains, mm
   c is the mean thickness of the grains, mm

Physically, the values a, b, and c are the dimensions of the smallest parallelepiped that fully encloses a seed with mean sizes.

With another formula, the *equivalent mean diameter d_e* of irregular shaped grains is given by the diameter of a sphere of identical volume (adapted from Sitkei, 1986):

$$d_e = \sqrt[3]{\frac{6w_s}{\pi \rho_s}} \qquad (3.17)$$

where:
   w_s is the weight of 1000 seeds/1000, g
   ρ_s is the mean true density (volumetric weight) of seed, g/mm³

Another important property of the seeds is the *mean projected area a_{ps}* on a surface, viewed from the perpendicular direction to a flat surface. This is best determined using image-processing software. The mean projected cross-sectional area can also be calculated based on the volume of the seed as follows:

$$a_{ps} = k_{ps} \left(\frac{w_s}{\rho_s} \right)^{2/3}$$ (3.18)

where k_{ps} is the body shape factor. For a sphere, $k_{ps} = 1.21$.

Sphericity sph_s of seed is used to characterize the similarity between the seed shape and a sphere and is quantified by the ratio of the equivalent mean diameter of the seed to the diameter of the smallest circumscribing sphere, which is the size a:

$$sph_s = \frac{d_{eg}}{a} 100$$ (3.19)

The *roundness* Θ_s of a seed expresses the measure of the sharpness of its edges and corners, and it is defined by the ratio of the largest projected area A_{ps} of the seed to the area of its smallest circumscribing circle, as follows:

$$\Theta_s = \frac{A_{ps}}{\frac{\pi a^2}{4}} = \frac{4A_{ps}}{\pi a^2}$$ (3.20)

High values of sphericity and roundness indicate that the seed's shape is close to a spherical one.

Porosity (*Por*) of grain (bulk seeds) is defined as the proportion of the space in the bulk grain that is not filled by the seeds. In other words, porosity is the ratio of the intergranular volume to the total volume occupied by the grain. This ratio depends on the mean true density of seeds (kernels) and the grain bulk density ρ_b as expressed by the following equation:

$$Por = \left(1 - \frac{\rho_b}{\rho_s} \right) 100$$ (3.21)

The porosity for clean grains ranges from approximately 25% to 70%. Porosity allows fluids to flow through a mass of particles referred to as a packed bed. Grain porosity influences seed susceptibility to breakage, drying rate, and resistance to fungal development.

Table 3.5 centralizes the values of physical properties: size, weight of 1000 seeds W_s (kg/m³), mean true density of seed (particle), and bulk density of selected crop grains. The ANSI/ASAE D241.4 standard contains further information on grain density, bulk density, and the mass–moisture relationship for grain storage.

Ponce-Garcia et al. (2008) showed that mechanical and *viscoelastic properties* of wheat kernels can be evaluated with the compression loading method by applying the theory of elasticity. All wheat grain samples showed a rupture when the loading force was around 18.5 N, independent of the wheat variety and moisture content. The wheat

TABLE 3.5

Physical Data of Selected Crop Grain

Plant Type	Length, mm			Width, mm			Thickness, mm			Moisture%, Wet Basis	Weight of Thousand Seeds, g	Mean True Density of Seed, g/mm³	Bulk Density, kg/m³
	Min	Mean	Max	Min	Mean	Max	Min	Mean	Max				
Wheat	6.4	7.1	7.86	3.45	3.52	3.66	2.72	3.05	3.37	9.8	30–48	1.22–1.26	772
Barley	6.4	8.45	10.7	2.62	3.6	4.73	1.57	2.65	3.68	9.7–10.4	35–50	1.18–1.45	618
Rye	6.1	7.11	8.0	3.8	4.65	5.15	2.1	3.18	3.4	9.7	35–40	1.23	721–741
Oats	7.9	10.1	11.3	2.45	2.85	3.21	2.13	2.21	2.33	10.3	30–45	1.03–1.18	412
Rice	6.12	7.23	8.31	1.70	2.15	2.97	1.40	1.88	2.52	11.9–12.4	17.5–31.5	1.12–1.52	579–805
Corn (maize)	9.79	11.8	13.8	6.19	8.02	9.85	3.53	4.43	5.33	9.0	380	1.19	652–721
Soybean	6.15	7.32	8.49	5.56	6.79	8.02	4.76	5.78	6.8	7.0	160	1.15	772
Sunflower	14.3	22.8	31.0	4.73	7.06	9.82	2.36	3.98	6.67	15.0–16.8	125–175	0.44–0.55	412–577
Sorghum	3.81	4.04	4.62	3.29	3.50	4.53	3.09	3.44	3.61	9.5–9.9	16–28	1.24	721–800
Canola (rape)	1.1	2.06	2.6	1.3	1.75	2.3				6.6	2.5–5.5	1.13–13.2	540–670
Beans (lima)	13.55	15.1	16.5	10.7	12.1	13.2	6.3	7.5	8.4	9.8	200–350	1.17	721–800

TABLE 3.6

Mean Values[a] of Modulus of Elasticity in Wheat Kernels, Measured in a Single Kernel

Wheat Variety	Kernel Thickness, mm	Compression Area, mm²	Compression Force, N	Final Stress, N/mm²	Modulus of Elasticity, N/mm²
Saturno	2.90 ± 0.16	1.83 ± 0.43	69.92 ± 8.75	40.00 ± 1.07	232.14 ± 55.41
Salamanca	2.97 ± 0.20	1.78 ± 0.28	75.13 ± 7.79	43.30 ± 9.09	244.65 ± 45.81
Cortazar	3.06 ± 0.14	1.56 ± 0.21	80.24 ± 5.83	50.05 ± 9.46	308.54 ± 29.92
Rayon	2.82 ± 0.14	1.47 ± 0.12	86.93 ± 5.82	63.23 ± 1.39	321.46 ± 32.54
Alter	3.00 ± 0.13	1.29 ± 0.13	92.58 ± 3.66	71.05 ± 8.88	438.65 ± 50.03
Sofia	3.03 ± 0.16	1.22 ± 0.17	88.74 ± 6.72	71.53 ± 1.45	454.14 ± 79.36
Rafi	2.97 ± 0.13	1.15 ± 0.16	90.94 ± 6.20	78.95 ± 1.24	485.80 ± 54.33

Source: Ponce-Garcia, N. et al., *Cereal Chemistry* 85(5):667–672, 2008.

[a] Means ± standard deviation values.

kernels have an elastic behavior, obeying Hooke's law, as shown by a linear relationship between load applied and deformation. Wheat with 9.3% moisture content has significant elastic behavior in comparison with wheat tempered at 22.5% moisture that shows plastic behavior. Table 3.6 shows mean values of modulus of elasticity in wheat kernels, measured in a single kernel, along with significant testing data, including compression force, compression area, and final compression stress values. The values of modulus of elasticity of wheat kernels range from 99.2 N/mm² for 22.5% moisture content to 394.8 N/mm² for 9.3% moisture content.

3.4.2 Mechanical Damage of Grains

The quality of harvested grains may be affected by damaging them due to mechanical forces exerted by combine harvester active elements. Usually, mechanical damage of grains is followed by rapid spoiling that affects the surrounding other grains during storage.

The type, shape, and degree of grain damage during harvesting depend on the grain species, variety, and physical and mechanical properties, as well as the type, frequency, and strength of mechanical actions exerted on the grain by the combine harvester active elements. The grains are damaged primarily during conveying and threshing processes. Thus, it is very important to design and operate a combine harvester so that the mechanical damage of grains is minimized.

The common components of cereal grains are the bran, germ (embryo), and endosperm. The structure of wheat grain is shown in Figure 3.4. Grain damage occurs in the form of rupture or cracking due to excessive stresses induced by impact, compression, drawing, and so forth, as well as due to heavy friction during conveying.

The most common types of grain defects caused by combine harvesters are *knocking off, skinning, distortion, cracking*, and *splitting*. Grains may have the germ knocked off or scalloped out due to header action. Skinning is the damage of the grain protective husk (of barley, oats), and it is caused by intense conveying and threshing actions. A skinned grain may display the following types of damage: awn skinning, germ exposure, pearling, split back, or split skirt. Awn skinning is the grain damage when more than a third of the

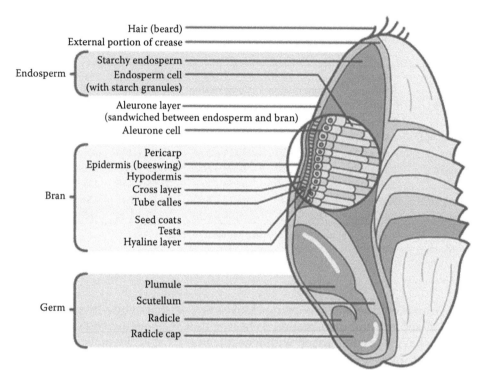

FIGURE 3.4
Structure of wheat seed. (Courtesy of HGCA, Kenilworth, England, and Nabim, London, England.)

husk from awn end to the grain center has been removed. The germ is exposed when the husk is removed from the germ end of the grain. Grain is pearled when the entire husk is removed.

Distortion of the grain concerns changes of the grain shape due to nonuniform distribution of a load on the grain. That causes the sliding of internal grain layers against each other or their local deformation because of differences in mechanical properties (e.g., allowable compression stress).

Cracking of the grain is considered when it is limited to cracking of the grain bran or surface layers due to high-impact forces or pressure. In such a case, the grain remains a single piece without falling apart into smaller parts. If the grain divides into several pieces, the damage is called splitting.

An individual kernel may have more than one defect. Certain damage may be assessed in different ways, depending on grain use: direct (animal feeding), direct processing, or storage. In the industrialized countries, grain damage is assessed based on government market standards. Thus, the presence of shrunken, broken, or sprouted grains at levels above the standard docks the grain grade and quality. Table 3.7 is an example of a primary grade determinant for wheat quality, Canada Western Red Spring (CWRS) variety, issued by the Canadian Grain Commission. This table also refers to other types of damage (e.g., natural stain, pink color) not produced by wheat combine harvesting, but possibly post-harvest favored.

The *Grain Inspection Handbook*, issued by the U.S. Department of Agriculture (2004), contains additional information on grain damage, sampling, and inspection.

TABLE 3.7

Primary Grade Determinant Table for Wheat Quality, CWRS Variety

Grade Name	Natural Stain %	Pink%	Sawfly, Midge %	Shrunken and Broken			Smudge and Black Point		Sprouted	
				Shrunken %	Broken %	Total %	Smudge %	Total %	Severely Sprouted %	Total %
No. 1 CWRS	0.5	1.5	2.0	4	5	7	0.30	10	0.10	0.5
No. 2 CWRS	2	5	5	4	6	8	1	20	0.20	1.0
No. 3 CWRS	5	10	10	4	7	9	5	35	0.30	3.0
No. 4 CWRS	5	10	10	4	7	9	5	35	0.5	5
CW feed	No limit	No limit	No limit	No limit	No limit	No limit within broken tolerances	No limit	No limit	No limit	No limit

Source: CGC, Wheat, in *Official Grain Grading Guide*, chap. 4, Canadian Grain Commission, Winnipeg, Manitoba, 2012, http://www.grain-scanada.gc.ca/oggg-gocg/04/oggg-gocg-4f-eng.htm.

3.5 Aerodynamic Properties of Plant Grains and MOG

Different weeds infest grain-producing areas throughout the world. Wacker (1993) described different ways to separate the weeds during harvesting, such as removing material other than grain (MOG) components by collecting them in a trailer, adjusting combine settings to collect the weed kernels, and destroying their viability during the harvest process. Thus, during combine harvesting, the weed kernels and MOG components must be separated from grains based on physical characteristics (dimension, sphericity, aspect ratio, and weight) and aerodynamic properties (drag coefficient and terminal velocity).

A body placed into a fluid flow is subject to the action of friction force with the fluid, as well as a *drag force* that resists the body's motion through the fluid due to asymmetrical pressure distribution on the body's surface. Unlike the friction force, which is independent of fluid flow velocity, the drag force depends on velocity. The drag force always decreases fluid velocity relative to the body that was placed in the fluid flow.

The drag force F_d (N) exerted on a given object can be derived from Bernoulli's equation of the pressure in a fluid and is given by the following equation:

$$F_d = \frac{1}{2} c_d \rho_f A_{sf} v_f^2 \tag{3.22}$$

where:
  c_d is the *drag coefficient* (dimensionless)
  ρ_f is the fluid mass density, kg/m³
  A_{sf} is the cross-sectional body area normal to the fluid flow direction, m²
  v_f is the relative velocity between the object and fluid flow, m/s

The drag coefficient value depends on the body shape orientation, state of its surface, flow direction relative to the body position, and Reynolds number *Re*. For bodies of well-defined shape (sphere, cylinder, flat disc, etc.), the drag coefficient values can be found in Figure 3.5.

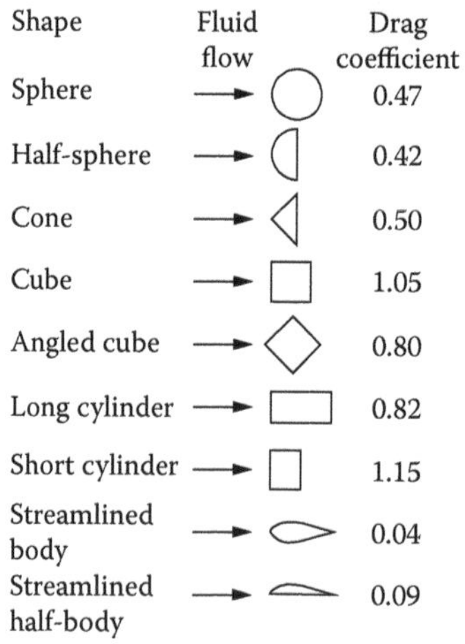

Shape	Fluid flow	Drag coefficient
Sphere	→ ○	0.47
Half-sphere	→	0.42
Cone	→	0.50
Cube	→ □	1.05
Angled cube	→ ◇	0.80
Long cylinder	→	0.82
Short cylinder	→ □	1.15
Streamlined body	→	0.04
Streamlined half-body	→	0.09

FIGURE 3.5
Drag coefficients of regular shapes and bodies placed in a fluid flow.

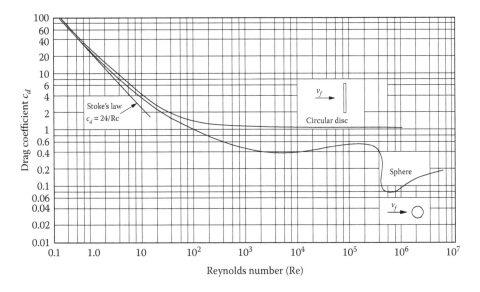

FIGURE 3.6
Drag coefficient variations with Reynolds number.

The variation of drag coefficient value with the Reynolds number is shown in a system of logarithmic coordinates in Figure 3.6. In the case of a very small Reynolds number, the Stokes' law is applied. Thus, for $Re < 1$, the inertia forces may be neglected and only the forces due to fluid viscosity can be considered. Above a Reynolds number of 1000, the drag coefficient of a circular disc (very thin) becomes a constant: 1.17.

Harvested grains of many species are irregularly shaped, not to mention the MOG components that are cut, fragmented, and deformed during different combine processes, such as cutting, threshing, separation, and cleaning. During cleaning, grains and MOG orient themselves randomly and rotate in the airflow of the cleaning unit. That means the projected area normal to the flow changes, making the cleaning process, or even carrying out experiments for gathering the required design data, difficult. Table 3.8 contains drag coefficient values for various crops.

TABLE 3.8

Drag Coefficient and Terminal Velocity Values for Various Crop Grains

Crop Grains	Drag Coefficient	Terminal Velocity, m/s (ft/s)	Reynolds Number, Re
Wheat	0.50	5.79–9.6 (19–31.5)	2200–2750
Barley	0.50	7.01–9.0 (23–29.53)	2280–2570
Rye		6.1–8.23 (20–27)	2130–2450
Oats	0.47–0.51	5.2–8.3 (17.1–27.2)	1750–2550
Soybean	0.45	7.92–14.5 (26–47.57)	6100–6300
Corn (maize)	0.56–0.7	9.7–11.4 (31.82–37.4)	5600–6680
Flax	0.5–0.52	4.6–4.7 (15.1–15.42)	836
Alfalfa	0.5	5.45 (17.89)	600
White kidney beans	—	5.5–8.5 (18–27.89)	—

Considering a fluid that moves vertically from bottom to top, the velocity of the main body of the fluid, in which a placed object remains in the same position, is called the *terminal velocity* of the object in that considered fluid. In this case, the drag force equals the weight of the body, that is,

$$W = \frac{1}{2} c_d \rho_f A_{sf} v_f^2 \tag{3.23}$$

Separating of grains from MOG during combine harvesting is achieved by using air blown by fans of different designs. Combine harvested grains and MOG components have different shapes and sizes. In a laminar flow, a particle, once settled, will generally maintain its orientation. So, what is the most probable particle orientation into the fluid body so that we can calculate or measure the cross-sectional particle area A_{sf} normal to the fluid flow direction?

Some professionals consider the minimum cross section area of the particle when placed in a laminar flow. However, the grains and wheat straw orient randomly and rotate about the axis along their length, while this axis tends toward a horizontal plane position (Jiang et al., 1984; Bilanski and Lal, 1965). Thus, the resultant force acting on the particle is the vectorial summation of the vertical force exerted by the fluid through the particle center of gravity, the horizontal force that tends to rotate the particle about a vertical axis, and the torque of a couple that tends to change the inclination of the particle's long axis.

For a certain grain with weight w_s, the terminal velocity v_{ts} in an ascending airflow is given by the following equation (Mohsenin, 1980):

$$v_{ts} = \sqrt{\frac{2w_s(\rho_s - \rho_a)}{c_d a_{sf} \rho_s \rho_a}} \tag{3.24}$$

where:
  a_{sf} is the cross-sectional area of a seed, normal to the fluid flow, m²
  ρ_a is the air density, kg/m³
  ρ_s is the grain (seed) density, kg/m³
  w_s is the weight of 1000 seeds/1000, kg

Table 3.8 contains terminal velocity values for various crops. Similar information about crop MOG components is given in Table 3.9.

3.6 Friction Coefficients of Plant Grains and MOG

During harvesting, the plant components (stems, leaves, chaff, grains, etc.) are subject to friction due to sliding against each other, as well as due to their motion against the combine harvester working devices: reel, cutting blades, conveyors, threshing rotor elements and concave, cleaning sieves, and so forth. Friction forces are very important for the design of machine elements, for preserving the quality of the harvested grains, and for preventing grain damage during combine processing.

TABLE 3.9

Drag Coefficient and Terminal Velocity Values of MOG
Components of Selected Crops

Crop MOG Components	Drag Coefficient	Terminal Velocity, m/s (ft/s)
Wheat straw		
6 mm long	0.84	5.15 (16.9)
25 mm long	0.80	4.25 (13.94)
75 mm long	0.90	3.0 (9.84)
250 mm long	0.91	2.7 (8.86)
Various lengths		2.2–6.8 (7.22–22.31)
Wheat heads, unthreshed		6.5–9.2 (21.33–30.18)
Wheat heads, threshed		1.6–4.5 (5.25–14.76)
Corn cobs		6–12.3 (19.69–40.35)
Corn stover		6.4–15.4 (21–50.52)

Although not a fundamental force, a *friction force* arises at the contact point or surface between two bodies as a summation of fundamental electromagnetic forces between the charged particles of the contacting surfaces of those bodies.

A *static friction force* is the friction force that arises between two objects that have no relative movement to one other. For example, it is the force that prevents an object from sliding down a ramp with a relatively small inclination angle. The static friction force is always exerted on a body in the direction opposite that of its tentative direction of movement.

A *dynamic friction force* arises between two objects that move in relation to each other. The dynamic friction force is always exerted on a body in the direction opposite that of the body's movement.

The law of friction between a force P parallel to the movement direction and required to start or maintain the movement of a body, which exerts a force N in a direction normal to the contact surface, is expressed in a simple form:

$$P = \mu N \tag{3.25}$$

where μ is the *static* μ_s or *dynamic friction coefficient* μ_d between the sliding body and the surface.

Equation 3.21 is known as Coulomb's law, and it is based on the following assumptions:

1. The friction force is independent of the dimensions of the contact surface between the bodies.
2. The friction force is independent of sliding velocity.
3. The friction force depends on the nature of the contact surfaces.

These statements are valid mostly for dry friction cases; when wet materials are involved, additional adhesion forces develop at the contact surface and the friction coefficient may be altered. This happens because the adhesion forces depend on the size and shape of the contact surface between the bodies. On the other side, the temperature rises at the contact surface, and the friction coefficient may also be affected due to temperature sensitivity of plant components.

The static friction coefficient μ_s of a body sliding on the surface of a certain material is given by the tangent of the angle at which the body starts to slide down the slope of

TABLE 3.10

Static Friction Coefficients of Selected Crop Grains

Crop Grains	Moisture Content, Wet Basis, %	Steel	Wood
Wheat	12.7	0.29–0.32	0.42–0.47
	16.4	0.42–0.49	0.50–0.52
Barley	12.7	0.29–0.30	0.43–0.47
	16.4	0.34–0.36	0.47–0.51
Rye	12.7	0.30	0.41
	16.4	0.37	0.47
Oats	12.7	0.27–0.30	0.40–0.44
	16.4	0.33–0.43	0.44–0.49
Triticale	12.7	0.39	0.39
	16.4	0.41	0.47
Rice	11.8–12.9	0.209–0.231	0.228–0.251
Soybean	8–16	0.17–0.29	0.30–0.43
Corn	5.15–22	0.38–0.57	0.36–0.59
White kidney beans	10–25	0.348–0.38	0.32–0.42

the considered material. The values of static friction coefficients of grains are shown in Table 3.10 for different crop species, moisture contents, and natures of the material.

Since the force to maintain the motion of a sliding object is less than or equal to the force required to initiate the motion, the *dynamic coefficient of friction* is less than or equal to the static coefficient of friction.

The dynamic friction coefficient is very sensitive to the moisture content of the grains and MOG, because the moisture wets the friction surface and the frictional force increases. This requires a greater power requirement to process wet crops.

Plant MOG components (fragments of stalks, leaves, chaff, etc.) are not isotropic materials; consequently, their friction coefficient depends on their orientation relative to the direction of movement. Figure 3.7 shows the variation of the friction coefficient of wheat straw on a straw layer with the moisture content for various angles of orientation. Thus, the mathematical model that describes this dependency is

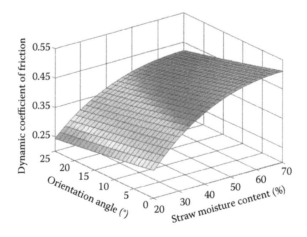

FIGURE 3.7

Dynamic friction coefficient of straw-on-straw layers for various angles of orientation.

$$\mu_{dM} = au_M e^{(-bu_M - c\beta)}$$

(3.26)

where:

μ_{dM} is the dynamic coefficient of wheat straw on straw
a is the coefficient
b and c are the exponents
u_M is straw moisture content
β is the angle of straw orientation relative to the movement direction

Equation 3.26 is a result of nonlinear regression analysis of the data shown in figure 276 (Sitkei, 1986). Thus, $a = 0.0195$, $b = 0.0141$, and $c = 0.0078$, while $R^2 = 0.99$.

For comparison, the dynamic (sliding) coefficient of straw on a galvanized steel surface is 0.30.

For more information on mechanical properties of agricultural materials, the practitioner is also referred to Sitkei (1986) and Stroshine (2004).

References

ANSI/ASAE. 1998. Density, specific gravity, and mass–moisture relationships of grain for storage. ANSI/ASAE D241.4. Washington, DC: American National Standards Institute.

Bilanski, W.K. and R. Lal. 1965. Behavior of threshed materials in a vertical wind tunnel. *Transactions of the ASAE* 8:411–416.

CGC. 2012. Wheat. In *Official Grain Grading Guide*, chap. 4. Winnipeg, Manitoba: Canadian Grain Commission. http://www.grainscanada.gc.ca/oggg-gocg/04/oggg-gocg-4f-eng.htm (accessed February 15, 2013).

Deger, G., M. Pakdemirli, F. Candan, et al. 2010. Strength of wheat and barley stems and design of new beams/columns. *Mathematical and Computational Applications* 15(1):1–13.

Gere, J.M. and S.P. Timoshenko. 1997. *Mechanics of Materials*. 4th ed. Boston: PWS Publishing Company.

Hopkins, R.B. 1970. *Design Analysis of Shafts and Beams*. New York: McGraw-Hill.

Jiang, S., W.K. Bilanski, and J.H.A. Lee. 1984. Analysis of the separation of straw and chaff from wheat by an air blast. *Canadian Agricultural Engineering* 26:181–187.

Mohsenin, N.N. 1980. *Physical Properties of Plants and Animal Materials*. New York: Gordon and Breach Science Publishers.

Parkinson, M.B., G.M. Roach, L.L. Howell, et al. 1996. Predicting the large-deflection path of tapered cantilever beams. DMI-9624574. Arlington, VA: National Science Foundation.

Ponce-Garcia, N., J.D.C. Figueroa, G.A. Lopez-Huape, H.E. Martinez, and R. Martinez-Peniche. 2008. Study of viscoelastic properties of wheat kernels using compression load method. *Cereal Chemistry* 85(5):667–672.

Sitkei, G. 1986. *Mechanics of Agricultural Materials*. Amsterdam: Elsevier

Song, H. and J.B. Litchfield. 1991. Predicting method of terminal velocity for grains. *Transactions of the ASAE* 34(1):225–231.

Stroshine, R. 2004. *Physical Properties of Agricultural Materials and Food Products*. R. Stroshine.

USDA. 2004. *Grain Inspection Handbook: Book II*. Washington, DC: U.S. Department of Agriculture.

Wacker, P. 1993. Bekämpfing von Unkräutern bei der Getreideernte [Fighting weeds while harvesting]. *Jahrbuch Landtechnik* 6(89):215–219.

Bibliography

Agri-Facts. 2007. Using 1,000 kernel weight for calculating seeding rates and harvest losses. Agdex 100/22-1. Edmonton, AB: Alberta Agriculture and Food.

Beighley, D.H. 2010. Growth and production of rice. In *Encyclopedia of Life Support Systems: Soils, Plant Growth, and Crop Production*. UNESCO–EOLSS.

Blahovec, J. 2011. Bending properties of plants. In *Encyclopedia of Agrophysics*, ed. J. Glinski, J. Horabik, and J. Lipiec. Dordrecht, Netherlands: Springer.

Boyd, N.S., E.B. Brennan, R.F. Smith, and R. Yokota. 2009. Effect of seeding rate and planting arrangements on rye cover crop and weed growth. *Agronomy Journal* 101(1):47–51.

Boz, H., K.E. Gercekaslan, M.M. Karaoglu, et al. 2012. Differences in some physical and chemical properties of wheat grains from different parts within the spike. *Turkish Journal of Agriculture and Forestry* 36:309–316.

Campbell, D.W. 1980. Modeling the combine harvesting. PhD dissertation, University of Saskatchewan.

CGC. 2000. Visual identification of small oilseeds and weed seed contaminants. *Grain Biology Bulletin*, 3:1–30.

Chandler Jr., R.F. 1979. *Rice in the Tropics: A Guide to the Development of National Programs*. Boulder, Colorado: Westview Press.

Charlet, L.D. and R.M. Aiken. 2005. Influence on planting date on sunflower stem weevil larval density. *Crop Management*, doi: 10.1094/CM-2005-0812-01-RS.

Chohan, M.S.M., M. Naeem, M. Khan, et al. 2004. Forage performance of different varieties of oats (*Avena sativa* L.). *International Journal of Agriculture and Biology* 6(4):751–752.

Crook, M.J. and A.R. Ennos. 1996. Mechanical differences between free-standing and supported wheat plants, *Triticum aestivum* L. *Annals of Botany* 77:197–202.

Degouet, C., B. Nsom, E. Lolive, and A. Grohens. 2007. Physical and mechanical characterization of soya, colza, and rye seeds. *Applied Rheology* 17(3):36546-1–36546-7.

Dillon, M.A. 2009. 2009 small grain research report. Center: Colorado State University.

Doehlert, D.C., M.S. McMullen, and J.-L. Jannink. 2005. Oat grain/groat size ratios: A physical basis for test weight. *Cereal Chemistry* 83(1):114–118.

Doehlert, D.C. and D.P. Wiessenborn. 2007. Influence of physical grain characteristics on optimal rotor speed during impact dehulling of oats. *Cereal Chemistry* 84(3):294–300.

Dunca, J. 2008. Mechanical properties of cereal stem. *Research in Agricultural Engineering* 54:91–96.

Engineering Toolbox. 2013. Density of some common materials. http://www.engineeringtoolbox.com/density-materials-d_1652.html (accessed January 6, 2013).

Esehaghbeygi, A., B. Hoseinzadeh, M. Khazaei, and A. Masoumi. 2009. Bending and shearing properties of wheat stem of Alvand variety. *World Applied Sciences Journal* 6(8):1028–1032.

Folkins, L.P. and M.L. Kaufmann. 1974. Yield and morphological studies with oats for forage and grain production. *Canadian Journal of Plant Science* 54:617–620.

Ghadge, P.N. and K. Prasad. 2012. Some physical properties of rice kernels; variety PR-106. *Journal of Food Process Technology* 3(8):1000175.

Ghasemi, A., M.M. Ghasemi, and M. Pessarakli. 2012. Yield and yield components of various grain sorghum cultivars grown in an arid region. *Journal of Food, Agriculture and Environment*, 10(1):455–458.

GTA. 2011. *Visual Recognition Standards Guide 2011/2012 for Grain Commodity Sampling and Assessment*. Sydney: Grain Trade Australia.

Gürsoy, S. and E. Güzel. 2010. Determination of physical properties of some agricultural grains. *Research Journal of Applied Sciences, Engineering and Technology* 2(5):492–498.

Hall, B. 2006. Fine tuning canola seeding rates and row widths. Guelph: Ontario Ministry of Agriculture, Food and Rural Affairs. http://www.omafra.gov.on.ca/english/crops/field/news/croptalk/2006/ct_0306a9.htm (accessed May 4, 2012).

Hanna, S., S.P. Conley, and J. Santini. 2008. Impact of fungicide application timing and crop row spacing on soybean canopy penetration and grain yield. *Agronomy Journal* 100:1488–1492.

Hansen, H.B., B. Moller, S.B. Andersen, et al. 2004. Grain characteristics, chemical composition, and functional properties of rye (*Secale cereal* L.) as influenced by genotype and harvest year. *Journal of Agriculture, Food and Chemistry* 52:2281–2291.

Hirai, Y., E. Inoue, K. Mori, and K. Hashiguchi. 2002. Analysis of reaction forces and posture of a bunch of crop stalks during reel operations of a combine harvester. In *Agricultural Engineering International: The CIGR Journal of Scientific Research and Development* IV, Manuscript FP 02 002.

Huitink, G. Harvesting grain sorghum. In *Grain Sorghum Handbook*, chap. 8. Fayetteville, AR: University of Arkansas, Cooperative Extension and Service Publications.

Huitink, G. and T. Siebenmorgen. Maintaining yield and grain quality. In *Rice Production Handbook*, chap. 12. Fayetteville, AR: University of Arkansas, Cooperative Extension and Service Publications.

Isik, E. and H. Unal. 2011. Some engineering properties of white kidney beans (*Phaseolus vulgaris* L.). *African Journal of Biotechnology* 10(82):19126–19136.

Jafari, S., J. Khazaei, A. Arabhosseini, et al. 2011. Study on mechanical properties of sunflower seeds. *Electronic Journal of Polish Agricultural Universities* 14(1):6.

Jayas, D.S. and S. Cenkowski. 2006. Grain property values and their measurements. In *Handbook of Industrial Drying*, ed. A.S. Mujumdar, 575–603. 3rd ed. Philadelphia: Taylor & Francis.

Jezowski, S., T. Adamski, M. Surma, et al. 2000. Variation of some physical and geometrical stem features in doubled haploids of barley. *International Agrophysics* 14:187–189.

Jian, W., Z. Jinmao, L. Quiqin, et al. 2006. Effects of stem structure and cell wall components on bending strength in wheat. *Chinese Science Bulletin* 51(7):815–823.

Jouki, M. and N. Khazaei. 2012. Some physical properties of rice seed (*Oryza sativa*). *Research Journal of Applied Sciences, Engineering and Technology* 4(3):1846–1849.

Kandel, H. 2010. Soybean production: Field guide for North Dakota and Northwestern Minnesota, A-1172. Fargo, ND: North Dakota State University, Extension Service. http://www.ag.ndsu.edu/extensionentomology/field-crops-insect-pests/Documents/soybean/a-1172-soybean-production-field-guide (accessed May 4, 2012).

Kaya, Y., G. Evci, S. Durak, et al. 2009. Yield components affecting seed yield and their relationships in sunflower (*Helianthus annuus* L.). *Pak. J. Bot.* 41(5):2261–2269.

Kelly, J.P. 2009. By-plant prediction of corn (*Zea mays* L.) grain yield using height and stalk diameter. BSc thesis, Oklahoma State University.

Kheiralipour, K., M. Karimi, A. Tabatabaeefar, et al. 2008. Moisture-depend physical properties of wheat (*Triticum aestivum* L.). *Journal of Agricultural Technology* 4:53–64.

Kibar, H. and T. Ozturk. 2008. Physical and mechanical properties of soybean. *International Agrophysics* 22:239–244.

Kim, S.H. and J.M. Gregory. 1989. Power requirement model for combine cylinders. ASAE Paper 891592. St. Joseph, MI: American Society of Agricultural Engineers.

Kukelko, D.A., D.S. Jayas, D. S. White, et al. 1987. Physical properties of canola (rapeseed) meal. *Canadian Agricultural Engineering* 30(1):61–64.

Lee, C. and J. Herbek. 2005. Estimating soybean yield. Lexington: University of Kentucky, Cooperative Extension Service.

Madsen, E. and N.E. Langkilde. 1987. *ISTA: Handbook for Cleaning of Agricultural and Horticultural Seeds on Small-Scale Machines.* Zurich: International Seed Testing Association.

Miu, P. 1995. Modeling of the threshing process in cereal combine harvesters [in Romanian]. PhD dissertation, Politehnica University of Bucharest.

Muir, W.E. and R.N. Sinha. 1987. Physical properties of cereal and oilseed cultivars grown in western Canada. *Canadian Agricultural Engineering* 51–55.

Mujumdar, A.S. 2007. *Handbook of Industrial Drying.* Boca Raton, FL: CRC Press/Taylor & Francis.

Navarro, S. and R.T. Noyes (eds.). 2001. *The Mechanics and Physics of Modern Grain Aeration Management.* Boca Raton, FL: CRC Press.

Oelke, E.A., E.S. Oplinger, H. Bahri, et al. 1990. Rye. In *Alternative Field Crops Manual.* Madison, WI: University of Wisconsin.

Oelke, E.A., E.S. Oplinger, and M.A. Brinkman. 1989. Triticale. In *Alternative Field Crops Manual.* Madison, WI: University of Wisconsin.

Oladokun, M.A.O. and A.R. Ennos. 2006. Structural development and stability of rice *Oryza sativa* L. var. *Nerica 1. Journal of Experimental Botany* 57(12):3123–3130.

Oplinger, E.S., E.A. Oelke, J.D. Doll, et al. 1990. Flax. In *Alternative Field Crops Manual.* Madison, WI: University of Wisconsin.

Parnell Jr., C.B., D.D. Jones, R.D. Rutherford, and K.J. Goforth. 1986. Physical properties of five grain dust types. *Environmental Health Perspectives* 66:183–188.

Pedersen, P. 2007. Soybean plant population. Ames: Iowa State University. http://extension.agron. iastate.edu/soybean/production_plantpopulation.html (accessed May 2, 2012)

Poole, N. 2005. *Cereal Growth Stages.* Lincoln, New Zealand: Grains Research and Development Corporation.

Rahman, M.S. 2007. *Handbook of Food Preservation.* Boca Raton, FL: CRC Press.

Ransom, J., T. Friesen, R. Horsley, et al. 2011. North Dakota barley, oat, and rye. Fargo, ND: North Dakota State University, Extension Service.

Razavi, S.M.A., S. Yeganehzad, and A. Sadeghi. 2009. Moisture dependent physical properties of canola seeds. *Journal of Agricultural Science and Technology* 11:309–322.

Robinson, A.P. and S.P. Conley. Plant populations and seeding rates for soybeans. West Lafayette, IN: Purdue University. https://www.extension.purdue.edu/extmedia/ay/ay-217-w.pdf (accessed May 3, 2012).

Sapra, V.T. and E.G. Hayne. 1974. Variations in yield characteristics in three populations of winter triticale. *Transactions of the Kansas Academy of Science* 76(1):18–24.

Shinners, K.J., G.C. Boettcher, D.S. Hoffman, et al. 2009. Single-pass harvest of corn grain and stover: Performance of three harvester configurations. *Transactions of the ASABE* 52(1):51–60.

Shoughy, M.I. and M.I. Amer. 2004. Physical and mechanical properties of Faba bean seeds. *Misr Journal of Agricultural Engineering* 23(2):434–447.

Shroyer, J., H. Kok, and D. Fjell. 1998. Seedbed preparation and planting practices. In *Grain Sorghum Production Handbook.* Manhattan, KS: Kansas State University.

Silk, W.K., L.L. Wang, and R.E. Cleland. 1982. Mechanical properties of the rice panicle. *Plant Physiology* 70:460–464.

Simetric Co. UK. 2011. Density of materials. Colchester, UK: Simetric Co. http://www.simetric.co.uk/ si_materials.htm (accessed May 1, 2012).

Simonyan, K.J., A.M. El-Okene, and Y.D. Yiljep. 2007. Some physical properties of Samaru sorghum 17. *Agricultural Engineering International: The CIGR Ejournal,* IX:1–15, Manuscript FP 07 008.

Soliman, N.S., M.A. Abd El Maksoud, G.R. Gamea, and Y.A. Qaid. 2009. Physical characteristics of wheat grains. *Misr Journal of Agricultural Engineering* 26(4):1855–1877.

Sykorova, A., E. Sarka, Z. Bubnik, et al. 2009. Size distribution of barley kernels. *Czech Journal of Food Science* 27(4):249–258.

Tamm, U. 2003. The variation of agronomic characteristics of European malting barley varieties. *Agronomy Research* 1:99–103.

Tarighi, J., A. Mahmoudi, and N. Alavi. 2011. Some mechanical and physical properties of corn seed (var. DCC 370). *African Journal of Agricultural Research* 6(16):3691–3699.

Tavakoli, H., S.S. Mohtasebi, and A. Jafari. 2009. Effect of moisture content, internodes position, and loading rate on the bending characteristics of barley straw. *Research in Agricultural Engineering* 55:45–51.

Taylor, B.N. 2001. *The International System of Units (SI).* NIST Special Publication 330. Gaithersburg, MD: National Institute of Standards and Technology.

Taylor, R.K. 1998. Harvesting grain sorghum. In *Grain Sorghum Production Handbook.* Manhattan, KS: Kansas State University.

Tiwari, V. 2010. Growth and production of oat and rye. In *Encyclopedia of Life Support Systems. Soils, Plant Growth, and Crop Production.* Paris: UNESCO–EOLSS.

Touming, L., L. Li, Y. Zhang, et al. 2011. Comparison of quantitative trait loci for rice yield, panicle length, and spikelet density across three connected populations. *Journal of Genetics* 90:377–382.

Tzvelev, N.N. 1976. *Cereals of the USSR*. Leningrad: Nauka.

Uhl, J.B. and B.J. Lamp. 1966. Pneumatic separation of grain and straw mixtures. *Transactions of the ASAE* 9(2):244–246.

USDA. 2010. Sorghum: Area, yield, production, and value. Agricultural statistics, table 1-62. Washington, DC: U.S. Department of Agriculture. http://www.nass.usda.gov/Publications/Ag_Statistics/index.asp (accessed May 6, 2012).

USDA. 2012. Crop production, 2011 summary. Washington, DC: U.S. Department of Agriculture National Agricultural Statistics Services.

Washmon, C.N., J.B. Solie, W.R. Raun, and D.D. Itenfisu. 2002. Within field variability in wheat, grain yields over nine years in Oklahoma. *Journal of Plant Nutrition* 25:2655–2662.

Wright, C.T., P.A. Pryfogle, N.A. Stevens, et al. 2005. Biomechanics of wheat/barley straw and corn stover. *Applied Biochemistry, Biotechnology*, 2005:5–19

4

Plant Cutting, Gathering, and Conveying Processes and Equipment

4.1 Introduction

Plant cutting, gathering, and conveying are among the main combine harvesting processes; they are performed by changeable headers coupled to a feeder-house, as required for harvesting particular crops or for some harvesting technology (e.g., plant cutting vs. grain stripping only). The types of headers were briefly described in the Chapter 1; they are the standard grain header (platform header), draper header (for wheat, rice, barley, oats, rye, triticale, and soybeans), corn header (for ears only or ears and stems), stripper header (for rice), sunflower header, and pick-up header (for beans). In the following, the header components are described along with mathematical models of their processes.

4.2 Grain Header

Ultimately, the main goal of combine harvesting is to retrieve from the field as much grain as possible. The grain header is the first combine equipment that handles and gathers the crop plants from the field.

Grain headers are used for harvesting cereals with small grains: wheat, barley, oats, rye, triticale, rice, soybean, flax, milo, and other cereal crops. The headers are built in series of four to six sizes (e.g., 6–12 m, 20–40 ft) to work with combines of different sizes.

The grain header (Figure 4.1a) is made of a platform (1) with a frontal, reciprocating-knife cutter bar (2); two lateral dividers of the plants (3); cross-mounted, equidistant crop-lifting devices (4); a reel (5) with slates and metal or plastic tines (fingers); and an intake auger (6), combined with a retractable-finger drum (7). All components, including power transmission (8), are assembled on the header frame that has a welded structure.

While the combine is driven in the crop field, the reel rotates and its bars with fingers enter vertically among the plants, which are then pushed toward the cutter bar, falling on the platform and being gathered to the header center by the intake auger. Then, the gathered material is pushed by the retractable-finger drum into the combine feeder-house, which is the physical coupling interface between the header and the combine. The header of modern combines is maintained at the operating height using a closed-loop control system that receives the input from a sensor arm. The most common sensor arm is in the form of a ground runner, which engages and rides over the soil surface (Figure 4.1b).

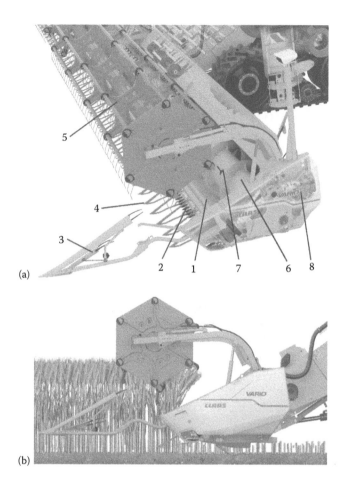

FIGURE 4.1
(a) Main functional components of a grain header. **(See color insert.)** (b) Lateral view of a grain header in the field. (Courtesy of CLAAS, Harsewinkel, Germany.)

A grain header may have some improved or additional components, as follows: extendable platform, flexible floating cutter bar, full-fingered intake augers, full-width retracting fingers, side cutter bars (Figure 1.28), different knife–guard combinations, headlights, or an extended platform for rapeseed harvesting. A flexible cutter bar can flex over field contours and ridges to better cut the plants that have pods close to the ground (e.g., soybean).

4.2.1 Plant Supporting and Orienting Reel

The reel, the first active component of a header that comes into contact with the plants in the field, has a pentagonal or hexagonal–prismatic structure spanning the entire header length. Although reel designs may differ among major combine manufacturers, a reel (Figure 4.2) is mainly composed of the following subassemblies: reel arms (1) whose position is adjusted hydraulically, driving and nondriving end shafts, a central tube (2) that has equidistant flanges (3), an eccentric end shield (4), bars with tines (fingers) (5), end

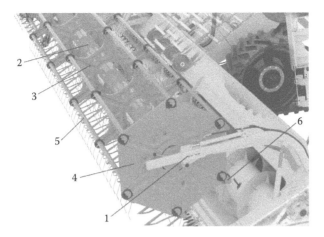

FIGURE 4.2
Main components of a reel.

cranks (6), and power transmission components (mechanical or hydraulic). The space between the flanges varies according to the length of the reel and the considered mechanical robustness of the reel structure versus the total weight. Figure 4.3 shows a bar with double tines; the helical spring between them makes the tines elastic, to minimize grain shattering loss and plants entangling with them.

The main functions of a reel are as follows: dividing the plants in the field in bunches, orienting them toward the cutter bar of the header, supporting them during the cutting process, and shunting all of them on the header platform or draper. Additional functional requirements must be satisfied by the reel: lifting the plants that are fallen on the ground and avoiding grain shattering due to the impact of tine bars on the plant ears.

The functional and constructional parameters of the reel process are shown in Figures 4.4 and 4.5. As a combine progresses into the grain field, the tine bar must vertically enter through the crop plants, and then the stalks are bent backward toward the cutter bar, which slices them at the desired height, up to just below the heads of ripe grains.

With reference to the fixed Cartesian coordinate system XOY in Figure 4.4, the absolute movement of a tine bar is the result of the combination of the translational movement of

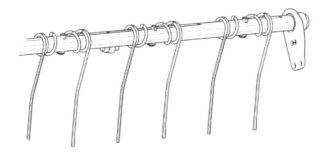

FIGURE 4.3
Reel bar with tines.

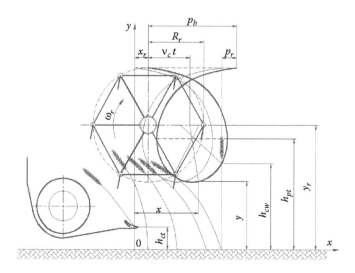

FIGURE 4.4
Motion parameters of conventional tined reel.

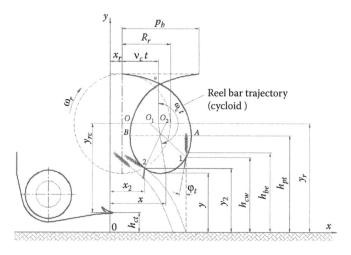

FIGURE 4.5
Prolate cycloid trajectory of a reel tine bar.

the combine harvester with its velocity v_c (m/s) and the rotation n_r (rpm) of the reel around its own axis.

In the analysis of the reel tine bar kinematics, the following assumptions have been made:

- The reel rotational speed is constant and taken to be positive in the clockwise sense of rotation.
- The velocity v_c of the combine harvester is constant and specified from throughput requirements.
- The tines have no rotational motion relative to the fixed coordinate system *XOY*.

Let's consider a reel bar at its initial position on the top of the reel (above its shaft). The reel shaft initial position is given by the coordinates x_r, y_r. At an arbitrary time t, the position of the center of the considered bar (we neglect the bar radius) is represented by the following parametric equations of a cycloid:

$$x = x_r + v_c t + R_r \sin \omega_r t$$

$$y = y_r + R_r \cos \omega_r t$$

(4.1)

where:
   x_r is the abscissa of the reel axis, m
   y_r is the ordinate of the reel axis, m
   R_r is the reel radius (to the tine bar center), m
   ω_r is the angular speed of the reel, m/s

The angular speed of the reel can be calculated with Equation 4.2 as follows:

$$\omega_r = \frac{\pi n_r}{30}, \, s^{-1}$$

(4.2)

Then,

$$y_r = h_{ct} + y_{rc}$$

(4.3)

where:
   h_{ct} is the plant cutting height, m
   y_{rc} is the vertical distance between the reel axis and the cutting bar plane, m

For reel kinematics analysis, the plant cutting height is considered known.

In a matrix format of homogenous coordinates, the displacement equations (4.1 and 4.3) can be written as

$$\begin{bmatrix} x \\ y \\ 1 \end{bmatrix} = \begin{bmatrix} x_r + v_c t & \sin \omega_r t & \cos \omega_r t \\ h_{ct} + y_{rc} & \cos \omega_r t & -\sin \omega_r t \\ 1 & 0 & 0 \end{bmatrix} \begin{bmatrix} 1 \\ R_r \\ 0 \end{bmatrix}$$

(4.4)

In the above equations, the following variables are considered unknown: x_r, y_{rc}, ω_r, and R_r.

Let us define the dimensionless kinematic index λ_r of the reel as the ratio

$$\lambda_r = \frac{\omega_r R_r}{v_c}$$

(4.5)

The shape of the cycloid depends on the λ_r value. Thus, if $\lambda_r < 1$, the reel bar trajectory is a *curtate cycloid*. In such a case, the reel bar not only does not bend the plants toward the cutter bar, but contrarily, it pushes the plants toward the front direction of the combine. If $\lambda_r = 1$, the reel bar has no influence on the plants, except that it may shatter the grains by striking some ears.

The reel function is accomplished only on a segment of the trajectory, on the arc between points 1 and 2, respectively, of the *prolate cycloid*, as shown in Figure 4.5. Along this trajectory arc, the bar velocity is $v_x < 0$. Thus, by deriving the first equation of the equation system (4.1) with respect to the time, the reel bar speed must be

$$v_x = v_c + \omega_r R_r \cos \omega_r t < 0 \tag{4.6}$$

Accordingly, the reel function is accomplished if $\lambda_r > 1$, because $\cos \omega_r t < 1$.

The interval limits of the bar action on the plant correspond to points A and B, where the tangents to the cycloid are vertical. That means, in these points, the horizontal component of bar speed $v_x = 0$, that is,

$$v_c + \omega_r R_r \cos \omega_r t_A = 0 \tag{4.7}$$

Thus, the angle the bar enters among the plants in the field is given by the equation

$$\omega_r t_A = -\arccos \frac{1}{\lambda_r} \tag{4.8}$$

This requires the ordinate of these points to be equal or slightly lower than the plants' height. In other words, the height of the reel axis is given by the equation

$$y_r \leq h_{pt} + \frac{R_r}{\lambda_r} \tag{4.9}$$

Accordingly, the vertical position of the reel axis above the cutting bar plane can be calculated with the following equation:

$$y_{rc} \leq h_{pt} - h_{ct} + \frac{R_r}{\lambda_r} \tag{4.10}$$

For an efficient action of the reel and to avoid excessive bending of the plants, as well as prematurely breaking their stems, a bar of the reel shall act at or above the height of the plant center weight. Mathematically, this can be expressed as follows:

$$y_{rc} \geq h_{cw} - h_{ct} + R_r \cos \omega_r t_1 \tag{4.11}$$

where h_{cw} is the height to the plant center weight (m).

Thus, the height y_{rc} of the reel axis should satisfy both Equations 4.10 and 4.11.

At the lower limit, with Equation 4.11, we can calculate the angle $\varphi_1 = \omega_r t_1$ and then the abscissa x_1 of point 1 as follows:

$$x_1 = x_r + v_c t_1 + R_r \sin \omega_r t_1 \tag{4.12}$$

Applying Equation 3.11 to the actual configuration shown in Figure 4.5, for most bent plants (located at $x = x_1$) we can write the following equation:

$$y = h_{cw}\sqrt{1 - \left(\frac{x_1 - x}{c_{bp}h_{cw}}\right)^2} \tag{4.13}$$

where c_{bp} is the plant bending coefficient, calculated with Equation 3.12.

With the equations above, we can simulate the bending of the plant when the reel bar moves from point 1 to point 2.

It is extremely important that the reel is low enough that the plant stem will be in contact with the plant on the entire cycloid arc between points 1 and 2. As a consequence, there is another functional condition that must be satisfied: the length of the bent plant from the ground to the last contact point (2) shall be less than or equal to the stem height h_{be}.

Mathematically, we express this as follows:

$$\int_0^{x_1 - x_2} \left(1 + \left[y'(x)\right]^2\right)^{1/2} dx \le h_{be} \tag{4.14}$$

where

$$y'(x) = \frac{dy}{dx} \tag{4.15}$$

Corresponding speed components of the reel bar can be found by differentiating the displacement matrix (4.4) with respect to time; thus, we get

$$\begin{bmatrix} v_x \\ v_y \\ 0 \end{bmatrix} = \begin{bmatrix} v_c & \omega_r \cos \omega_r t & -\omega_r \sin \omega_r t \\ 0 & -\omega_r \sin \omega_r t & -\omega_r \cos \omega_r t \\ 0 & 0 & 0 \end{bmatrix} \begin{bmatrix} 1 \\ R_r \\ 0 \end{bmatrix} \tag{4.16}$$

where v_y is the reel bar velocity on the Y axis. The 3×3 square matrix is the velocity matrix; the null components of the third row correspond to the velocity on the Z axis that obviously does not apply to the reel bar in this study.

Let us introduce a dimensionless parameter R_λ as the ratio

$$R_\lambda = \frac{v_c}{\omega_r} \tag{4.17}$$

that is, the advance of the combine per radian of reel rotation.

That means

$$\lambda_r = \frac{R_r}{R_\lambda} \tag{4.18}$$

From Figure 4.5, it can be inferred that

$$\omega_r t_A + \omega_r t_B = 2\pi \tag{4.19}$$

Combining Equation 4.19 with the reel bar motion equations (4.1), we get the following equation (Oduori et al., 2012):

$$\frac{x_A - x_B}{2R_r} = \sqrt{1 - \left(R_\lambda / R_r\right)^2} - \frac{R_\lambda}{R_r}\cos^{-1}\left(R_\lambda / R_r\right) \tag{4.20}$$

This may be simplified to

$$\frac{x_A - x_B}{2R_r} = \left[1 - \left(R_\lambda / R_r\right)\right]^{1.55} \tag{4.21}$$

This equation serves to help us choose the right kinematic index of the reel for a crop with a certain plant height.

Let us note the number of the reel bar with z_r. The distance between two homologous, successive points of the trajectory of a bar is called the *reel pitch* p_r. Its size is given by Equation 4.22:

$$p_r = \frac{2\pi R_r}{\lambda_r} \tag{4.22}$$

The distance between two homologous points of the trajectories of two consecutive bars is called the *bar pitch* p_b and can be calculated as follows:

$$p_b = \frac{2\pi R_r}{\lambda_r n_b} \tag{4.23}$$

where n_b is the number of reel bars.

While the reel bar follows its trajectory between points 1 and 2, the contact point between the row plants being bent and the reel bar changes between the points corresponding to h_{cw} and h_{be}. One can combine Equations 4.14 and 4.16 to find the speed of the bar relative to the plant stem over the cycloid arc between points 1 and 2.

Let us note with l_t the length of the tine whose vertical inclination is given by the angle φ_t (Figure 4.5). The coordinates x_t and y_t of the tine tip are given by the following matrix:

$$\begin{bmatrix} x_t \\ y_t \\ 1 \end{bmatrix} = \begin{bmatrix} x_r + v_c t - l_t \sin\varphi_t & \sin\omega_r t & \cos\omega_r t \\ h_{ct} + y_{rc} - l_t \cos\varphi_t & \cos\omega_r t & -\sin\omega_r t \\ 1 & 0 & 0 \end{bmatrix}\begin{bmatrix} 1 \\ R_r \\ 0 \end{bmatrix} \tag{4.24}$$

The angle φ_t can be adjusted with the help of an eccentric end shield mounted on a lateral side of the reel that rotates the crank of each bar. Once the angle φ_t is adjusted, the corresponding speeds, velocities, and accelerations of a reel bar and tine tip at each trajectory point are equal since the tines translate without rotating. The value of the reel kinematic index shall be chosen so that when the bar and tines penetrate the lodged and tangled crop foliage, the grain loss through ear shattering will be avoided or minimized.

Ideally, the tine orientation should be changed over a full bar cycloid to satisfy the following requirements:

- Avoid grain shattering by entering among the crop plants in a vertical position
- Lift and rank the plants when feeding the plants to the cutter bar and header platform
- Release the cut plants on the header platform by retracting from the plants in a vertical position

There must be a correlation between the reel and cutter bar actions. According to Figure 4.6, when the leading bar follows its cycloid path between points 1 and 2, the cutter bar moves between points C_1 and C_2. This is a common working phase of the reel and cutter bar. The plant that was bent the most is not cut during this time, simply because its stem is not within the C_1C_2 interval. After escaping from the bar influence, it relaxes a bit while being cut by the cutter bar within the C_3C_4 interval due to the action of the lagging reel bar.

Within the C_2C_3 distance, the cutter bar works in two phases: an idle phase and a cutting phase of relaxed plant stems, which are subjected to the wave propagation of plant bending by the consecutive bar. An efficient process of the reel is obtained by choosing all functional parameters so that the idle-phase duration of the cutter bar tends to zero.

When designing the header, the process of the reel needs to be optimized for harvesting a relatively large range of crops, whose physical properties were given in Chapter 3. Another thing must also be taken into consideration: the bending wave propagation through the plants not being in direct contact with the reel bar. Ideally, modeling of plant bending waves toward the reel or away from the reel due to wind influence would be the best improvement of the mathematical models described above.

In the following, we suggest a mathematical model of wave propagation through the plants due to the reel bar action. The neighboring plants that are behind (relative to the combine sense of direction) are also pushed by the reel toward the cutting bar because of the transfer of bent plant momentum via pressure and friction drag. A similar effect is the wave propagation through a field crop caused by vortices. As a turbid medium, the plants attenuate the wave propagation.

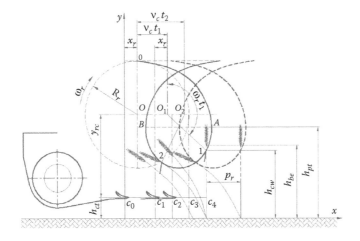

FIGURE 4.6
Correlation of reel and cutter bar actions.

In a manner consistent with the *Beer–Lambert exponential extinction law*, we can assume that the deflection f of plants in the X direction from their roots' vertical direction decreases exponentially with the distance measured from the reel bar current position, according to the following law:

$$f = f_0 e^{-abx} \tag{4.25}$$

where:

    f is the deflection of plants not in contact with the reel bar, from their vertical position, m
    f_0 is the deflection of plants in contact with the reel bar, from their vertical position, m
    a is the plant density attenuation coefficient
    b is the plant foliage absorbing coefficient

The author considers that this is a new field where additional knowledge must be acquired and incorporated into the models described in this chapter.

In modern combine harvesters, the diameter of the reel varies between 1.016 m (40 in.) for the grain headers (with auger) and 1.65 m (65 in.) for the draper headers. The reel rotation varies hydraulically or mechanically between 5 and 85 rpm for harvesting a large range of cereals. The values of the kinematic index of the reel vary with the crop and crop conditions. It is recommended that $\lambda_r = 1.15–1.5$.

To vary the angular speed, a reel can be powered mechanically using a double or multiple V-belt, or hydraulically. The height of the reel above the cutter bar can be adjusted by rotating the reel arms with an angle within the range of –11° to +10° from its baseline using hydraulic cylinders placed at the ends of the reel. Positioning the reel in the horizontal direction relative to the cutter bar is also done with lateral hydraulic cylinders mounted on the reel arms, as seen in Figure 4.1a.

4.2.2 Plant Cutting: Cutter Bar

4.2.2.1 General

During combine harvesting, depending on the crop and header type, the plants may be cut by one of the following cutter types:

- Reciprocating cutter bar (grain header or draper header)
- Rotary blade cutters (corn header and sunflower header)
- Cutting discs (corn header)

A reciprocating cutter bar cuts the plant stalks between a moving and a resting blade, while fast-moving rotary blade cutters make a free cut where the countersupport is ensured by the moment of inertia of the stalk.

A corn header section may just *wiredraw* the stalk of the corn plant between two stripping rotors, while a stripper (for rice, oats) does not cut the plant stems, the grain being detached by combing the ears and stripping the grain only.

The standard reciprocating cutter bar (or *sickle bar*) is composed of three main components: the carrying frame, the cutting blade system, and the driving mechanism. The cutting blade system consists of a single or double reciprocating multiknife (multiteeth) bar; a single reciprocating multiknife bar works in conjunction with stationary

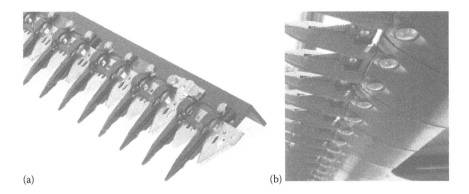

FIGURE 4.7
(a) Top view of a standard cutter bar (sickle bar). (Courtesy of Gebr. Schumacher. Eichelhardt, Germany.)
(b) Bottom view of a cutter bar assembled to the header platform.

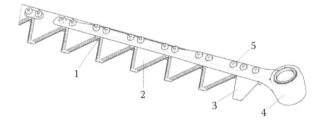

FIGURE 4.8
Single reciprocating multiknife bar.

guard sickles acting also as a carrying frame (Figure 4.7a and b). The multiknife bar (Figure 4.8) is composed of a flat bar (1), on which the multiteeth knives (2) and the end blade (3) are assembled with the fasteners (5). The head assembly (4) connects with the crank mechanism so that the multiknife bar reciprocates through the guard fingers. A multiteeth (serrated) blade, typically the cutting tool, is a trapezoid-shaped tempered-steel plate. The guard fingers may have cutouts with small riveted knives (counter-knives), or the steel fingers' edges are sharpened to act as counterknives. The fingers divide the plants and support them during the cutting process; they also guard the knives against rocks or soil unevenness. A required clearance between the knives and counterknives or guard fingers is maintained by equally spaced, adjustable clips fastened to the counterbar.

4.2.2.2 Cutter Bar Working Process

The stem cutting process of the reciprocating cutter bar can be decomposed in three phases, as follows:

1. Separation of the plant stems and their slippage along the fingers toward the knives
2. Forward and lateral inclination of the plants between the knives
3. Cutting of the stems and release of the knives from cut material

During the first phase, the fingers divide the crop plants in strips by inclining them laterally, while the cutting bar mounted on the combine header moves forward.

During the second phase, the stems come progressively into contact with the cutting edges of the knives and counterknives (finger edges). During phase 3, the stems are cut through a shearing mode. The geometry and design of the knives depend on the forces required to cut the plant and the desired cutting area covered by the absolute motion of the knives; they are studied in correlation with the combine speed and the lateral speed of the knives, which results from the reciprocating motion of the multiknife bar relative to the guard fingers.

Figure 4.9 shows the forces acting on the counterknife (finger) by a plant stalk during the phase 1 contact. The plant that was laterally moved by the finger exercises the normal force N on it. Due to friction, the finger pushes the plant stalk with force F_f. The plant stalk will be cut if the plant slides along the finger edge into the working area of the cutter bar. This condition is met if the resistant force T, due to stalk bending resistance and inertia, is higher than the friction force, that is, $T > F_f$. This means

$$R\cos\alpha_f > \mu_f R\sin\alpha_f \tag{4.26}$$

where μ_f is the friction coefficient of the plant with the finger (counterknife).

If φ_f is the friction angle of the plant with the finger, then

$$\mu_f = \tan\varphi_f \tag{4.27}$$

Combining the last two equations, we get the required condition so that the plant stalk slides along the finger edge:

$$\cos(\alpha_f + \varphi_f) > 0 \tag{4.28}$$

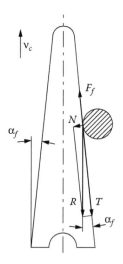

FIGURE 4.9
Forces acting on the counterknife (finger) by a plant stalk.

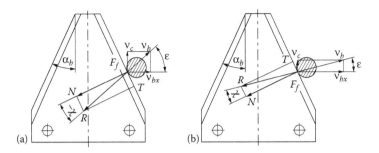

FIGURE 4.10
Forces developed by the knife (blade) on a plant stalk during the contact.

During the second phase, the plants come progressively into contact with the cutting edges of the blades. Figure 4.10 shows the forces developed by a knife (blade) on a plant stalk during the contact. Due to the reciprocating motion of the cutter bar, the value and direction sense of the blade speed v_{bx} change continuously. If the cutter bar is driven using an eccentric crank–connecting rod (crank–conrod) mechanism whose crank angular speed is ω_c and length r_c, the blade speed v_{bx} is given by the following equation:

$$v_{bx} = r_c \sin \omega_c t \tag{4.29}$$

where t is the time.

The instantaneous absolute speed of a point of the cutting blade relative to the ground is v_b, with the direction given by the angle ε, whose value changes as the speed v_{bx} value changes.

During the contact with the cutting blade, the plant stalk develops a resistance force R due to the stalk bending, as well as due to plant inertia. Its component T tends to move the stalk along the cutting edge, while the friction force F_f opposes the movement of the plant relative to the blade edge. If φ_b is the friction angle of the plant stalk with the cutting blade, then we can write

$$T = \tan \varphi_b \tag{4.30}$$

Change in the direction of the force R is given by the value and sign of the angle χ, as shown in Figure 4.10a and b. Consequently, there are two distinct situations, as follows:

1. Angle $\chi < 0$. In this case, the force T acts toward the blade base. In such a case, congestion of the plant stalks at the blade base occurs; this will negatively impact the cutting process during the next phase. Taking into account that

$$\chi = \varepsilon - \alpha_b \tag{4.31}$$

   provided that the plant stalks do not slip toward the cutting blade base, the following condition applies:

$$\varepsilon \leq \alpha_b + \varphi_b \tag{4.32}$$

This is equivalent to

$$\frac{v_c}{v_{bx}} \leq \tan\left(\alpha_b + \varphi_b\right) \tag{4.33}$$

Since the maximum value of $\sin \omega_c t = 1$, the following condition must be met to prevent sliding the plant stalks to the base of the cutting blade:

$$v_c \leq \frac{\omega_c r \left(\tan \alpha_b + \tan \varphi_b\right)}{1 - \tan \alpha_b \tan \varphi_b} \tag{4.34}$$

2. Angle $\chi > 0$. In this case, the force T acts toward the front of the cutting blade; thus, there is a risk of the plant escaping and remaining uncut. Taking into account that, in this case,

$$\chi = \alpha_b - \varepsilon \tag{4.35}$$

to prevent the plants from escaping uncut from the cutter bar, the following condition applies:

$$\varepsilon \geq \varphi_b - \alpha_b \tag{4.36}$$

The last inequality is equivalent to the following one:

$$v_c \geq \frac{\omega_c r \left(\tan \alpha_b - \tan \varphi_b\right)}{1 + \tan \alpha_b \tan \varphi_b} \tag{4.37}$$

According to the above-stated conditions, when harvesting a crop whose plant stalk's friction angle with the cutter blade is φ_b, the combine harvester speed v_c should be chosen within the interval given by the following inequalities:

$$\frac{\omega_c r \left(\tan \alpha_b - \tan \varphi_b\right)}{1 + \tan \alpha_b \tan \varphi_b} \leq v_c \leq \frac{\omega_c r \left(\tan \alpha_b + \tan \varphi_b\right)}{1 - \tan \alpha_b \tan \varphi_b} \tag{4.38}$$

The equation above implies correlation of design parameters r and α_b, and functional parameter ω_c with the plant stalk's friction angle φ_b when combine speed is specified. The friction angle depends on crop parameters such as stalk diameter and rigidity, stalk moisture at harvesting time, plant density, and cutting height relative to the ground. Cutting quality depends on the angles α_b and α_f, sharpening angles of the blade and counterknife, thickness of the cutting edge, clearance between the blade and the counterknife mounted on the finger, and cutting speed.

Optimum values of the blade sharpening angles are 19°–25°. Recommended values of cutting edge thickness are 25–30 μm (0.0001–0.0012 in.). Satisfactory cutting of the cereals is ensured as long as the edge thickness of the cutting blade is less than or equal to 0.125 mm (0.005 in.).

Clearance between the cutting blade and counterknife influences not only the quality of cutting, but also the wear of the cutting edges and cutting power consumption. A good quality of cutting implies a lower resistance of the stalk to cutting than its resistance to bending within the space defined by the clearance between the knife and counterknife. When the cutter bar is made of smooth-edged knives, the clearance is 0.1–0.3 mm (0.004–0.012 in.). When the cutter bar is made of serrated knives, the clearance between a knife and counterknife should be a maximum 0.3 mm (0.012 in.).

In the following, we analyze the forces (shown in Figure 4.11) that act on the plant stalk during the cutting process to determine the values of design angles α_b and α_f. The stalk is retained between the active cutting edges if the sum of the forces on the X and Y directions satisfies the following two conditions:

$$\sum F_x = 0 \tag{4.39}$$

$$\sum F_y \leq 0 \tag{4.40}$$

That is,

$$N_1 \cos \alpha_b + F_1 \sin \alpha_b - N_2 \cos \alpha_f - F_2 \sin \alpha_f = 0$$

$$N_1 \sin \alpha_b - F_1 \cos \alpha_b + N_2 \sin \alpha_f - F_2 \cos \alpha_f \leq 0$$

$$F_1 = N_1 \tan \varphi_b$$

$$F_2 = N_2 \tan \varphi_f \tag{4.41}$$

By solving Equation 4.41, we get the following condition for the plant stalk to be retained between the knife and counterknife:

$$\alpha_b + \alpha_f \leq \varphi_b + \varphi_f \tag{4.42}$$

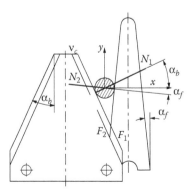

FIGURE 4.11
Forces acting on the stalk during cutting.

The values of friction angles $\varphi_b + \varphi_f$ depend on crop and stalk moisture and the blade type: if smooth-edged or serrated blade. In wheat and rye, when the blades are smooth-edged, these values are $\varphi_b + \varphi_f = 20°–30°$ when the plant's moisture is 14%, and $\varphi_b + \varphi_f = 25°–35°$ when the moisture is 22%. Thus, the values of $\alpha_b = 18°–29°$.

The cutter bar can also be made of two reciprocating-knife bars whose movement directions are always opposite to each other. This construction offers a great advantage: a better dynamic balancing. It is more expensive though.

4.2.2.3 Driving Mechanisms of Cutter Bar

The reciprocating motion of the multiknife bar relative to the fingers is generated by different mechanisms that convert the rotational motion of a shaft to a harmonic oscillatory motion of the multiknife bar that acts as a slider. In combine harvester headers, one of the following mechanisms is used to drive the cutter bar:

- Crank–conrod–slider mechanism (two-dimensional (2D) or three-dimensional (3D) configuration, with concentric or eccentric slider)
- Crank–conrod–rocker mechanism
- Wobble mechanism

The *crank–conrod–slider mechanism* with 2D configuration is well known from an engineering point of view and clearly explained in other engineering books, both mathematically and in terms of design. In the following, we analyze an eccentric crank–conrod–slider mechanism. In such a mechanism, the direction of motion of the slider does not go through the center of the crank rotation, but is eccentrically shifted at the distance e_s (Figure 4.12). The rotation of the crank with the length r_c can be represented by its angular speed ω_c and angular acceleration ε_c. The length of the connecting rod (conrod) is l_r.

For mathematical modeling of this mechanism, the following functional assumptions are made:

- The crank rotates with a uniform angular speed.
- The unique degree of freedom $\theta = \omega_c t$ is measured from the conrod position corresponding to the outer dead center (ODC) of the mechanism. The t is the time variable.

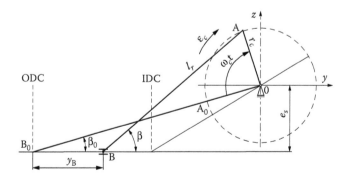

FIGURE 4.12
Crank–conrod–slider diagram. IDC, internal dead center; ODC, outer dead center.

- The cutter bar is connected to the slider joint and moves along the y_B direction. The movement origin corresponds to point B.
- Both the crank and the conrod are considered infinitely rigid bodies.

The slider (cutter bar) displacement y_B during the time t is given by the following equation:

$$y_B = (r_c + l_r)\cos\beta_0 - r_c \cos(\theta - \beta_0) - \sqrt{l_r^2 - [e_s + r_c \sin(\theta - \beta_0)]^2} \qquad (4.43)$$

The value of the angle β_0 at the outer dead position is determined geometrically from the following equation:

$$\beta_0 = \arcsin\frac{e_s}{r_c + l_r} \qquad (4.44)$$

Let us make the following notations:

- The crank angular position θ^* measured from the negative X axis is

$$\theta^* = \theta - \beta_0 \qquad (4.45)$$

- The stroke/conrod length ratio ς is defined as

$$\varsigma = \frac{r_c}{l_r} \qquad (4.46)$$

- The eccentricity ratio ϑ is defined as

$$\vartheta = \frac{e_s}{l_r} \qquad (4.47)$$

Thus, Equation 4.43 becomes

$$y_B = (r_c + l_r)\cos\beta_0 - r_c \cos\theta^* - l_r\sqrt{1 - (\vartheta + \varsigma\sin\theta^*)^2} \qquad (4.48)$$

Let us develop the square root expression in a power series based on the following formula:

$$(1+x)^n = 1 + nx + \frac{1}{2}(-1+n)nx^2 + \frac{1}{6}(-2+n)(-1+n)nx^3 + 0[x]^4, \ x \in [0,3] \qquad (4.49)$$

Because $n = 1/2$ is small, we retain the first two terms only, and the square root expression can be written as

$$\left[1-\left(\vartheta+\varsigma\sin\theta*\right)^2\right]^{1/2} = \left[1-\left(\vartheta^2+2\vartheta\varsigma\sin\theta*+\varsigma^2\sin^2\theta*\right)\right]^{1/2}$$

$$= 1-\frac{1}{2}\left(\vartheta^2+2\vartheta\varsigma\sin\theta*+\varsigma^2\sin^2\theta*\right) \tag{4.50}$$

That leads to the following form of Equation 4.48:

$$y_B = l_r\left[\left(\varsigma+1\right)\cos\beta_0 - \varsigma\cos\theta*-1+\frac{1}{2}\left(\vartheta^2+2\vartheta\varsigma\sin\theta*+\varsigma^2\sin^2\theta*\right)\right] \tag{4.51}$$

By deriving it with respect to time, the equation of the slider (cutter bar) linear speed is

$$\dot{y}_B = r_c\omega_c\left[\sin\theta*+\vartheta\cos\theta*+\frac{1}{2}\varsigma\sin\left(2\theta*\right)\right] \tag{4.52}$$

Consequently, the acceleration of the slider is given by the following equation:

$$\ddot{y}_B = r_c\omega_c^2\left[\cos\theta*-\vartheta\sin\theta*+\varsigma\cos\left(2\theta*\right)\right] \tag{4.53}$$

Because the stroke/conrod length ratio $\varsigma = 0.04$–0.1, it can be considered that the cutter bar movement is similar to a harmonic oscillatory one. In his PhD thesis, C. Maglioni (2009) has extensively developed the sensitivity analysis for this mechanism type, including the dynamic analysis.

The *crank–conrod–rocker mechanism* works in a way similar to that of the crank–conrod mechanism. The main role of the rocker is to satisfy geometry-related requirements of mechanical design, although this equates to introduction of a few other mechanical parts, and the overall functional reliability of this mechanism is lower than the reliability of the crank–conrod mechanism.

The *wobble mechanism* (Figure 4.13) uses a crankshaft (1), which is a crankshaft with a bent-axis crank (2) and element (3), to oscillate the pivot (4) and the sickle yoke shaft (5). The pivot (4) axis is perpendicular to the element (3) of the crankshaft. The output shaft (6), which is an integral part of the yoke shaft (5), rotates the crank (7) whose length is l.

The wobble mechanism has the functional advantage of three kinematic rotation couplings with concentric axes. The extreme points of rotating elements 3 to 5 move within separate spherical surfaces with a common center O.

Let us consider that point A of crank 2 rotates with a constant angular speed ω_c within a plane that is parallel to the *YOZ* plane. At the same time, point B of pivot 4 has an oscillatory motion within the plane *XOZ*.

The space x the cutter bar moves is given by the following equation:

$$x = l\left(\sin\alpha - \sin\zeta\right) \tag{4.54}$$

Element 3 has the length l_y, and the length of the pivot radius OB is l_p.

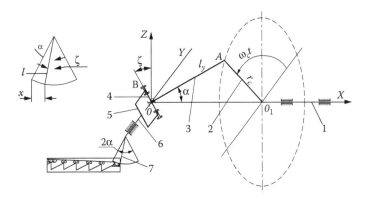

FIGURE 4.13
Kinematics of a wobble mechanism.

The equations of the vectors $\bar{l}_y$ and $\bar{l}_p$ are as follows:

$$\begin{bmatrix} \bar{l}_y \\ \bar{l}_p \\ 0 \end{bmatrix} = \begin{bmatrix} \bar{i} \\ \bar{j} \\ \bar{k} \end{bmatrix} \cdot \begin{bmatrix} l_y \cos\alpha & l_y \sin\alpha \sin\omega_c t & l_y \sin\alpha \cos\omega_c t \\ -l_p \sin\zeta & 0 & l_p \cos\zeta \\ 0 & 0 & 0 \end{bmatrix}$$

(4.55)

Because the axis of pivot 4 is perpendicular to element 3, we can write

$$\bar{l}_y \cdot \bar{l}_p = 0$$

(4.56)

This leads to the following equation:

$$\tan\zeta = \tan\alpha \cos(\omega_c t)$$

(4.57)

By introducing all of the above into Equation 4.54, we get the equation of the space on which the cutter bar moves as follows:

$$x = l\sin\alpha \left[1 - \frac{\cos\omega_c t}{\cos\alpha \left(1 + \tan^2\alpha \cos^2\omega_c t\right)^{1/2}} \right]$$

(4.58)

Consequently, the equations of the cutter bar speed and acceleration are

$$v_x = \omega_c l \tan\alpha \frac{\sin\omega_c t}{\left(1 + \tan^2\alpha \cos^2\omega_c t\right)^{3/2}}$$

(4.59)

$$a_x = \omega_c^2 l \tan\alpha \cos\omega_c t \frac{1 + 3\tan^2\alpha - 2\tan^2\alpha \cos^2\omega_c t}{\left(1 + \tan^2\alpha \cos^2\omega_c t\right)^{5/2}}$$

(4.60)

The speed of a cutter bar driven by a wobble mechanism is lower at the middle of the stroke than the speed of a cutter bar driven by a crank–conrod mechanism. The difference in speed is proportional to the angle α, whose values are within the 15°–30° range. The maximum value of the acceleration of a cutter bar driven by a wobble mechanism is also lower than the maximum acceleration of a cutter bar driven by a crank–conrod mechanism.

The crankshaft is usually belt driven, and overload protection is ensured in the form of controlled-drive belt slippage. The bent part of the crankshaft, the bearing over element 3, the bearing housing, and the yoke can be fully enclosed by a box with oil bath lubrication. Figure 4.14 shows a view with 90° cross sections of such a wobble box.

4.2.2.4 Knife Cutting Speed, Pattern, and Area

The cutting speed of a knife has an important influence on the plant cutting process quality and power consumption; that is why we study this in conjunction with the combine speed, and also derive the cutting pattern and area of the field covered by the absolute movement of the knife. It is very difficult to know the instantaneous displacement of the knife inside the plant stem. Thus, it is more useful to study it based on the absolute displacement of the knife. To cut the plants, the cutting speed has to be higher than or at least equal to the minimum cutting speed, which is established experimentally for each crop. It is enough to satisfy this condition at the beginning and at the end of the cutting stroke, which is smaller than the movement stroke of the knife.

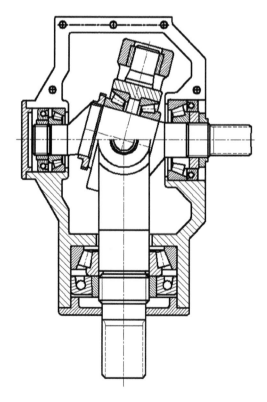

FIGURE 4.14
Cross sections of a wobble mechanism box.

If we simplify the knife movement to a harmonic oscillatory one, the space and speed equations can be written as follows:

$$x = r_c \left(1 - \cos \omega_c t \right) \tag{4.61}$$

$$v_x = \omega_c r_c \sin \omega_c t \geq v_{\min} \tag{4.62}$$

The cutting speed depends on the cutter bar type and the type of driving mechanism. If v is the recommended cutting speed of the crop stems, then the minimum angular speed ω_c of the crank is given by

$$\omega_c \geq \frac{v}{\sqrt{x_s \left(S_c - x_s \right)}} \tag{4.63}$$

where:
$S_c = 2r_c$ is the cutter bar stroke, in m
x_s is the position of the knife at the beginning or end of the cutting process

Because the angular speed of the input shaft is

$$\omega_c = \frac{\pi n}{30} \tag{4.64}$$

the rotation n per minute of the input shaft can be calculated as follows:

$$n \geq \frac{30v}{\pi \sqrt{x_s \left(S_c - x_s \right)}} \tag{4.65}$$

In cereals, $v = 1.5–2$ m/s, and in corn or other plants with a thick stem, $v = 2.5$ m/s.
When the cutter bar is driven by a crank–conrod mechanism, the mathematical law that describes the relative speed of a knife along the cutter bar is given by the equation of a circle with the radius

$$r_c = S_c / 2 \tag{4.66}$$

as shown in Figure 4.15.
If the graph is drawn at ω_c scale, then the speed at any abscissa x of the knife position is

$$v_x = \omega_c r_c \sin \omega_c t = \omega_c \overline{AB} \tag{4.67}$$

During the cutting process, the plant stems are cut along the cutting edges of the counterknives mounted on the cutter bar fingers. The knife displacement over a certain surface of the field results in bending the plants toward the counterknife, which has a rectilinear motion as the combine harvester moves in the field with the speed v_v, which we assume to

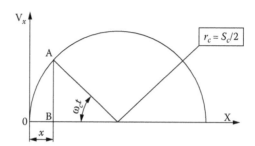

FIGURE 4.15
Knife speed variation along the stroke.

be constant for the purposes of this study. Due to a combination of the reciprocating rectilinear motion of the knife and the forward motion of the combine harvester, each knife follows an S-curve, which helps in defining the knife cutting pattern and effective cutting area (Figure 4.16).

In Figure 4.16, the following notations have been used:

S is the knife stroke, m
p is the knife pitch, m
p_0 is the finger pitch, m
y_c is the feedrate per knife cutting edge, kg/(sm)
$2b$ is the width of the ledger plate (counterknife) mounted on the finger, m
a_k is the length of the active part of the knife, m

$$S = p = p_0 = 2r_c \tag{4.68}$$

During the time $t = \pi/\omega$, the crank rotates 180° and point O moves to point O_1. Thus, the equation of the point O trajectory is

$$x_O = r_c \left[1 - \cos\left(\pi y_O / y_c \right) \right] \tag{4.69}$$

In Figure 4.16, one can distinguish a few surfaces that are covered by the knife and counterknife, as follows:

- Plants under the finger are pushed laterally and forward due to the friction with the finger and the counterknife.
- Surface F_1 is covered by the knife when the plants are cut by a knife cutting edge with the counterknife along line a_1b_1.
- Surface F_2 is covered by both cutting edges of the knife during the forward and return strokes.
- Surface F_0 is not covered by any knife edge.
- Surface F_1 represents 60%–72% of the total surface.

The plants from the surface F_0 are practically pushed forward and cut around point d_2 during the return stroke of the knife; the agglomeration of the stems around this point causes a rapid wearing out of the cutting edge of the knife in that particular location.

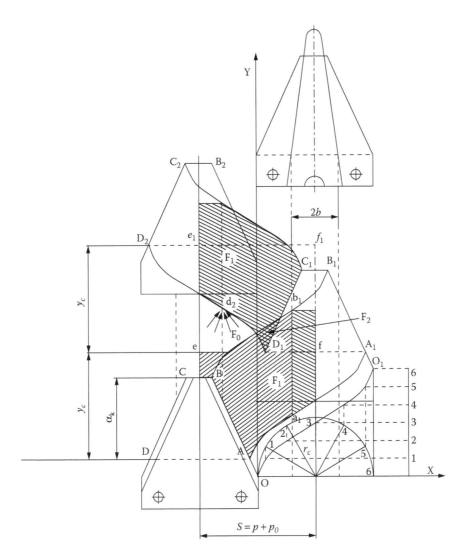

FIGURE 4.16
Knife cutting pattern and area.

Because some of the plants are inclined more than the others until their stems are cut, the height of the stubble is not uniform.

4.2.3 Plant Collecting: Auger and Draper

Tremendous gains in combine harvester capacity and efficiency have placed ever-increasing demands on the grain header, whose header auger has to gather the cut material and forward it to a retractable-finger drum, and then through the feeder-house into the combine. Gathering an ever-larger volume of material is very important, although the impact of design and correct setting of the auger is often downplayed.

The *header auger* spans the whole length of the header to drag the cut plants in. The cross auger is made of two similar sections that are arranged symmetrically with the retractable-finger drum (Figure 4.17). Each section is made of a steel barrel (1) and a helical high spiral (2). The spirals have opposite wrapping angles, so that the material is collected and transported from the lateral sides to the center of the header. The barrel diameter d_a has values within the 200–300 mm range; the outside diameter D_a = 400–500 mm, while the spiral pitch p_a = 200–400 mm.

The feedrate q_a of the auger can be calculated with the following formula:

$$q_a = \frac{\omega \left(D_a^2 - d_a^2 \right) p_a \psi \rho_p}{8}, \text{ kg/s} \tag{4.70}$$

where ρ_p is the volumetric mass of cut material (kg/m³) and ψ is the fill factor, $0 \leq \psi \leq 1$. This factor characterizes how much the space between the helical spirals is filled with cut material. Usually, $\psi = 0.05$–0.35.

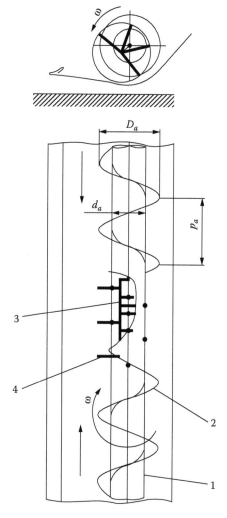

FIGURE 4.17
Header auger with retractable-finger drum.

The *retractable-finger drum* (Figure 4.17) takes the material delivered by the auger sections and pushes it through the bottom section into the feeder-house. It is made of a crank-shaft (3) with a fixed but adjustable position and the fingers (4), which are articulated on the crankshaft and slides through spherical bushings mounted on the auger barrel. These bushings are mounted on the barrel on a helical path at an angle of 60°–90°. The distance between the paths of two adjacent fingers is 25–30 mm (1–1 5/16 in.). The drum has a length of 1–1.5 m (40–59 in.) and a diameter equal to the auger barrel diameter d_a.

The most important parameters of the retractable-finger drum are the drum angular speed, which is equal to the auger angular speed; the instantaneous speed of the tip of the finger; and the variation law of the fingertip path outside the auger barrel.

The speed of the fingertip must be lower than the speed of grain detachment from the ears. Then, the finger has to be outside of the barrel between the points corresponding to the start of material gathering from the auger and leaving the material at the entrance of the feeder-house section, so that no material will be recirculated.

The geometric and functional parameters of a retractable-finger drum are shown in Figure 4.18a and b. The drum radius is $r_a = d_a/2$. The crank of the crankshaft has the length r_c, and its orientation is described by the adjustable angle α. The finger has the length l_f and is articulated with the crank in joint A.

The finger shown in Figure 4.18a is in the process of gathering the cut material, while the finger shown in Figure 4.18b is in the position of leaving the material at the feeder in the feeder-house. While the drum rotates with the angular speed ω_a, the finger rotates with the angular speed ω_f. At the moment when the finger overlaps the crank ($\varphi_a = 0°$) and when the finger is in the crank extension position ($\varphi_a = 180°$), $\omega_f = \omega_a$. Since the angles φ_a and φ_f vary during the same time, we can write

$$\frac{\varphi_a}{\omega_a} = \frac{\varphi_f}{\omega_f} \tag{4.71}$$

For a complete rotation of the drum, the angle β_f between the finger direction and radial direction is given by the following equation:

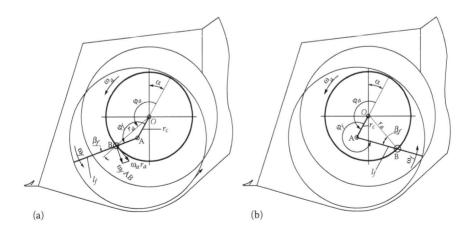

(a)   (b)

FIGURE 4.18
(a, b) Geometry and kinematics sketch of a retractable-finger drum.

$$\beta_f = |\varphi_a - \varphi_f| \tag{4.72}$$

Thus, depending on the finger position, from the triangles OAB, we can write

$$r_a = \text{AB} \cdot \cos\beta_f + r_c \cos(\pi - \varphi_a)$$
$$r_a = \text{AB} \cdot \cos\beta_f + r_c \cos(\varphi_a - \pi) \tag{4.73}$$

Since

$$\cos(\pi - \varphi_a) = \cos(\varphi_a - \pi) = -\cos\varphi_a \tag{4.74}$$

we can write

$$r_a = \text{AB} \cdot \cos\beta_f - r_c \cos\varphi_a \tag{4.75}$$

From the speed triangle, the following equation can be developed:

$$\omega_a r_a \cos\beta_f = \text{AB} \cdot \omega_f \tag{4.76}$$

In the OAB triangle, by applying Pythagoras's generalized theorem, we get

$$\text{AB}^2 = r_a^2 + r_c^2 - 2r_a r_c \cos(\pi - \varphi_a) \tag{4.77}$$

From Equations 4.75 through 4.77, the following equation for the finger angular speed can be developed:

$$\omega_f = \frac{\omega_a r_a (r_a + r_c \cos\varphi_a)}{r_a^2 + r_c^2 - 2r_a r_c \cos\varphi_a} \tag{4.78}$$

The length of the finger l_{fo} outside of the drum can be calculated with the following formula:

$$l_{fo} = l_f - \sqrt{r_a^2 + r_c^2 - 2r_a r_c \cos\varphi_a} \tag{4.79}$$

and has values within the following range: $l_{fo} = [l_f - r_a, l_f - r_c]$.

Due to progressive withdrawal of the fingers in the back of the drum (relative to the combine movement direction), the material is released and taken up by the chain conveyor.

A *draper header* uses fabric or a rubber conveyor apron instead of a cross auger. A draper header may have two or three belts; two cross (lateral) belts gather and transport the cut material to the middle of the header, while the center in-feed belt or a retractable-finger drum pushes the material toward the chain conveyor or other type of conveyor for feeding the material into the threshing system.

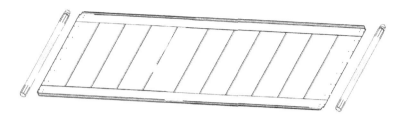

FIGURE 4.19
Cross belt of a draper header.

A belt (Figure 4.19) is composed of a drive roller, an idle roller, the fabric or a rubber flat belt with transversal scrapers, and a tensioning system to maintain the belt tensioned against slippery over the rollers. A draper roller is a cylindrical roller with sealed, nongreasable bearings. The belt is usually guided on its lateral sides by different forms of guides, such as V-shaped guides. The cross belt width is 1.016–1.067 m (40–42 in.) and runs at a speed within the 0–4.4 m/s (0–173 in./min) range. The center in-feed belt width is 1.25–2 m (49–78.75 in.) and runs at a speed within the 0–3.9 m/s (0–135.5 in./min) range.

Figure 4.20 shows a sketch of a cross belt with its geometric and functional parameters. Because the cross belt gathers and transports, with a constant speed v_B, the cut material from one end of the header to its center, the belt load varies proportionally with the belt length L_B.

Let us assume that the combine speed v_c is constant and the speed direction is oriented from the reader to this page. If m_{p0} is the mass of the cut plants per square meter, then the specific throughput q_{p0} (kg/m·s) of the material falling down per unit of belt length is given by the following equation:

$$q_{p0} = m_{p0}v_c \tag{4.80}$$

The total throughput q_{py} (kg/s) of cut material flowing in the X direction (opposite direction of the combine) over the length y of the belt is

$$q_{py} = q_{p0}y \tag{4.81}$$

The direction of the material flow is changed from the X sense of direction to the Y direction. Thus, we can write that at coordinate y, the throughput on the Y direction is

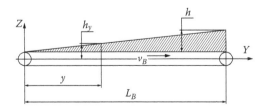

FIGURE 4.20
Sketch of the cross belt transportation process.

$$q_{py} = h_y l_{cp} v_B \rho_{bp} \varepsilon \qquad (4.82)$$

where:

l_{cp} is the average length of the cut plants, m
v_B is the belt speed, m/s
ρ_{bp} is the volumetric mass of the bulk plants, kg/m³
ε is the filling coefficient, with values between 0 and 1

This coefficient depends on the cut plant's orientation on the belt, as well as on the plant's compactness on the belt.

The height h_y of the material at coordinate y is given by the following formula, deducted from combining Equations 4.80 through 4.82:

$$h_y = \frac{m_{p0} v_c y}{v_B l_{cp} \rho_{bp} \varepsilon} \qquad (4.83)$$

Therefore, the height of cut plants over the belt varies from 0 to a maximum value at the coordinate $y = L_B$.

At $y = L_B$, the total throughput q_p of cut plants is given by the following equation:

$$q_p = q_{p0} L_B \qquad (4.84)$$

If the result is multiplied by 2, we get the total throughput of cut material that has to be fed into the combine through the center in-feed belt.

4.3 Stripper Header

4.3.1 Introduction

The stripper header was originally conceived by Keith Shelbourne (Shelbourne Reynolds, 2013) in the mid-1980s as an alternative solution to the classical grain header and an approach to the threshing process without cutting the plants from the field. The basic concept of the stripping process is illustrated in Figure 4.21. A rearward rotating rotor (1), carrying comb-like stripping elements (2) with keyhole-slotted teeth, strips the grains and fragments of ears from the crop as the combine moves through the crop. The rotor fingers comb the stalks and tear off the ears. By stripping, the ears, pedicels, ear fragments, and chaff are pulled apart by the stripping fingers. As a result, the mixture of grains, chaff, and ear fragments move backward along the upper hood (3) and drop into a conveying auger (4) due to the inertial force and the airflow-carrying force. Plants standing in the field are channeled into the stripping area by first being bent toward by the leading bar (guide nose) (5).

Today, Shelbourne Reynolds (2013) manufactures stripper headers of 3–9.8 m width that equip combine harvesters of different makers, such as CLAAS, John Deere, New Holland, Massey Fergusson, and others. Figures 1.40 through 1.42 show a schematic and views of the Shelbourne Reynolds stripper header.

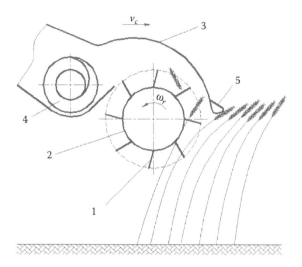

FIGURE 4.21
Basic concept of grain-stripping process.

There are some advantages to using a stripper header instead of a classical grain header:

- Up to 20% of the straw enters the combine harvester; combine harvester performance depends heavily on the grain-to-straw ratio.
- Under normal harvesting conditions (e.g., crop maturity, moisture), the combine throughput increases by over 50% to more than 100% at equal grain loss levels (Klinner et al., 1987a,b).
- In wheat, up to 80% of the grains are threshed by the stripper.
- The combine harvester has reduced power consumption.

Grain stripping has also disadvantages. When grains are harvested with a stripper, the straw is left on soil. The straw may be incorporated into the soil at 0.15–0.2 m (6–8 in.) depth by plowing and then becomes an organic fertilizer. Thus, the yield in wheat and barley production increases during the first year after plowing. During grain stripping, the stripper header losses increase significantly with crop maturity. Some research studies show that up to 75% of the total field grain losses are due to the stripping process (Glancey, 1997).

4.3.2 Grain-Stripping Process

The functional and constructional parameters of the stripper process are shown in Figure 4.22. We denote by R_r the radius of the rotor, which rotates with the constant angular speed ω_r. The fingers are mounted in an inclined position relative to the radial direction, so that R_f is the corresponding radius to the fingertip.

Let us denote by h_m the minimum height of the crop plants just below the ears, while h_M is the maximum height of the crop plants, including the ear (note the difference in definitions). The stripper working height h_r shall be between h_m and h_M.

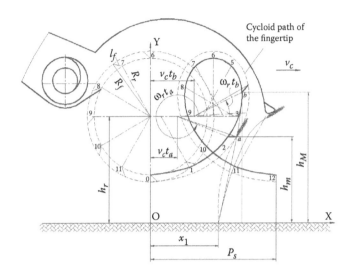

FIGURE 4.22
Functional and constructional parameters of the stripper process.

As a combine progresses into the grain field with the constant speed v_c, the guide nose of the stripper initially bends the plants forward so that if they escape, the finger will catch the plant stem.

The stripper finger shall be in a horizontal position (point a) when the smallest plants are approached at the height h_m. At the position corresponding to point b at the height h_M, the stripping process of the highest plants is completed.

With reference to the fixed Cartesian coordinate system XOY in Figure 4.22, the absolute movement of a stripper finger is the result of a combination of the translational movement of the combine harvester with a velocity v_c and the rotation of the stripper rotor around its own axis. Thus, a point at the finger base will describe a cycloid.

Let us define the dimensionless kinematic index λ_r of the stripper as the ratio

$$\lambda_r = \frac{\omega_r R_r}{v_c} \tag{4.85}$$

The shape of the cycloid depends on the λ_r value. Thus, if $\lambda_r < 1$, the stripper finger trajectory is a *curtate cycloid*. In such a case, the stripper not only does not strip the plants, but contrarily, it pushes the plants toward the front of the combine. The stripper rotor function is accomplished only on the first segment of the trajectory, on the arc between points 2 and 5, respectively, of the *prolate cycloid*, as shown in Figure 4.22.

The parametric equations of the cycloid are as follows:

$$x = v_c t + R_r \cos \omega_r t$$
$$y = h_r + R_r \sin \omega_r t \tag{4.86}$$

The angular speed of the reel can be calculated with Equation 4.87 as follows:

$$\omega_r = \frac{\pi n_r}{30}, \, s^{-1} \tag{4.87}$$

where n_r is the rotor rotation (rpm).

At point a, the equations of the finger-base path are

$$x_a = v_c t_a + R_r \cos \omega_r t_a$$
$$y_a = h_m = h_r - R_r \sin \omega_r t_a \tag{4.88}$$

From the second equation, we get the angle when the finger plate should be in a horizontal position:

$$\sin \omega_r t_a = \frac{h_r - h_m}{R_r} \tag{4.89}$$

From the first equation of the system (4.88), we get

$$\cos \omega_r t_a = \frac{x_a - v_c t_a}{R_r} \tag{4.90}$$

If we replace the term $\cos \omega_r t_a$ in the second equation of the system (4.88), we obtain the cycloid equation

$$y_a = h_r - R_r \sqrt{1 - \left(\frac{x_a - v_c t_a}{R_r}\right)^2} \tag{4.91}$$

This is equivalent to the following equation:

$$y_a = h_r - \sqrt{R_r^2 - (x_a - v_c t_a)^2} \tag{4.92}$$

When the finger approaches the lowest-height plants under the ears, the bent stalk curve is tangent to the path in point a. From a geometric stand point of view, they have a common tangent; mathematically, the first derivatives of the finger trajectory and the stalk curve are equal. In the following, we will apply this concept.

Thus, by applying Equation 3.11, we can write the stalk curve equation:

$$y_a = h_{be\min} \sqrt{1 - \left(\frac{x_a - x_1}{c_b h_{be\min}}\right)^2} \tag{4.93}$$

where the coefficient of bending c_b can be calculated with Equation 3.12 and $h_{be\min}$ is the minimum height of crop stalks below the ears.

By equalizing the first derivatives of Equations 4.92 and 4.93, one can determine the stripper height or combine speed.

Another approach assumes positioning of the stripper rotor low enough so that the fingers begin stripping at the point of the vertical-upward velocity; this means that the plants are tangent to the cycloid and in a vertical position (Yuan and Lan, 2007). That implies a higher h_s coordinate of the rotor axis; this may become a problem when crops are lodged or laid closed to the ground.

The distance between two homologous, successive points of the trajectory of a finger row is called the *stripper pitch* p_s (Equation 4.94):

$$p_s = \frac{2\pi R_r}{\lambda_r} \tag{4.94}$$

The distance between two homologous points of the trajectories of two consecutive finger rows is called the *row pitch* p_r and can be calculated as follows:

$$p_r = \frac{2\pi R_r}{\lambda_r n_r} \tag{4.95}$$

where n_r is the number of consecutive finger rows.

In terms of grain loss, while the rotor is gradually lifted, the loss increases due to splashing the grain at a lower elevation than the guide nose level or due to an improper geometric configuration of the upper hood.

The designing of a stripper is ultimately an optimization problem to correlate dimensional and functional parameters when a variety of crops are considered. A good suggestion is to develop a genetic algorithm. Readers may want to adapt the genetic algorithm discussed in Chapter 5 as an example for optimization of an axial threshing unit process.

4.4 Corn Header

4.4.1 Introduction

An introduction to corn headers was given in Section 1.5.3. For a more detailed description, some parts of that section are mentioned here as well.

Corn headers can be classified based on their main functions, as follows:

- *Corn headers* for collecting the ears only, while the stalks are pulled down
- *Chopping corn headers* that collect the ears and chop the stalks whose fragments are spread on the field
- *Integral corn headers* that collect the ears and chop and collect the stalk fragments, to be left on the soil as a continuous swath or to be collected into a trailer that moves in parallel to the combine

Combine harvesters are equipped with row crop corn headers for harvesting 6 to 24 rows at a time (Figure 1.31). These corn headers are built in series of 6-row, 8-row, 12-row, 16-row, and so forth. The number of harvested rows is determined by the combine class following capacity and engine power requirements, for example, 165–215 kW (220–290 hp) for 6–8 rows, 225–345 kW (300–460 hp) for 8–12 rows, and so forth. The distance between cornrows can be as tight as 0.38 m (15 in.), although 0.762 m (30 in.) is most common.

The components of a corn header are shown in Figure 4.23: a structural welded frame (1), outer fenders and shields (2) and outer snout with point (3) (on left hand; same on the

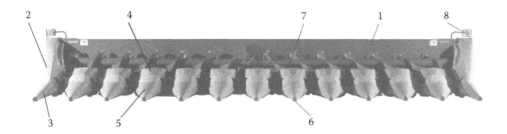

FIGURE 4.23
Components of a corn header.

right hand), crop/row dividers composed of a narrow row center snout (5) with point and a center divider (4), a wear tip (6), the gathering auger (7), and lateral lights (8). Under every two adjacent row dividers, there is mounted a row unit composed of stalk rolls with augers (9), snapping plates (10), and ear gathering chains (11) with pins (Figure 4.24). The driving gearbox of the stalk rolls is not shown. The crop/row dividers match accurately the row width (some models may have an adjustable row spacing design). Most modern combines have an automatic header height control, terrain contour-following capability, and a flexible sensing wand–based row steer function, as well as single-unit coupling systems for hydraulic and electric connections with the machine. For crops with tangled corn stalks or lying down on the soil, a very good solution is represented by hydraulic-driven spiral augers mounted under outer shields (Figure 4.25). Such an auger pushes the stalks into the header. A good example is Polytin corn headers (developed by VH Manufacturing, Rock Valley, IA), whose lateral augers are 0.2 m (8 in.) in diameter and 0.91 m (3 ft) long.

The *chopping corn headers* have the same working components as the corn headers and one of the following additional assemblies:

- A pair of rotary knives under each pair of stalk rolls
- A combination of stalk rolls with knives that replaces the stalk rolls
- Transversal rotary knives

A chopping corn header may have a two-blade or three-blade *rotating cutter* (Figure 1.35), which is driven right from the row unit gearbox. Such a cutter chops the stalks that are

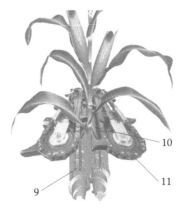

FIGURE 4.24
(See color insert.) Functional components of a corn header row section.

FIGURE 4.25
Spiral auger mounted under the outer shield.

pulled down by the stalk rolls. The chopped material is spread on the field, and later it is incorporated into the soil during plowing or tillage operations.

The *stalk rolls with knives* (Figure 4.26) sever the stalks while being pulled down. Such a stalk roll with knives manufactured by Cressoni is made of a shaft (1), helical spirals (2), radial multiblades (3), and four longitudinal knives (4).

Geringhoff (2013) manufactures chopping corn headers whose row units use a combination of a regular stalk roll and a shaft with transversal cutting blades. Such innovative design solutions eliminate a subsequent stalk-shredding operation, though some stalks may break and enter the machine with their ears.

A corn header that is equipped under the row units with a *transversally positioned rotary cutter* and an auger may be classified as a chopping corn header if the cut material is spread on the soil, or as an integral corn header if the spiral auger gathers the cut material, which could be further pushed by means of a fan into a collecting unit that is pulled close to the combine harvester. The rotary cutter chops the stalks and simultaneously throws the chopped stalks, leaves, and weeds back to the auger, which collects this material mixture and can leave it on the field as a continuous swath to be later picked up and baled.

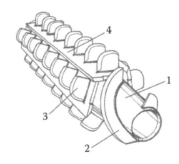

FIGURE 4.26
Stalk roll with knives and transversal blades. (Courtesy of Cressoni, Volta Mantovana, Italy.)

4.4.2 Row Unit Process and Working Elements

The row unit process is comprised of three phases: grasping the stalks between the stalk rolls, stalk drawing, and corn ear detachment.

The feeding of each row unit of a *corn header* is ensured by poly snouts and by tapered-finger gathering chains (Figure 4.24) that drive the plant stalks between the cantilevered stalk rolls placed underneath the deck plates. As the combine harvester moves through the crop, the poly snouts are positioned between the rows and guide the corn stalks to the stalk rolls. If the corn is down or logged, the snouts lift the stalks and guide them further toward the stalk rolls.

The ear-snapping process is ensured by pulling down the stalks by the stalk rolls (Figure 4.27), while the ears are retained on top of the snapping plates and pushed backward further by the chains, toward the gathering auger. The detaching process of the ears from the stalks is based on the fact that the tensile strength of the stalk is greater than the tensile strength of the ear peduncle. However, this difference in tensile strength may be affected by the way the stalks are handled; that depends on the roll and plate design, settings, and chain speed versus combine speed.

When the ear reaches the snapping plates, its peduncle breaks and the stalk is further pulled down. It is clear that the adjustable distance between the snapping plates should be wide enough to avoid butt shelling, but not too wide to allow the ears to escape. In fact, this distance should be narrower than the diameter of the smallest ears of the crop. The

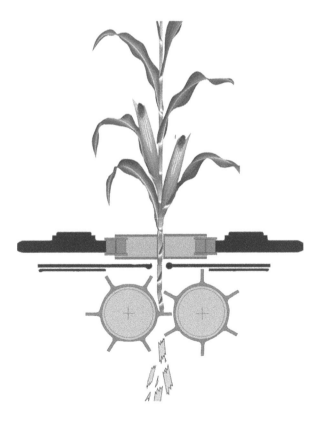

FIGURE 4.27
Row unit process.

gathering chains that operate just above the snapping plates move the ears rearward to the cross auger, which conveys them to the combine feeder.

The row units can be classified in two categories:

1. Row units with distributed-function rolls
2. Row units with multiple-function rolls

The *row units with distributed-function rolls* are made up of two types of components: rolls that pull the stalk down and components for snapping (detaching) the ear from the stalk. The process of such a unit has been described above and is illustrated in Figure 4.27, where the snapping plates detach the ear from the stalk. Another example is given in Figure 4.28, where a shaft equipped with disc cutters cuts the stalk into small fragments that will later be incorporated into the soil through plowing. Further examples have two pairs of rolls: one pair of rolls pull the stalk down, while the other pair retains the ear after its peduncle is broken.

The *row units with multiple-function rolls* are usually made of two rolls that, through an adequate design, pull down the stalk and, at the same time, detach the ear from the stalk.

Regardless of their function, each roll has on its front end a taper auger whose design differs among the corn header manufacturers. Over the rest of the length, the rolls may have a cam profile or helical ribs, or they could be grooved rolls. Each design involves some advantages and disadvantages. The cylindrical rolls with helical ribs can detach the corn ears without damaging them, but they develop a lower stalk pulling force. The grooved rolls and the rolls with a rectangular cross section have a better pulling capacity, but they can damage the ears to a higher degree.

The distance between the rolls can be adjusted within some constructional limits.

The stalks are guided between the rolls by chains, which also convey the detached ears rearward to the transversal auger. The chains are driven by tooth gears. The distance between the drive sides of the chains is also adjustable to accommodate a large variety of corn plant physical dimensions and properties.

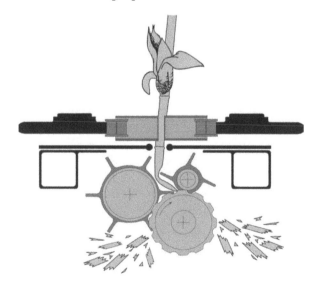

FIGURE 4.28
Row unit with multiple disc cutters on a shaft. (Courtesy of Geringhoff, Ahlen, Germany.)

4.4.3 Stalk Grasping and Drawing Process and Roll Design

Figure 4.29 illustrates the process of axial grasping of the stalks by the stalk rolls. Due to advancement of the combine through the crop, and due to stalk rigidity as well, a stalk is grasped by the taper auger of the rolls' front end.

To get the corn stalks grasped by the rolls, there are two design requirements, which can be explained and mathematically described as follows:

1. The stalk is not to be rejected (pushed far away) by the taper auger front end of the rolls; that means

$$\gamma_t < \varphi_{sr} \tag{4.96}$$

   where:
   γ_t is the half of the roll taper angle, °
   φ_{sr} is the angle of dynamic friction between the stalk and the roll surface

   This is equivalent to

$$\tan \gamma_t < \mu_{sr} \tag{4.97}$$

   where μ_{sr} is the coefficient of dynamic friction between the stalk and the roll surface.

2. The distance between the rolls' front ends should be calculated with the following formula:

$$a_r > d_a + d_{pt} + \frac{d_{pt}}{\cos \alpha_r} \tan \gamma_t \tag{4.98}$$

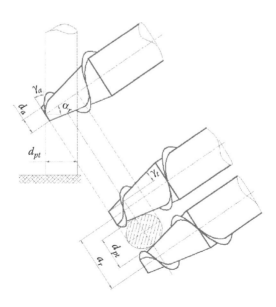

FIGURE 4.29
Schematic of the process of axial grasping of the corn stalks.

That is,

$$a_r > d_a + \frac{d_{pt}\left(1+\cos\alpha_r\right)}{\cos\alpha_r}\tan\gamma_t \tag{4.99}$$

Depending on the manufacturer, the auger on the tapered front end of the stalk roll may have the spiral of a constant or variable height, as well as a constant or variable pitch. The auger forces the stalk to move rearward when the following condition is met:

$$\gamma_a \le \alpha_r \tag{4.100}$$

During the second phase, the stalk undergoes a drawing process. The dynamics of the rolls' action on the stalk during drawing is illustrated in Figure 4.30. The resultant force N is the summation of radially distributed elementary forces exercised on the corn stalk by each roll. Figure 4.30 shows the diagram of these forces as a curvilinear triangle. Due to the contact with the stalk, a resultant friction force F_f develops in the tangential direction at the contact point. The forces N and F_f can be decomposed into their components in the Y and Z directions. Thus, a stalk is drawn between the rolls if

$$F_{fz} > N_z \tag{4.101}$$

That is equivalent to the following inequality:

$$\tan\alpha_d < \mu_{sr} \tag{4.102}$$

where α_d is the stalk drawing angle. This angle is a fraction of the grasping angle α_g, and typically it is

$$\alpha_d = \frac{1}{3}\alpha_g \tag{4.103}$$

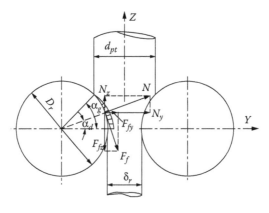

FIGURE 4.30
Dynamics of the roll action during stalk drawing.

The component forces N_y and F_{fy} squeeze the stalk, producing the relative deformation ψ_d, as defined by Equation 4.104:

$$\psi_d = \frac{d_{pt} - \delta_r}{d_{pt}} = 1 - \frac{\delta_r}{d_{pt}} \tag{4.104}$$

The value of the relative deformation varies within the 0.6–0.9 range. The stalk cross section remains constant if $\psi_d > 0.9$; in such a case, there is a small possibility of breaking the stalk during the drawing process.

While a stalk is drawn by the stalk rolls, the combine harvester moves with the speed v_c that is assumed to be constant. When the stalk drawing process starts, the row unit chains, whose speed of the driving side is v_{ch}, pull the stalk rearward (Figure 4.31).

The maximum stalk tensile resistance is preserved if the stalk is drawn in the vertical direction with the speed v_z. Thus, the length of the stalk engaged with the rolls is high. The functional condition of the rolls to draw the stalk vertically can be mathematically expressed as follows:

$$v_c = v_{ch} \cos \alpha_r = v_z \left(1 - \varepsilon_d\right) \sin \alpha_r \tag{4.105}$$

In Equation 4.105, we denoted by ε_d the coefficient of slippage of the stalk relative to the rolls. Depending on the corn variety and stalk and leaf moisture, this coefficient takes values within the 0.05–0.3 range.

If $v_c < v_z (1 - \varepsilon_d) \sin \alpha_r$, the stalk load of the rolls decreases, while if vice versa, the corn stalks will be bent forward and there is a risk of blocking the rolls from rotation due to the load increase.

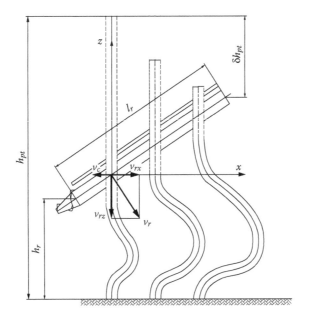

FIGURE 4.31
Schematics of stalk roll length calculation.

The main design parameters of the stalk rolls are the diameter D_r and the length l_r, while the most important functional parameter is the angular speed v_r. The stalk drawing condition expressed by Equation 4.102 can be written

$$\frac{\sqrt{1-\cos^2 \alpha_d}}{\cos \alpha_d} < \mu_{sr} \tag{4.106}$$

That is equivalent to the following inequality:

$$\cos \alpha_d > \frac{1}{\sqrt{1+\mu_{sr}^2}} \tag{4.107}$$

Taking into account Equation 4.103, and the fact that the stalk grasping angle formula is

$$\cos \alpha_g = 1 - \frac{\psi_d d_{pt}}{D_r} \tag{4.108}$$

we get the following formula for calculating the stalk roll minimum diameter:

$$D_r \geq \frac{\psi_d d_{pt}}{1-\cos 3\alpha_d} \tag{4.109}$$

The trigonometric formula for $\cos 3\alpha_d$ is

$$\cos 3\alpha_d = 4\cos^3 \alpha_d - 3\cos \alpha_d \tag{4.110}$$

Introducing Equations 4.107 and 4.110 into Equation 4.109, Equation 4.109 becomes

$$D_r \geq \frac{\psi_d d_{pt}}{1-\dfrac{4-3\left(1+\mu_{sr}^2\right)}{\left(1+\mu_{sr}^2\right)^{3/2}}} \tag{4.111}$$

This is the minimum diameter of the cylindrical roll without additional helical spirals. In practice, the stalk roll diameter varies between 0.06 and 0.08 m (2.375 and 3.125 in.), and with additional spirals this may go up to 102 mm (4 in.).

The length l_r of the stalk rolls can be determined by considering the stalk drawing dynamics, as illustrated in Figure 4.31. Thus, the stalk rolls shall draw the maximum δh_{pt} length of the corn plants while the combine harvester moves through the crop with the velocity v_c over a distance that is equal to the horizontal projection of the stalk roll length. Mathematically, this condition can be expressed as follows:

$$\frac{l_r \cos \alpha_r}{v_c} \geq \frac{h_{pt} - h_r - l_r \sin \alpha_r}{v_r \left(1-\varepsilon_d\right)\cos \alpha_r} \tag{4.112}$$

Consequently,

$$l_r \geq \frac{v_c \left(h_{pt} - h_r\right)}{v_r \left(1 - \varepsilon_d\right)\cos^2 \alpha_r + v_c \sin \alpha_r} \tag{4.113}$$

The length of the stalk rolls of the existent equipment varies between 0.63 and 1.10 m (24.500 and 43.375 in.).

The minimum angular speed ω_r of the stalk rolls should satisfy the following requirement: to avoid a high load, each roll should completely rotate at least one turn while the combine runs over a distance that is equal to the average space s_{pt} between two consecutive plants on the row. Mathematically, that can be expressed as

$$\frac{s_{pt}}{v_c} \geq \frac{2\pi}{\omega_r} \tag{4.114}$$

Practically, the angular speed of the stalk rolls varies between 50 and 100 s^{-1}, which corresponds to a rotation range of 475–950 rpm.

4.4.4 Corn Ear Detachment Process

The ear detachment process starts when the bottom end of the ear comes into contact with the snapping plates because the adjustable distance between the snapping plates is smaller than the minimum ear diameter. At this stage, the stalk and ear are subjected to a tensile force created by the stalk rolls. The ear peduncle breaks, and then the ear is conveyed rearward by the chains toward the transversal auger.

The dynamic action of the stalk rolls and snapping plates is illustrated in Figure 4.32. The maximum tensile force developed by the stalk rolls is

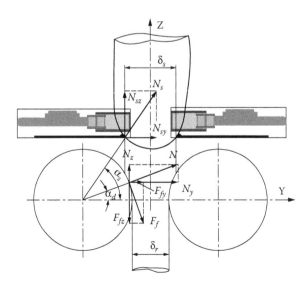

FIGURE 4.32
Dynamics of the roll and snapping plate action during ear detachment.

$$F_d = 2\left(F_f \cos\alpha_d - N\sin\alpha_d\right) = 2N\left(\mu_{sr}\cos\alpha_d - \sin\alpha_d\right) \tag{4.115}$$

where:
  α_d is the stalk drawing angle, degrees
  μ_{sr} is the coefficient of dynamic friction between the stalk and the roll surface

The snapping plates oppose a resistance force R_s that is equal to the tensile breaking force of the ear peduncle, and its formula can be expressed as follows:

$$R_s = 2N_{sz} = 2N_s\sin\alpha_s \tag{4.116}$$

where α_s is the ear-snapping (bearing) angle (°).
  Using the symbols from Figure 4.32, one can write a formula for the bearing angle:

$$\cos\alpha_s = 1 - \frac{\delta_s - \delta_r}{D_r} \tag{4.117}$$

The ear will be detached when $F_d > R_s$. Solving this inequality by using Equations 4.115 through 4.117 helps determining the required force to detach a corn ear or determine the stalk roll diameter.

The clearance β_s between the snapping plates shall be smaller than the diameter of the smallest ears of the corn crop. Some corn headers have the deck plate position hydraulically adjusted. A smaller clearance allows excess thrash (debris) into the combine. To avoid this, the opening between the snapping plates is practically tapered, set wider at the rear than at the front by about 3.18–4 mm (0.125–0.160 in.).

The gathering chains and the stalk rolls of each row unit are driven through a row unit gearbox. Some manufacturers offer a variable-speed drive that varies in sync with the combine speed. The speed of the chains is 14–22 m/s (255–402 ft/min).

The flight-to-trough clearance of the ear auger should be tight enough to prevent small ear shelling. This clearance is 12.5–15 mm (0.5–0.6 in.). Auger speed should be correlated with the machine speed as well.

Corn headers require about 6 kW (8 hp) per row; this is about twice as much as the power needed for grain headers.

4.4.5 Stalk Cutting/Chopping Processes

Following the ear detachment, the drawn stalks, except those drawn and cut by rolls with knives (Figure 4.33) or with radial blades (Figure 4.26), may be further processed in one of the following ways:

- The stalks, in a randomly inclined position relative to the ground, can be later incorporated into the soil as is by plowing, or after a presowing process performed with disc harrows.
- The stalks may be cut/chopped immediately after getting down from the stalk rolls; this process is performed by separate *rotary cutters with blades* placed underneath each row unit of the corn header.

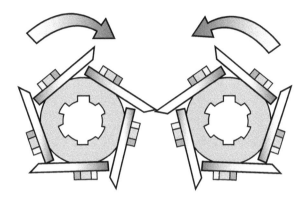

FIGURE 4.33
Stalk rolls with 5 in. phase knives. (Courtesy of Cressoni, Volta Mantovana, Italy.)

- The stalks, in a randomly inclined position relative to the ground, can be cut/chopped by a *transversal rotary cutter* and then the cut material may be
  - Spread on the ground to be later incorporated into soil by plowing
  - Left on the ground as a continuous swath to be bailed later
  - Collected with an auger and pushed by a blower into a trailer that is pulled in parallel with the combine harvester

Every stalk chopper must satisfy a series of functional specifications, such as

- Performing a free cutting of standing or inclined plants
- Ensuring the length of cut material within a certain range, for example, less than 100 mm (4 in.)
- Ensuring a high uniformity of the length of cut material (fragments of stalks and leaves), for example, minimum 80%

Each row unit has a rotary cutter that is placed under the stalk rolls and rotates in a plane, which is parallel to the snapping plates.

In this subsection, we analyze the process performed by the rotary cutters with blades.

A rotary cutter with blades may have different design configurations, such as

- A disc with two to five blades.
- A pair of cutters that rotate in opposite directions.
- A counterknife or not. The counterknife, which is placed under the cutter, usually consists of a plate with sharp edges of an open V-shape.

The cutting process kinematics and dynamics of a rotary cutter with blades and counterknife are shown in Figures 4.34 and 4.35, respectively.

In Figure 4.34, the cutter's axis speed relative to the ground is the combine speed v_c. Let us note the number of the blades with z_b.

With reference to the fixed Cartesian coordinate system XOY relative to the ground, shown in Figure 4.34, the absolute movement of each point of the cutting edge of a

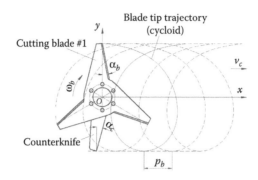

FIGURE 4.34
Cutting process kinematics of a rotary cutter with blades.

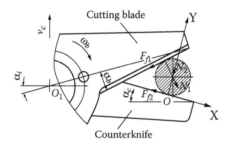

FIGURE 4.35
Cutting process dynamics of a rotary cutter with blades.

blade is the result of a combination of the translational movement of the combine harvester with the velocity v_c (m/s) and the rotation n_c (rpm) of the cutter around its own axis.

In the analysis of the cutter blade kinematics, the following assumptions have been made:

- The cutter rotational speed ω_b is constant and taken to be positive in the clockwise sense of rotation.
- The velocity v_c of the combine harvester is constant and specified from throughput requirements.

Let's consider at radius R the extreme point of the cutting edge of blade 1 at its initial vertical position, as shown in Figure 4.34. At an arbitrary time t, the position of this point is represented by the following parametric equations of its cycloid trajectory (represented by a continuous, thin line):

$$x = v_c t + R \sin \omega_b t$$

$$y = R\left(1 - \cos \omega_b t\right)$$

(4.118)

where:
    x is the abscissa of considered point, m
    y is the ordinate of considered point, m

R is the radius of considered point, m
ω_b is the angular speed of the cutter's blade, m/s

The angular speed of the cutter can be calculated with Equation 4.119 as follows:

$$\omega_b = \frac{\pi n_b}{30}, \, s^{-1} \tag{4.119}$$

where n_b is the cutter's rotation, rpm.

The distance between two successive trajectories of the same blade point is called the *cutter pitch* p_c. Its size is given by Equation 4.120:

$$p_c = \frac{2\pi R}{\lambda_c} \tag{4.120}$$

The distance between two homologous points of the trajectories of two consecutive blades is called the *blade pitch* p_b and can be calculated as follows:

$$p_b = \frac{2\pi R}{\lambda_c z_b} \tag{4.121}$$

where:
z_b is the number of cutter blades
λ_c is the kinematic index of the cutter motion

This index is defined as the following ratio:

$$\lambda_c = \frac{\omega_b R}{v_c} \tag{4.122}$$

The speed of a considered point is given by the first derivative of Equation 4.118:

$$v_x = v_c + \omega_b R \cos \omega_b t$$
$$v_y = \omega_b R \sin \omega_b t \tag{4.123}$$

The absolute speed v relative to the ground of a considered point can be calculated with the following formula:

$$v = \sqrt{v_x^2 + v_y^2} = v_c \sqrt{\lambda_c^2 + 2\lambda_c \cos \omega_b t + 1} \tag{4.124}$$

A drawn stalk will be cut if the v_x speed component is oriented to the $-X$ direction when $\cos \pi = -1$; mathematically, this can be written as

$$v_c \le \omega_b R \tag{4.125}$$

Consequently, the kinematic index shall satisfy the following condition:

$$\lambda_c > 1 \tag{4.126}$$

There is another functional condition for the cutter with blades: while the combine harvester moves through the crop over a distance s_{pt} that is equal to the average space between two consecutive plants in a row, the upper portion of a drawn stalk $(h_{pt}-h_c)$ is completely cut in fragments of specified length l_f, where h_{pt} is the average height of the plants (m) and h_c is the height of the cutter to the ground (m).

Thus, the number of rotations n_s required to cut the upper portion of the drawn stalk is

$$n_s = \frac{h_{pt} - h_c}{z_b l_f} \tag{4.127}$$

The above-mentioned functional condition can be mathematically formulated in terms of the time as follows:

$$t = \frac{2\pi n_s}{\omega_b} = \frac{h_{pt} - h_c}{(1-\varepsilon_d) v_r \cos\alpha_r} \tag{4.128}$$

From the combination of Equations 4.127 and 4.128, one can calculate the required angular speed of the cutter with blades:

$$\omega_b = \frac{2\pi(1-\varepsilon_d) v_r \cos\alpha_r}{z_b l_f} \tag{4.129}$$

The design angular speed of a cutter with blades varies from 18 to 26 s⁻¹.

In the following, we analyze the cutting process dynamics by considering the forces (represented in Figure 4.35) that act on the plant stalk, to determine the values of design angles α_b and α_c. The stalk is cut only if it not expelled outside of the cutting edges. Thus, the stalk is retained between the active cutting edges if the sum of the forces on the X and Y directions satisfies the following two conditions:

$$\sum F_x \leq 0 \tag{4.130}$$

$$\sum F_y = 0 \tag{4.131}$$

That is,

$$N_1 \sin(\alpha_b + \alpha_c + \alpha_i) \leq F_{f1} \cos(\alpha_b + \alpha_c + \alpha_i)$$
$$N_1 \cos(\alpha_b + \alpha_c + \alpha_i) - F_{f1} \sin(\alpha_b + \alpha_c + \alpha_i) = 0$$
$$F_{f1} = N_1 \tan\varphi_b \tag{4.132}$$
$$F_{f2} = N_2 \tan\varphi_c$$

where:

α_i is the angle stalk grasping, degrees
φ_b is the friction angle of the plant with the blade
φ_c is the friction angle of the plant with the counterknife

By solving Equation 4.132, we get the following condition for the plant stalk to be retained between and cut by the blade and counterknife:

$$\alpha_b + \alpha_c + \alpha_i \leq \varphi_b + \varphi_c \tag{4.133}$$

The values of friction angles $\varphi_b + \varphi_c$ depend on stalk and leaf moisture and on the blade type: smooth-edged or serrated blade. In wheat and rye, when the blades are smooth-edged, these values are as follows: $\varphi_b + \varphi_c = 20°–35°$ when the plant moisture is 15%, and $\varphi_b + \varphi_c = 25°–40°$ when the moisture is 25%. If the angle of stalk grasping increases, the retention of the stalk between the blade and the counterknife worsens.

In practice, the blade angle α_b takes values within the 15–40° range and the outside diameter of the blades is 0.175–0.25 m (6.875–10 in.). The counterknife angle $\alpha_c = 10$–12°. The cutting edges of the blades and counterknife are sharpened at an angle of 18–28°.

Many technical solutions can be applied to perform a single-pass corn harvesting process that will collect the chopped stalks directly into a wagon that is pulled by the combine or into a trailer that is pulled by a tractor. Such technology offers the advantage that the chopped corn stover does not touches the ground; therefore, it is not contaminated with soil. Then, the stover can be used as feedstock for cellulosic ethanol. In such a case, the corn stalk fragments should have a length smaller than 3 cm (1.2 in.). Besides the problem of transportation of big volumes of corn residue, there are other problems regarding the capacity and profitability of a biorefinery, biomass storage needs, and how much corn stover should be collected from the field, since this represents a returning of organic matter to the soil and soil protection from winter erosion effects.

4.5 Sunflower Header

Sunflower seeds are physiologically mature when the back of the flower head is yellow (seeds have maximum weight), but their harvesting starts when the bracts surrounding the head turn brown. Harvesting can also start earlier to avoid bird damage and reduce loss from plant lodging and seed shattering. When sunflower seeds are delivered, the accepted standard moisture is 9%.

The detaching process of the sunflower heads aims at cutting the stalk underneath the flower head by a cutting bar or rotary cutters. This process is performed by dedicated sunflower headers (Figures 1.36, 1.38, and 4.36). Readers should note that the sunflower may be harvested with a modified grain head as well, but the percentage of seeds lost due to shattering and gathering is usually higher than the percentage of seeds lost from a specialized sunflower head.

A schematic of a sunflower head working process is shown in Figure 1.37. The seed pans separate the plants of adjacent rows and collect the seeds fall down due to an inherent shaking effect. While the combine moves through the crop, the plants are bent

FIGURE 4.36
Frontal view of a sunflower head and combine. (Courtesy of CLAAS.)

forward by an adjustable front plate, getting between the fingers, and then the sunflower heads are detached due to the action of the snapping roll placed under the seed pans. The gradient of the seed pans, reel position, and rotation speed can be adjusted to accommodate the most diverse crop conditions during harvesting and to maximize the throughput of the machine. The reel pushes the sunflower heads toward the collecting auger, which conveys them to the header center and then to the feeder-house conveyor. Due to their rigidity and low moisture content, sunflower stalks do not bent elastically as other plants (wheat, barley, oats), and therefore developing a theory of sunflower head detachment is a difficult task.

For modeling and designing a sunflower head, the theory outlined in this chapter of the processes of grain header and corn header components may be of great use.

References

Geringhoff. 2013. Geringhoff corn headers. Ahlen, Germany: Geringhoff. http://www.geringhoff.eu/en/products/corn-headers.html (accessed July 18, 2013).

Glancey, J.L. 1997. Analysis of header loss from pod stripper combines in green peas. *Journal of Agricultural Engineering Research* 68:1–10.

Klinner, W.E., M.A. Neale, and R.E. Arnold. 1987a. A new stripping header for combine harvesters. *Agricultural Engineer* 42(1):9–14.

Klinner, W.E., M.A. Neale, R.E. Arnold, et al. 1987b. A new concept in combine harvester headers. *Journal of Agricultural Engineering Research* 38:37–45.

Maglioni, C. 2009. Analysis of reciprocating single blade cutter bars. PhD dissertation, University of Bologna.

Oduori, M.F., T.O. Mbuya, J. Sakai, and E. Inoue. 2012. Kinematics of the tined combine harvester reel. *Agricultural Engineering International Journal* 14(3):53–60.

Shelbourne Reynolds. 2013. Stripper header. Suffolk, England: Shelbourne Reynolds. http://www.shelbourne.com/3/products/1/harvesting/31_ stripper-header (accessed June 29, 2013).

Yuan, J. and Y. Lan. 2007. Development of an improved cereal stripping harvester. *Agricultural Engineering Journal: The CIGR Ejournal* IX, Manuscript PM07 009.

Zeleke, E. 2012. Design and modification of appropriate reel mechanism to harvest tef crop. PhD dissertation, Addis Ababa University.

Bibliography

Gangu, V. 1996. Research on the grain harvesting process by ear raking. PhD dissertation, Transilvania University of Brasov.

Kenney, P.M. 2009. An initial study to determine a friction-factor model for ground vegetation. PhD dissertation, University of Toledo.

Miu, P. 2000. Calculation and construction of agricultural harvesting machinery. Course notes, Politehnica University of Bucharest, Romania.

Neculaiasa, V. and I. Danila. 1995. *Working Processes and Harvesting Machinery* [in Romanian]. Iasi, Romania: A92 Publisher.

Niewenhof, P. 2003. Modeling of the energy requirements of a non-row sensitive corn header for a pull-type forage harvester. PhD dissertation, University of Saskatchewan.

Raupach, M.R., J.J. Finnigan, and Y. Brunet. 1996. Coherent eddies and turbulence in vegetation canopies: The mixing-layer analogy. *Boundary-Layer Meteorology* 78:351–382.

Smith, W. and J. Betran. 2004. *Corn: Origin, History, Technology, and Production*. Hoboken, NJ: John Wiley & Sons.

Van Delden, S.H., J. Vos, A.R. Ennos, and T.J. Stomph. 2010. Analysis lodging of the panicle bearing cereal teff. *New Phytologist* 186:696–707.

FIGURE 1.7
(b) Cutaway of a modern conventional combine 3D model. (Courtesy of CLAAS.)

FIGURE 1.10
Longitudinal section view of a conventional combine harvester. (Courtesy of Same Deutz-Fahr.)

FIGURE 1.22
Section view of Case IH Axial-Flow AFX combine harvester. (Courtesy of Case IH.)

FIGURE 1.25
Cutaway of a Lexion CLAAS combine with a hybrid threshing–separating system. (Courtesy of CLAAS.)

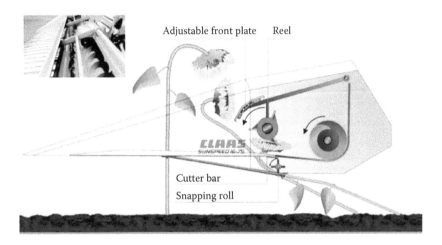

FIGURE 1.37
Schematic of a sunflower head working process. (Courtesy of CLAAS.)

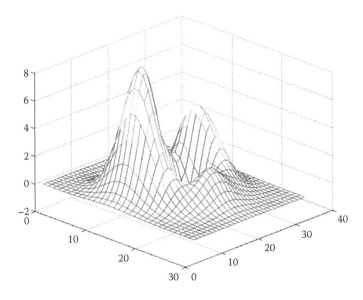

FIGURE 2.17
Global and local optima of a two-dimensional function.

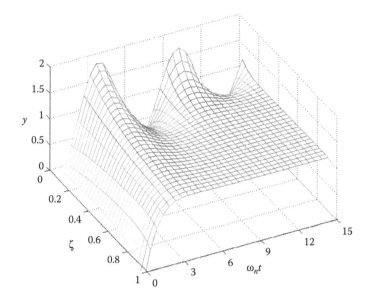

FIGURE 2.28
Transient responses of the second-order underdamped system for different damping ratio values.

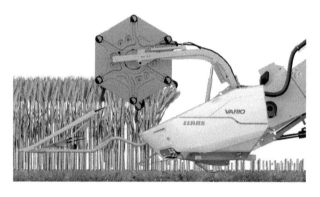

FIGURE 4.1
(b) Lateral view of a grain header in the field. (Courtesy of CLAAS, Harsewinkel, Germany.)

FIGURE 4.24
Functional components of a corn header row section.

FIGURE 5.1
Illustration of threshing and separating processes. (Courtesy of CLAAS, Harsewinkel, Germany.)

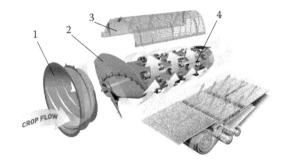

FIGURE 5.4
Material flow in the AXIAL-FLOW threshing unit. (Courtesy of Case IH, Racine, WI.)

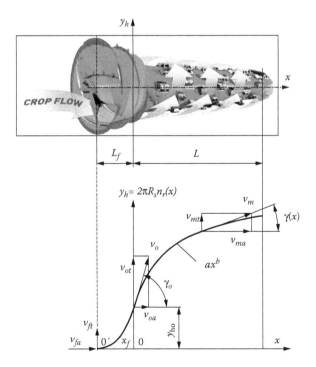

FIGURE 5.9
Schematics of material kinematics in an axial threshing unit.

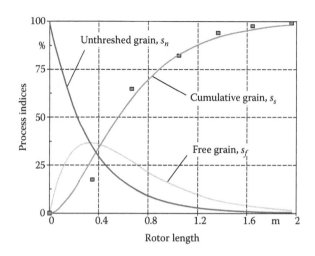

FIGURE 5.21
Graphs of process indices of a different axial threshing unit.

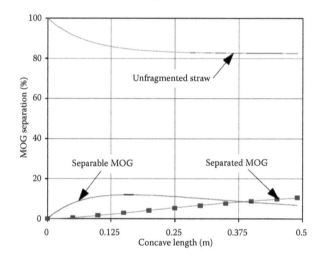

FIGURE 5.36
MOG processing in a tangential unit (feedrate 6.2 kg/s, cylinder speed 29.6 m/s).

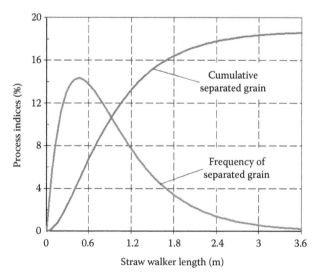

FIGURE 6.7
Straw walker process indices.

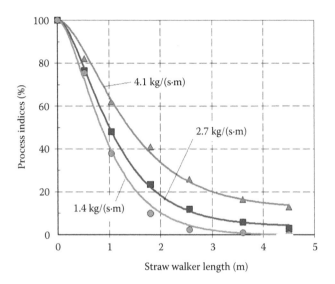

FIGURE 6.8
Agreement of model-predicted data with experimental data published by Beck (1999). (Data from Beck, F., Simulation der Trennprozesse im Mähdrescher [Simulation of separation processes in combine harvesters], PhD dissertation, Hohenheim University, Forschritt-Berichte VDI, Reihe 14(92), 1999, figure 63.)

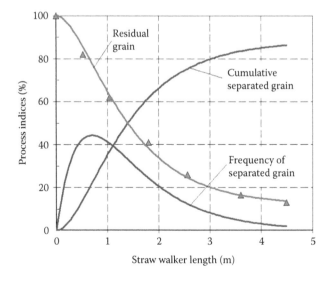

FIGURE 6.9
Indices of grain separation process on straw walkers (experimental data published by Beck, 1999). (Data from Beck, F., Simulation der Trennprozesse im Mähdrescher [Simulation of separation processes in combine harvesters], PhD dissertation, Hohenheim University, Forschritt-Berichte VDI, Reihe 14(92), 1999, figure 63.)

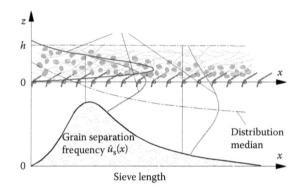

FIGURE 7.24
Transformation of grain distribution $u(z, x)$ within the MOG layer, over the sieve length, to the grain separation frequency under the sieve.

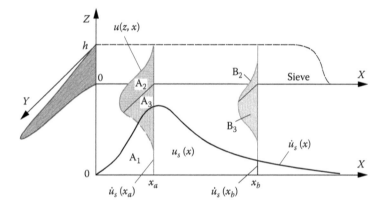

FIGURE 7.25
Transformation of grain distribution $u(z, x)$ within the MOG layer to the grain separation frequency distribution $\dot{u}_s(x)$ under the sieve.

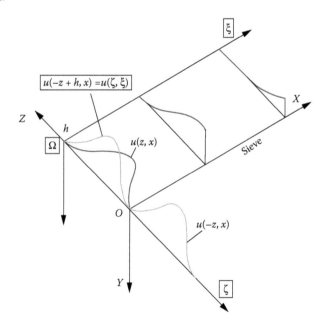

FIGURE 7.26
Transformation of grain distribution $u(z, x)$ to the distribution $u(\zeta, \xi)$.

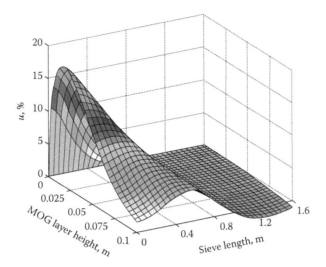

FIGURE 7.27
Predicted grain distribution within MOG layer over the sieve length.

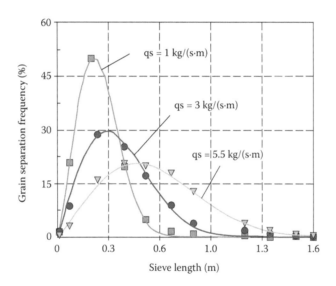

FIGURE 7.28
Prediction of grain separation over the sieve (vs. Figure 7.22).

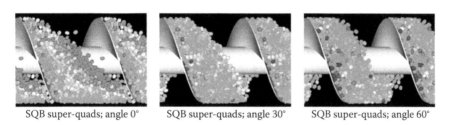

SQB super-quads; angle 0°   SQB super-quads; angle 30°   SQB super-quads; angle 60°

FIGURE 8.2
Influence of the auger inclination angle on the particle flow pattern. (From Owen and Cleary, 2009a and b.)

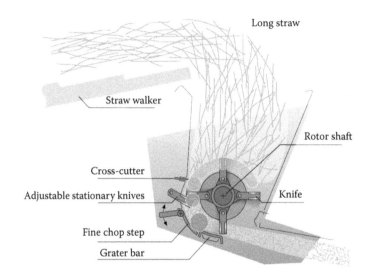

FIGURE 9.2
Schematic of the chopping process. (Courtesy of CLAAS, Harsewinkel, Germany.)

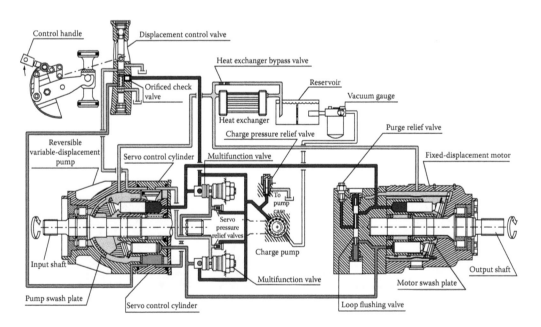

FIGURE 11.13
Diagram of a hydrostatic transmission with a variable-displacement pump and a fixed-displacement motor. (Courtesy of Sauer Danfoss, Neumünster, Germany.)

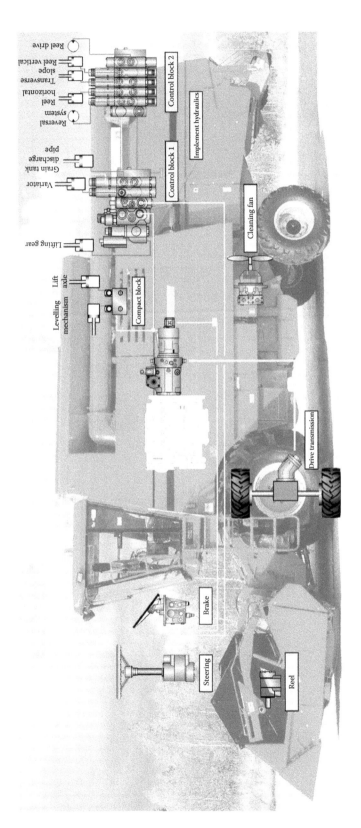

FIGURE 11.15
Hydrostatic and implement hydraulics in a combine harvester. (Courtesy of Bosch-Rexroth Group, Lohr am Main, Germany.)

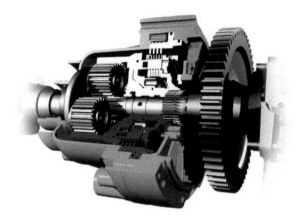

FIGURE 11.16
Continuous variable transmission for an axial threshing unit. (Courtesy of Case IH, Racine, WI.)

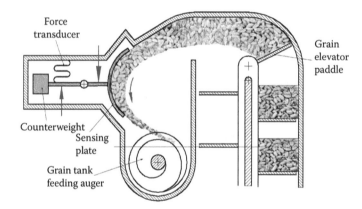

FIGURE 13.2
Design impact force–based grain yield sensor.

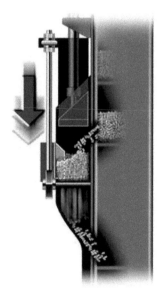

FIGURE 13.7
Grain moisture sensor installation. (Courtesy of CLAAS, Harsewinkel, Germany.)

FIGURE 14.1
Top view of a cab. (Courtesy of CLAAS Harsewinkel, Germany.)

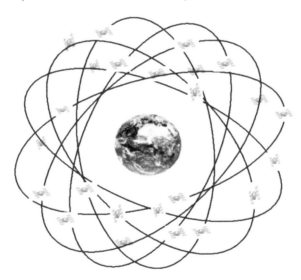

FIGURE 15.1
GPS constellation of 24 satellites (6 orbital planes, each with 4 satellites).

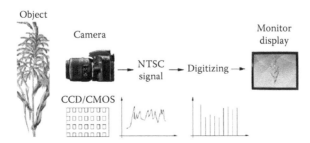

FIGURE 15.8
Image processing by an analog-to-digital (A/D) converter sampling the National Television System Committee (NTSC) signal.

5

Cereal Threshing and Separating Processes: Threshing Units

5.1 Introduction

The grain *threshing process* consists of grain detachment from the flowery cover (ears, panicles, cobs, pods, etc.) until they can move freely, due to a series of phenomena caused by impact and centrifugal forces, friction generated through a mechanical action, and friction among different components of the ears. The grain *separating process*, which follows closely the threshing process, consists of free-moving grain migration through the material-other-than-grain (MOG) mixture, followed by their directed passage through the openings of a separating metallic surface.

These sequential processes take place in a *threshing unit* or a *series of threshing units* of various shapes, sizes, and designs.

The threshing and separating processes are well illustrated in Figure 5.1. A rotating cylinder or rotor (1), equipped with active elements such as rasp bars (2), coupled with a separating body (3) (concave or cage), exercises impact forces on the plants fed into the threshing space that will ultimately separate them through the openings of the concave/cage, while most of the MOG leaves the unit at its rear.

The combine threshing unit is the most important assembly from the points of view of working processes and the power requirement. The ideal threshing unit is one that produces a perfect threshing of maximum crop throughput, with optimum grain separation, while preserving the natural shape and quality of grains and minimizing grain loss. The alternative of building threshing units of various designs has surpassed their thorough research; thus, combine manufacturers are not unanimous on the shape, composition, and dimensions of threshing units. Over the years, however, concepts have become clearer, and similar trends can be noticed in the design of threshing units.

Threshing units can be classified based on the following criteria:

- Material path: *Tangential* or *helical*, corresponding to tangential or axial threshing units.
- Material feeding direction: *Tangential* or *axial*. Axial threshing units may be fed with material either tangentially or axially.
- Number of cylinders (rotors): One, two, or multiple cylinders, or a combination of cylinders and rotors.

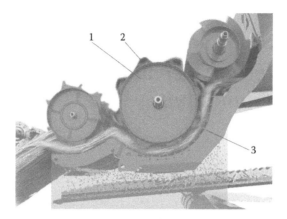

FIGURE 5.1
(See color insert.) Illustration of threshing and separating processes. (Courtesy of CLAAS, Harsewinkel, Germany.)

Figures 1.11 through 1.14 and the introduction to Chapter 1 represent a good overview of different threshing units' construction. In the following two sections we describe their construction in detail.

5.2 Construction of Tangential Threshing Units

A tangential threshing unit (Figure 1.11a) is mainly composed of a cylinder, a concave (a meshed grill), and a mechanism for adjusting the clearance between the cylinder and concave. Such a mechanism can be manually, hydraulically, or electrically driven (Figure 5.2). The threshing space is defined by the curved space that exists between the outside surface described by the cylinder rotation and the inside surface of the concave.

The cylinder is mainly composed of a driven shaft, stamped flanges that are equidistantly distributed over the shaft length, and rasp bars mounted parallel to the cylinder generator using special-purpose hardware.

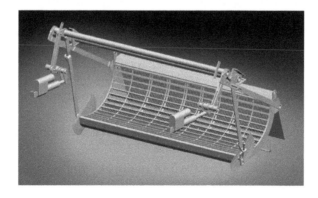

FIGURE 5.2
Electrical actuators and mechanism for adjusting the concave clearance. (Courtesy of SAME Deutz-Fahr, Treviglio, Italy.)

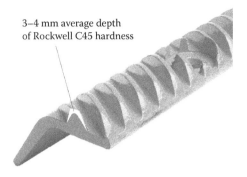

3–4 mm average depth
of Rockwell C45 hardness

FIGURE 5.3
Cylinder's rasp bar segment: design example.

The general shape of a rasp bar has serrations on its surface, and an example is shown in Figure 5.3. The rasp bars represent the active elements of a tangential threshing unit. The serrations or rasps have the following roles: intensifying and concentrating the impact forces of the bars on material to be threshed, increasing the working surface for a better wiredrawing of the material through the rasps, and moving the material axially while it is uniformly distributed within the threshing space. The bar serrations are inclined at a 45°–60° angle relative to the longitudinal axis of the cylinder. They can be left-hand or right-hand oriented. To prevent a continuous axial movement of the material to one end of the cylinder, the rasp bars are mounted in an interpolated orientation; thus, the material is well distributed over the length of the threshing unit while it is subjected to the alternate direction of impact forces created by the bars.

The rasp bars are usually hardened and may also be nonchromed or chromed. Hardened and nonchromed rasp bars have a gently contoured surface design that wears to an even, smooth profile. They are recommended for use in threshing corn, popcorn, soybeans, malting barley, edible beans, or any other seed crop that needs to be delivered at a high-quality level. Hardened and chromed rasp bars have a lifetime that is double that of the nonchromed ones because their sharp edges do not wear off fast. Such rasp bars are recommended for general use in cereal crops, where long-lasting threshing performance is the primary concern. There are also smooth bars whose surface has no serrations. Such bars perform a less aggressive threshing action, and they are recommended in food corn threshing.

The concave (Figure 5.2) has the shape of a curved grill, which wraps the cylinder over a certain center angle whose value evolved over time from 85° to 145°. The concave may be composed of more sequential sections corresponding to the feeding zone, threshing and separating zone, and discharge zone of the MOG. The threshing and separating zone of a concave is mainly composed of lateral supports, intermediate curved bars (oriented perpendicularly to the material movement direction), and reinforcing rods that are oriented along the material movement direction. The bars represent the main active elements of the concave. Their thickness is 8–12.7 mm (0.315–0.5 in.), their depth has values of 38–60 mm (1.500–2.375 in.), and their length equals the length of a tangential threshing unit, which could be between 0.7 and 1.7 m (27.5 and 67 in.), if not more in certain circumstances.

Improvement of tangential threshing and separating processes imposed decreasing the rod diameter from 8–9 mm (0.315–0.354 in.) to 3.5–4.5 mm (0.138–0.177 in.). The bars and rods form a grill through whose openings most of the threshed grains and a substantial part of MOG components get separated. The ratio of the concave opening surface (which represents the active surface of a concave) relative to the radial projection of the concave surface is within the 0.55–0.75 range.

The concave sections must be interchangeable as required by harvesting of different crops whose grain and stem sizes, as well as their mechanical and aerodynamic properties, are different.

The clearance between the cylinder and concave is bigger where the material enters in the threshing space, and it is smaller at the exit. This distance is to be measured radially from the point of the outside diameter of the cylinder to the top surface of a bar. The ratio between the values of the clearance at the entrance and the exit of the threshing space is 3–5.5 in cereals and beans and 1.8–2.5 in corn.

In certain crops and favorable harvesting conditions, the grain separation through the concave can be as high as 90%, but a proportion of 80% grain separation is common in a tangential threshing unit with one cylinder. This is a limiting factor of material throughput that can be processed by such a unit. A more efficient way to increase the material throughput is by using a *series of tangential threshing units*. A tangential threshing unit with two cylinders in series has two advantages: an increased separation area (up to 65%–75% more), that is, threshing efficiency (15%–25% higher), and the possibility of using a less energetic action on the material, which leads to less grain damage, that is, better process quality. This explains the design of a CLAAS threshing unit composed of multiple cylinders to replace the straw walkers.

For threshing of rice, the *cylinder* is designed *with teeth* instead of rasp bars. The concave also has a row of teeth in the feeding section (Figure 1.11b). The teeth have a convex profile that allows a relatively soft treatment of the material. Overall, a threshing unit with teeth is more aggressive than a conventional unit with rasp bars; it lends itself better to the threshing of the uneven developed plants, full of weeds, and with high moisture content. The throughput of the teeth threshing unit is usually 2–2.5 higher than that of a threshing unit with rasp bars of comparable dimensions; however, the grain damage is very high in the first case.

Section 1.4.1 discusses other technical solutions offered by modern combine harvesters. The specifications of tangential threshing unit components of modern combine harvesters are given in Table 1.1. Table 5.1 displays typical cylinder settings for a variety of crops.

TABLE 5.1

Typical Settings of a Conventional Rasp Bar Cylinder in a Variety of Crops

Crop	Peripheral Speed, m/s (ft/s)	Concave Clearance	
		Front, mm (in.)	Rear, mm (in.)
Barley	27–34 (88.5–111.5)	10–18 (0.39–0.71)	3–10 (0.12–0.39)
Beans	7–20 (23–65.5)	20–35 (0.79–1.38)	10–18 (0.39–0.71)
Corn (maize)	10–20 (33–65.5)	25–30 (0.98–1.18)	15–20 (0.59–0.79)
Oats	27–35 (88.5–115)	12–20 (0.47–0.79)	3–10 (0.12–0.39)
Peas	7–18 (23–59)	20–30 (0.79–1.18)	10–18 (0.39–0.71)
Rapeseed	15–24 (49–79)	20–30 (0.79–1.18)	10–20 (0.39–0.79)
Rice	20–30 (65.5–98.5)	14–18 (0.55–0.71)	3–6 (0.12–0.24)
Rye	25–35 (82–115)	12–20 (0.47–0.79)	3–10 (0.12–0.39)
Wheat	24–35 (79–115)	12–20 (0.47–0.79)	4–10 (0.16–0.39)

5.3 Construction of Axial Threshing Units

The limited grain separation efficiency and throughput of straw walkers in conventional combines has finally led to the integral transfer of their function to the axial threshing units.

An axial threshing unit (Figure 5.4, AXIAL-FLOW®) is mainly composed of the feeding section (cage 1 pairs with the rotor feeding zone (2)) and the threshing and separating section (cage 3 with a concave underneath, not shown, pairs with the rotor's threshing and separating zones (4)). The cage in the feeding section has two- or three-start helical ribs that continue into the threshing and separating zones to induce a helical path motion to the material processed in the threshing space. The feeding section of the rotor has an impeller made of three blades (a three-start helical ridge) that are mounted on the rotor cone. The impeller increases the speed of the material from 2.5 to 36 m/s in less than 3 s. The rotor is equipped with short rasp bars arranged in three spirals around it.

In comparison with tangential threshing units, the axial ones offer the following advantages:

- Lower percentage of grain damage (Jalnin, 1982; Wacker, 1985, 1988, 1990).
- Functional versatility: They work much better in all cereals, soybean, and corn.
- Lower grain losses (Arnold, 1979; Wacker, 1988, 1990).
- Better work quality in wet crop conditions (Vogt, 1982).
- With a 16%–20% larger power requirement, the working capacity of an axial unit is much larger (50%–90%) (Serii and Kosilov, 1986).
- A 17%–30% larger working capacity than that of a tangential unit (Eimer, 1980; Vogt, 1982).

A rotary threshing unit needs an increased power and yields a higher degree of MOG fragmentation and separation through concave and cage openings that leading to overcharging the cleaning unit of the combine.

In terms of the direction of material feeding, the axial threshing unit could be fed tangentially (Figure 1.23), axially (Figure 5.4), or oblique-axially (Figure 1.26). The material is

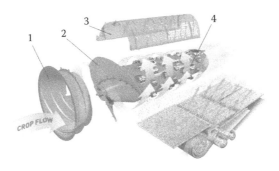

FIGURE 5.4
(See color insert.) Material flow in the AXIAL-FLOW threshing unit. (Courtesy of Case IH, Racine, WI.)

fed by the conveyor of the feeder-house or by the tangential threshing unit that is placed in front of the axial unit (Figure 1.24). The axial threshing units may have one or two rotors. The rotor, like the cylinder, needs a variable rotation for threshing and separation of different crop grains.

While the cage is fixed, the concave underneath the cage is composed of three or four interchangeable, adjustable sections for processing different crops in various field conditions. The specifications of axial threshing units of modern rotary combine harvesters are given in Table 1.2.

5.4 Performance Indices of a Threshing Unit

The assessment of the performance of a threshing unit must take into consideration the design and geometry of the threshing unit, processed crop type, maturity, and its condition due to environmental factors, material throughput, process quality indices, and power requirements to perform the threshing and separating processes. These processes, although sequential for individual grains, take place quasi-simultaneously when a proportion of processed grain is considered. Hence, it is difficult to quantify them separately, unless the approach is based on a stochastic analysis.

The performance indices of a threshing unit (composed of one cylinder or one rotor), accompanying or deriving from the threshing and separating processes, are shown in Figure 5.5. If more units (cylinders in a series or rotors in parallel) compose a threshing assembly, these indices should be calculated individually for each unit: the output from one unit may become the input for the next unit, and so on. Additionally, the required idle power of the cylinder, as well as the specific power consumption (kW/kg/s) per material throughput, characterizes the energy performance of that threshing unit.

Let us make the convention that the vegetal material entering into a threshing unit (measured in kg/s or lb/s) is composed of grains, s (100%), and MOG, pp (100%); the MOG may also include a negligible quantity of weeds or traces of the plants from the previous crop on the considered field, whose overall influence on the threshing and separating processes may be inconsequential or quite important.

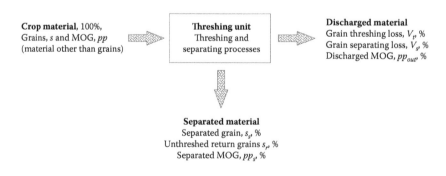

FIGURE 5.5
Performance indices of a threshing unit.

Let us consider a threshing unit of a certain design and dimensional specifications that performs *stationary threshing and separating processes*, that is, stochastic processes whose marginal distributions (means, variances) do not change when shifted in time. Thus, the performance of the threshing unit is characterized by the following *performance indices*:

- Total material throughput, q
- Cumulative percentage (or proportion), s_s, of the grains, relative to s, that separate through the concave, grates, and cage openings, as the case may be
- Percentage of unthreshed grains, s_n, which are either separated or discharged—hence the threshing loss, V_t
- Percentage of threshed and unseparated grains (free grains that are discharged)—hence the separating loss, V_s
- Grain separation efficiency, eff_s
- Cumulative percentage, pp_s, of separated MOG through the concave, grates, and cage openings
- Percentage of fragmented, separable MOG, pp_f, relative to the pp 100% throughput
- Percentage of damaged grains, s_b
- Specific power consumption, P_{ts}, that is, the required power per unit of material throughput
- Idle power, P_o, of the cylinder/rotor that needs to be overcome

These indices depend on the following:

- Crop: Type, variety, maturity, crop yield variation, fraction of green content (weed-to-MOG ratio)
- Vegetal material properties: MOG compressibility modulus, MOG bulk density, grain and MOG moisture contents, coefficients of friction of material with various surfaces, stem structure and shape
- Feeding-dependent parameters: Material cutting height (grain/MOG ratio), feeding direction (tangential, axial), material velocity, and throughput variation
- Threshing unit: Type, design, functional parameters, such as cylinder (rotor) speed, material throughput, and concave clearance

Figure 5.6 shows a qualitative comparison between performance indices of a tangential versus an axial threshing unit, as well as the influences of certain functional parameters, material moisture content, and green content.

There is a "sweet spot" of settings for the threshing unit; identification of these settings should begin during the product development phase of the threshing unit by thoroughly using process mathematical modeling theory, simulation methods, and optimization algorithms. Throughout this chapter, we attempt to analyze and assess the published mathematical models of threshing and separating processes, followed by a unitary and comprehensive modeling approach developed by the author. Then, a *genetic algorithm (GA)–based optimization technique* for the design and functional parameters of a threshing unit is described, and the *MATLAB program script* is laid out in the Appendix.

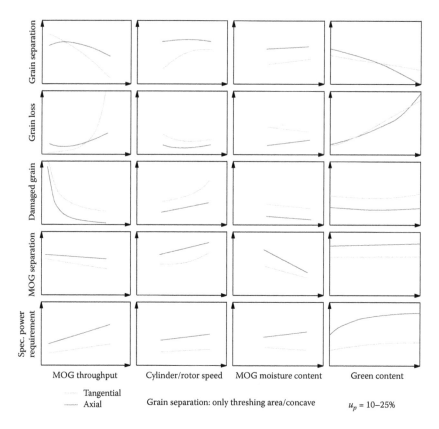

FIGURE 5.6
Qualitative comparison of performance indices of tangential and axial threshing units. (From Wacker, P. Untersuchungen zum Dresch- und Trennvorgang von Getreide in einem Axialdreschwerk [Researches on threshing and separation process in an axial threshing unit], PhD dissertation, Hohenheim University, Forschungsbericht Agrartechnik der MEG 117, 1985.)

5.5 General Assumptions in Theory, Mathematical Modeling, and Simulation

Self-propelled combine harvesters have already reached the limiting width of roads, so cost-effective continuous improvement of combine harvester performance is most likely to be obtained through mathematical modeling, simulation, and optimization of both processes and working units.

Mathematical modeling of threshing and separating processes offers the following advantages:

- Comprehensive understanding of fundamental relationships of the physical process parameters
- Quantification of nonmeasurable dynamic process parameters
- Programming keys for dynamic process simulation and optimization of design and functional parameters of the threshing unit

- Accurate prediction of unit performance over a large range of parameter variation
- Drastic reduction of testing time and costs
- Calculation elements for development of an automated control system for the combine harvester to allow it to achieve the overall expected performance when operating in the field
- Advanced knowledge for further design improvement and new unit development
- Improved decision-making process for future development of the combine harvester manufacturing and application business

Much experimental work has been carried out on threshing units, and much effort and inventive approaches have been extended in the analysis of experiment results. Thus, it is very important to accentuate a list of applicable general assumptions that allowed theory development and evolution as the concepts took shape. Here are general hypotheses (Miu, 1993; Miu and Kutzbach, 2008a,b) that make sense of the development and application of the theory of threshing and separating processes:

- Vegetal material to be processed is homogenous; that is, the ears (panicles, cob, and pods) are uniformly distributed within the MOG mass. The interpretation of their distribution is based on a comparative scale relative to their physical size.
- The threshing unit is fed with a constant feedrate of vegetal material and a constant grain/MOG ratio for a period of time much larger than the stationary operation time of the threshing unit. This assumption is required because it is about the study of data collected during a time period when the process is considered stationary; that is, its statistical properties do not change over time.
- Effective threshing of the grains (grain removal from the ears, panicles, etc.) starts at the beginning of the feeding zone of a threshing unit. In other words, we ignore quantifying a certain proportion of the grains that may have been shattered or detached from their flowery cover by the header and feeder-house conveyors.
- Material to be threshed moves through the threshing space (space bounded by the cylinder/rotor and concave/grates or cage) as a continuous stratum.
- The mass of material is continuously distributed in the threshing space, and its density is a continuous function of position and time.
- In a tangential unit, the material is homogenous in any section of the threshing space whose plane includes the longitudinal axis of the threshing unit, that is, in any section perpendicular to the material flow direction in the threshing space bounded by the cylinder and concave.
- In an axial threshing unit, the material is homogenous in any cross section of the threshing space, that is, perpendicular to the longitudinal axis of the unit.

5.6 Modeling of Material Kinematics

5.6.1 Material Kinematics in a Tangential Threshing Unit

Regardless of the type and design of a threshing unit, the threshing and separating processes occur due to the mechanical action of active elements (rasp bars, teeth, and ribs) and the static bars and rods of the concave/grates, and helical ribs inside the cage. The grains and MOG fragments that directly make physical contact with the rasp bars get higher tangential velocities than those that come in contact with the concave parts.

Thus, the material kinematics (position, speed, and acceleration) through the threshing space varies, and it is influenced by the following:

- Material properties: Type, variety, geometry, shape, and size of the grains and stem, moisture content, and grain/MOG ratio.
- Threshing unit design: Feeding direction and zone geometry, design of rasp bars/ teeth and helical ribs inside the cage, concave wrap-up angle, concave and cage opening design, and opening ratio.
- Process parameters: Material feedrate, rotor speed, and concave clearance.

Figure 5.7 shows the draft of the velocities of vegetal material between the cylinder and the concave. Notice that the material in the immediate vicinity of the cylinder rasp bar gets a velocity whose value is close to the peripheral velocity of the cylinder. Then, the velocities of the following material layers decrease based on the profiles shown in Figure 5.7.

Let us consider a cross section through a tangential threshing and the schematic of the material feeding as represented in Figure 5.8. The cylinder with the radius R and z rasp bars rotates with the angular speed ω; thus, the peripheral velocity v_c of the cylinder is

$$v_c = \omega R \tag{5.1}$$

The height of a rasp bar measured radially is δ_R.

The concave wraps up the cylinder over the angle α. Its total length is l_c, and the variable x is associated with the length of the concave, that is, $x \in [0, l_c]$. The clearance between the cylinder and concave top surface varies between δ_i and δ_e.

The material is fed into the threshing space with the velocity v_f at an angle γ_1.

Let us assume that in the feeding zone the density of the material as pressed by the feeder-house conveyor does not appreciably change. When the cylinder rotates within the time t, with an angle

$$\phi = \frac{2\pi}{z} \tag{5.2}$$

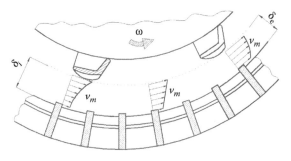

FIGURE 5.7
Draft of material velocity in a tangential threshing unit.

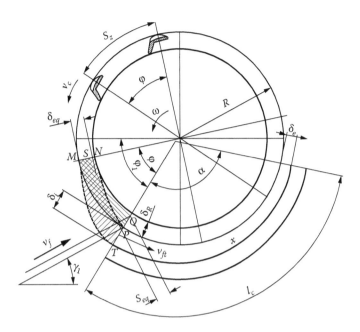

FIGURE 5.8
Schematic of tangential unit feeding with material.

the material must get into the space between two consecutive rasp bars on the distance δ_R. This means the space bounded by the MNPT cross section. Since this is difficult to evaluate, let us define an equivalent section MPQS with a similar area and length S_z. It means that the material will get in the space between two consecutive bars on an equivalent depth δ_{eq}:

$$\delta_{eq} = \frac{A_{MNPT}}{S_z} \tag{5.3}$$

Consequently, the following equation can be written for the time t:

$$t = \frac{S_{eq}}{v_f} = \frac{S_z}{v_c} \tag{5.4}$$

where S_{eq} is the real distance traveled by the material.
The curvilinear distance S_z value is given by the following equation:

$$S_z = \frac{2\pi R}{z} \tag{5.5}$$

This distance can be calculated as follows:

$$S_{eq} = \frac{\delta_{eq}}{\cos(\phi_1 - \gamma_1)} \tag{5.6}$$

where ϕ_1 is the mounting angle of the concave.

As a consequence, we obtain Equation 5.7 to calculate the value of the required velocity of the material to be fed into a tangential threshing unit:

$$v_f = v_c \frac{c_{eq}\delta_R z}{2\pi R \cos(\phi_1 - \gamma_1)} \tag{5.7}$$

Geometric calculations show that the coefficient $c_{eq} = 0.65$–0.8; the maximum value corresponds to the highest allowable velocity of uniform material feeding. The literature recommends values of material feeding velocities greater than 6.1 m/s (20 ft/s). At low feeding velocities (2.5–3 m/s), even if the feeding conveyor is uniformly fed with material by the header, the threshing unit is fed nonuniformly because of material partitioning by conveyor slats.

Within a reasonable approximation, we assume that the cross section size of the threshing space varies linearly from the entrance clearance δ_i to the exit clearance of the concave δ_e over the length l_c of the concave (Figure 5.8). Applying the properties of the theorem of Thales, we can write the following equality:

$$\frac{\delta_i - \delta_e}{l_c} = \frac{\delta_i - \delta}{x} \tag{5.8}$$

where δ is the concave clearance corresponding to the current position x along the concave length. Thus, at any given position x, the concave clearance can be calculated as follows:

$$\delta = \delta_i - \frac{x}{l_c}(\delta_i - \delta_e) \tag{5.9}$$

At limit,

$$\text{If } x = 0, \delta = \delta_i \text{ and}$$
$$\text{if } x = l_c, \delta = \delta_e$$

If we calculate the value of the tangential component v_{ft} of material feeding velocity at the entrance of the threshing space based on the feeding zone geometry, we propose the following formula for calculation of the material velocity v_t within the threshing space:

$$v_t = v_{ft} + v_c \left(\frac{x}{l_c}\right)^{\frac{1}{\varepsilon_1}} e^{-\varepsilon_2 \delta} \tag{5.10}$$

where:

ε_1 is the coefficient of material speed slippage relative to the rasp bar speed, $\varepsilon_1 < 1$

ε_2 is the coefficient of concave clearance resistance to the material movement through the threshing space

The values of these coefficients vary with material properties and design of the threshing unit.

5.6.2 Material Kinematics in an Axial Threshing Unit

There is a direct connection between the engineering design, the movement of material through the threshing space, and the overall performance of an axial threshing unit.

Until recently, the design process of an axial threshing unit was performed through a combination of the designer's experience and intuition that require a thorough knowledge of the threshing and separation process characteristics and the detailed specifications of similar products. The process of threshing unit design should always begin with a high-level description of the unit's desired behavior.

Published studies on material kinematics in axial threshing units are not numerous. Notable, though, are the contributions of Gasparetto et al. (1989) and Wacker (1985), who reported on the influence of a few parameters (feedrate, rotor speed, material moisture, and spiral angles) on material movement.

This section contains a further development of the study of the movement of vegetal material through the feeding zone and threshing and separating zones of axial threshing units, on which the author (Miu, 2002a, 2004; Miu and Kutzbach, 2007) has previously elaborated.

The kinematics equations are based on a nonlinear law governing the nonuniform movement of the material on an uneven helical path between the rotor and the concave–cage assembly.

The kinematics of the material through the threshing space is mainly influenced by

- Material properties: Type, variety, stem size, MOG moisture content, grain-to-MOG ratio
- Threshing unit design: Feeding direction, design of helical spirals (ribs) inside the cage, concave and cage openings' design
- Process parameters: Material feedrate, rotor speed, concave clearance

In developing the theory, the following specific assumptions have been made:

1. At the beginning of feeding zone of the threshing unit, the material feeding velocity has no tangential component.
2. Once introduced into the threshing space, the material moves as a continuum stratum.
3. Material velocity is a continuous function of material position over the length of the threshing space.
4. The radial reactive forces due to grain separation do not influence the material movement.

Let us consider that the velocity of the feeding material at the entrance of the feeding zone is v_f with the components v_{fa} and v_{ft} (Figure 5.9).

At the end of the feeding zone (i.e., the beginning of the threshing zone), the fed material velocity v_o has a tangential component v_{ot} and an axial component v_{oa}. Thus, as a vector, we can write

$$\bar{v}_o = \bar{v}_{oa} + \bar{v}_{ot} \tag{5.11}$$

The scalar value of the feeding velocity is

$$|\bar{v}_o| = \sqrt{|\bar{v}_{oa}|^2 + |\bar{v}_{ot}|^2} \tag{5.12}$$

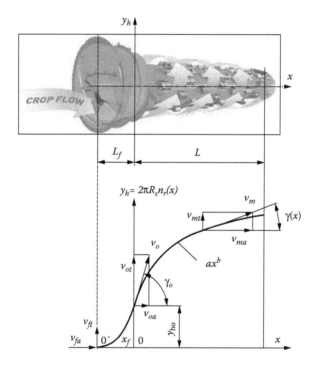

FIGURE 5.9
(See color insert.) Schematics of material kinematics in an axial threshing unit.

In Figure 5.9, the curve represents the flattened length of the material trajectory from the beginning of the unit feeding zone to the end of the discharge zone. The shape of the trajectory with respect to x is ultimately determined by the magnitude of material speed $v_m(x)$ as well as the angle $\gamma(x)$, where x is the current position over the rotor length ($x \geq 0$). The material speed $v_m(x)$ depends mainly on the feeding velocity and rotor speed, while the angle $\gamma(x)$ depends on the unit design (rasp bar angle, spiral angle, etc.).

The material moves through the threshing unit on a helical path with variable pitch. This movement is completely defined if the following equations are known:

$$\theta_m = \theta_m(t)$$
$$x = x(t) \tag{5.13}$$

where:

　　t　is the time
　　θ_m　is the current rotation angle of the material around the longitudinal axis of the threshing unit

By eliminating the time within Equations 5.13, it follows that

$$\theta_m = \theta_m(x) \tag{5.14}$$

Let $n_r(x)$ be the number of revolutions with radius R_s that have been completed by the material over the length x, where $n_r(x)$ is not necessarily an integer value. This means that

$$\theta_m(x) = 2\pi n_r(x) \tag{5.15}$$

The radius R_s can be expressed as

$$R_s = R + \frac{\delta}{2} \tag{5.16}$$

where:
    δ is the weighted average clearance between the rotor and concave, respectively, the rotor and cage
    R is the rotor radius

The variation of angle $\theta_m(x)$ is nonlinear and can be described by a function that

1. Has a *zero* at $x = 0$ (assumption 1)
2. Is continuous over the length x, that is, is differentiable and satisfies the previous assumption
3. Has no inflection points or extremes, to satisfy assumption 2

Consequently, we introduce the following relationship:

$$\theta_m(x) = \frac{a}{R_s} x^b \tag{5.17}$$

where:
    a is the coefficient that helps carry out the above-mentioned last condition
    b is the exponent

Thus, the number of material revolutions in the threshing and separating zones is

$$n_r(x) = \frac{a}{2\pi R_s} x^b \tag{5.18}$$

Let l_t be the length of the flattened curve $f(x) = ax^b$, which is the length of the material trajectory in the threshing and separating zones. According to the mathematical theory (Brohnstein and Semendyayev, 1985), the length of trajectory can be written as follows:

$$l_t = \int_0^L \left(1 + \left[f'(x)\right]^2\right)^{1/2} dx \tag{5.19}$$

Hence, the path length of the material over the interval [0, L] is given by equation

$$l_t = \int_0^L \left(1 + \left[abx^{b-1}\right]^2\right)^{1/2} dx \tag{5.20}$$

which must be computed by numerical methods using a ready available software such as MATLAB.

From geometrical interpretation of the first derivative of the function $f(x)$, we get the angle $\gamma(x)$ at any point x:

$$\tan \gamma(x) = f'(x) = abx^{b-1} \tag{5.21}$$

Hence, the relation between the components of material velocity is

$$v_{mt} = v_{ma}abx^{b-1} \tag{5.22}$$

where:
    v_{mt} is the tangential velocity of the material
    v_{ma} is the axial velocity of the material

The absolute velocity of material v_m on its path can be computed as follows:

$$v_m = v_{ma}\sqrt{1+\left(abx^{b-1}\right)^2} \tag{5.23}$$

The value of angle γ can be used to compute the conditional probability of kernel passage through the concave and cage openings, and to design these subassemblies (Miu et al., 1997).

The axial pitch p_{am} of a material trajectory is given by the following equation:

$$p_{am} = 2\pi R_s \frac{v_{ma}}{v_{mt}} = 2\pi R_s \frac{1}{abx^{b-1}} \tag{5.24}$$

The value of the material trajectory pitch is very useful for designing both the rotor and the cage in either tangential or inclined feeding sections of the threshing unit to ensure that fed material does not return after one complete revolution.

The above-given equations did not take into account the speed v_0 of the material at the beginning of the threshing zone, which is greater than the material feeding speed v_f due to the fact that the material movement in the feeding zone is accelerated abruptly by the impeller blades (three-start helical ridge), feeding auger, and so forth, depending on threshing unit design. Thus, the material is accelerated along the length L_f of the feeding zone, to which the variable x_f has been associated. Taking into account Equation 5.18, we can write that the total length of the material trajectory from the beginning of the feeding zone to the discharge end is

$$y_h = 2\pi R_s n_r(x) = y_{ho} + ax^b \tag{5.25}$$

We propose the following law that describes the flattened trajectory of the material within the feeding zone:

$$y_o = a_f\left(e^{b_f x_f} - 1\right) \tag{5.26}$$

where the coefficients a_f and b_f have positive values.

At $x_f = L_f$, we get

$$y_{ho} = a_f \left(e^{b_f L_f} - 1 \right) \tag{5.27}$$

In other words, we have the following definition of the function y_h:

$$y_h = \begin{cases} a_f \left(e^{b_f x_f} - 1 \right), & \text{for} \quad 0 \le x_f \le L_f \\ a_f \left(e^{b_f L_f} - 1 \right) + a x^b, & \text{for} \quad -x_f \le x \le L \end{cases} \tag{5.28}$$

The values of the coefficients a_f and b_f can be computed from experimental data that was not available to the author at the time of writing this book. The model parameters b and a depend on material properties, and on the design and functional parameters of the threshing unit. The parameter $b \le 1$ due to slippage between the material and the rotor, as well as a higher angle $\gamma(x)$ compared to the theoretical angle of trajectory, as determined by spirals and rasp bars or other active elements of the rotor. Because b is usually smaller than 1, the graphs of the functions $f(x) = ax^b$ are curves without any points of inflection or extremes.

Figure 5.10 shows the graphs of the number of revolutions performed by two different materials versus rotor length (Miu and Kutzbach, 2007). The functions are monotonic, increasing over the rotor length.

Because the factor $a > 0$, it stretches the y coordinates, as can be seen in Figure 5.11. The graph of the number of revolutions in Figure 5.12 displays a similar pattern.

MOG feedrate represents an important parameter influencing the trajectory length (Figure 5.13). The measured data is marked with symbols. The line represents the corresponding predicted data. With increasing MOG feedrate, the friction between material and the rotor/cage increases, and transportation of the material becomes more effective.

The variation of material velocity components over the rotor length versus rotor speed is shown in Figure 5.14. It can be noted that material sliding against the rotor is variable over the rotor length.

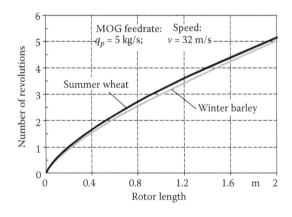

FIGURE 5.10
Number of revolutions of different materials.

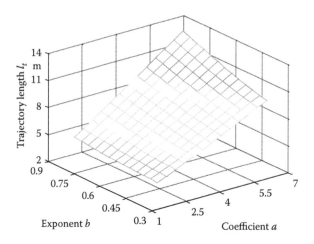

FIGURE 5.11
Trajectory length.

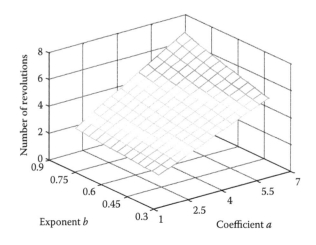

FIGURE 5.12
Number of material revolutions in the threshing space.

5.7 Modeling of Grain Threshing and Separating Processes

5.7.1 Literature Review of Mathematical Modeling of Grain Processing in Tangential/Axial Threshing Units

In this section different approaches in the published theories are mentioned; then, in Section 5.8, the theory the author has further elaborated will be presented.

Under optimal conditions, the cumulative grain separation in a threshing unit has the following values (Wacker, 1985; Serii and Kosilov, 1986; Miu, 1995):

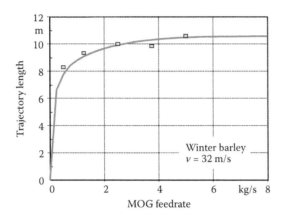

FIGURE 5.13
Influence of MOG feedrate on trajectory length.

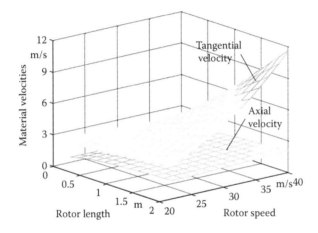

FIGURE 5.14
Tangential and axial velocities in barley.

- Up to 85% for a tangential unit composed of one cylinder
- Up to 90% for a tangential unit composed of two cylinders in series
- More than 99% for axial threshing units

Most scientific published works refer to equations for grain separation over the concave of a tangential threshing unit or over the rotor length of an axial one, as shown below. In the following, the variable x will be the current position associated with the threshing length (Miu, 1994, 1995; Miu and Kutzbach, 2008a,b), that is,

- The length l_c of the concave inside arc, $x \in [0, l_c]$, for a tangential threshing unit (Figure 5.8)
- The length L of threshing and separating zones, $x \in [0, L]$, for an axial unit (Figure 5.9)

Simple models use a single exponential function that represents the cumulative separation function s_s over the threshing length for a tangential threshing unit (Vasilenko, 1954; Arnold, 1964; Filatov and Chabrat, 1969); for an axial threshing unit (Wacker, 1985),

$$s_s(x) = 1 - e^{-\lambda x} \tag{5.29}$$

$$r_s(x) = e^{-\lambda x} \tag{5.30}$$

where:

r_s is the remaining grain to be separated

λ is the constant separation coefficient; that is, threshing and separation processes occur at a constant rate over the threshing length

These models offer the advantage of simple calculations with variable accuracy, depending on the current position x within the threshing length. A single separation coefficient cannot incorporate all influences of unit design specifications and functional parameters. The coefficients for threshing, segregation, and separation operations are embedded within the separation coefficient λ. Furthermore, Wacker (1985) proposed two different coefficients, λ_t and λ_s, for threshing and separating zones, respectively. Moreover, he found that second-order polynomials could quantify the influence of rotor speed v and MOG throughput q_p on the separation coefficients λ as follows:

$$\lambda = a_v + b_v v + c_v v^2 \tag{5.31}$$

$$\lambda = a_q + b_q q_p + c_q q_p^2 \tag{5.32}$$

where a, b, and c are experimental coefficients that depend on material properties and threshing unit design. The values of such coefficients corresponding to the influence of the rotor speed are given in Table 5.2. Figures 5.15 and 5.16 show the influence of rotor speed and MOG throughput, respectively (Wacker, 1985).

Maertens and De Baerdemaeker (2003) showed that Rusanov's equation, Equation 5.33, fits best to an experimental data set obtained from test runs with a New Holland combine. Rusanov (2001, 2002) and Vetrov (2002) further discuss this equation.

TABLE 5.2

Values of Experimental Coefficients Corresponding to the Influence of Rotor Speed

Material	q_p t/h	u_s %	u_p %	λ_t a_v	b_v	c_v	λ_s a_v	b_v	c_v
Summer wheat	17.5	18.0	37.5	−1.809	0.340	−0.0057	−0.786	0.226	−0.0036
Winter wheat	15.3	16.0	39.5	−4.489	0.446	−0.0070	−6.607	0.574	−0.0086
Summer barley	8.9	16.0	36.5	−9.164	0.758	−0.0120	−0.815	0.202	−0.0027
Winter barley	13.2	15.5	28.5	−7.694	0.627	−0.0090	−6.660	0.529	−0.0078

Source: Wacker, P., Untersuchungen zum Dresch- und Trennvorgang von Getreide in einem Axialdreschwerk [Researches on threshing and separation process in an axial threshing unit], PhD dissertation, Hohenheim University, Forschungsbericht Agrartechnik der MEG 117, 1985.

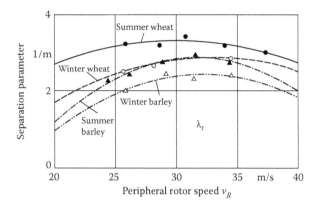

FIGURE 5.15
Influence of rotor speed v on separation coefficient λ_t in wheat and barley.

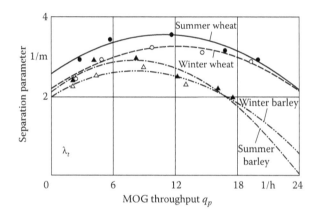

FIGURE 5.16
Influence of MOG throughput q_p on separation coefficient λ_t in wheat and barley.

Rusanov (1971) proposed the following exponential function to describe the remaining grain to be separated in a tangential unit:

$$r_s = e^{-\mu x^\alpha} \tag{5.33}$$

where the exponent $\alpha = 0.9$ is based on experimental data.

Caspers (1973) developed an exponential function with a third-degree polynomial exponent as follows:

$$s_s = k\left(1 - e^{-\left(a_1 \cdot x + a_2 \cdot x^2 + a_3 \cdot x^3\right)}\right) \tag{5.34}$$

where:

k is the total grain mass
a_1, a_2, a_3 are the coefficients that depend on unit design and material properties

For grain separation in a tangential unit, Lo (1978) developed an exponential function whose exponent is expressed as a fourth-degree polynomial of threshing length, Equation 5.35:

$$S_s = 1 - e^{-\left(a_1 \cdot x + a_2 \cdot x^2 + a_3 \cdot x^3 + a_4 \cdot x^4\right)} \tag{5.35}$$

Trollope (1982) developed an interesting set of differential equations of the threshing process in a tangential unit starting from the mass balance in a volume element of threshing space. The nature of the equations does not allow finding of the exact solution, but approximations of it. He made the assumption that the pressure p_0 exerted on the material is constant within the threshing space. Hence, the grain separation could be described by the following function:

$$S_s = q_s \left(1 - \frac{kp_0}{kp_0 - c} e^{-c\psi(\theta)}\right) \tag{5.36}$$

where the parameter φ depends on the threshing length, rotor speed, and material velocity, while the parameter c depends on the throughput. The main issue is that the pressure exerted on the material is not constant, although this should be a desideratum to maximize the process efficiency.

Other researchers have quantified the grain separation using the difference of two exponential functions, by taking into consideration the threshing/migration and separation operations. The coefficients in such functions are also influenced by crop properties and functional and design parameters of threshing units.

The first such example is Equation 5.37, developed by Alferov and Braginec (1972):

$$S_s = 1 - e^{-\lambda_1 x} - S_n \frac{1}{\lambda_1 - \lambda_0}\left(e^{-\lambda_0 x} - e^{-\lambda_1 x}\right) \tag{5.37}$$

where S_n is the mass of unthreshed grain.

Klenin and Lomakin (1972) have developed a similar equation, as follows:

$$S_s = 1 - \frac{S_n}{k_2 \mu}\left(k_2 e^{\mu x} - e^{k_2 x}\right) - \left(1 - S_A\right)e^{\mu x} \tag{5.38}$$

Trollope (1982) also developed an equation with two coefficients:

$$S_s = 1 + \frac{1}{kp_o - c}\left(ce^{-kp_o \psi} - kp_o e^{-c\psi}\right) \tag{5.39}$$

The grain separation model developed by Huynh et al. (1982) uses stochastic concepts to quantify the threshing process tangential threshing unit. This forerunner model represents a good approach for the threshing process description. The difficulty lies in measuring or estimating the time for which the grains remain within the threshing space, as well as the material velocity, which is variable over the threshing length (Miu, 2002a).

Analysis of the research results and models that are presented in the literature leads to the following remarks (Miu, 1995, 2004; Miu and Kutzbach, 2000, 2008a,b; Kutzbach, 2003; Maertens and Baerdemaeker, 2003):

- Most of the research has had the initial aim of knowledge and understanding of the threshing and separating processes.
- All presented functions represent valuable approaches for modeling threshing and separation processes.
- Most previous work has been done on conventional combines having rasp bar cylinders for threshing.
- Many studies and research have reported on variables that affect separation only.
- The influences of crop properties and functional and design parameters have been reflected in graphs, but only a few related mathematical relationships have been published.
- The models are partial; that is, they describe only particular aspects of the threshing–separating process. Most models are composed of a single equation and usually describe grain separation or the to-be-separated remaining grain.
- The models have been developed using different assumptions for certain types of threshing units and for particular test run conditions, so none of the models are universal.

Certain assumptions made in these previous efforts, as well as approach methods developed by mathematicians, have been used by the author in developing and validating the theories presented in this book.

5.7.2 Conjoint Mathematical Model of Grain Processing in Tangential/Axial Threshing Units

In the following, the author's unitary theory and accompanying mathematical models quantify the threshing and separating processes in both tangential and axial threshing units so that only coefficient values need to be adjusted for different crops. The mathematical models are further embedded in a GA for unit design and process optimization, allowing optimum setting of the threshing units as well as enhancing the decision-making act in design and new product development.

In mathematically modeling the grain processing in threshing units, the author considered the assumptions outlined in Section 5.5. In addition to that, to understand and apply this model, the following assumption must be noted: the threshing process—by which the grains are removed and separated from the ear—begins

- At the entrance in a tangential unit, respectively
- At the beginning of the threshing section of an axial threshing unit

This assumption can be eliminated; this shall imply a precise correction of the initial percentage of unthreshed grain (e.g., 97.8% instead of 100%) if a certainly known proportion of grains have already been removed from their flowery cover by the forefront conveyors or unit impeller.

In a threshing unit, which is composed of a cylinder and a concave, respectively, a rotor and a cage, the grains are detached from the ears, and most of them separate through the openings of the concave, or grates.

The threshing and separating process can be divided into the following three sections:

1. Detachment of the grain from the ears (the grain becomes *free* grain in the threshing space)
2. Motion of free grain kernels through the straw mat
3. Passage of free grain kernels through the openings of the concave or grates

The probability that grains will reach the separation surface is the same over the separation length as is the probability of free grain passage through the openings of the separation surface (Huynh et al., 1982; Mailander, 1984; Miu, 1994, 1995; Miu and Kutzbach, 2008a,b).

The probabilistic laws that respectively describe the above-mentioned events have been identified as follows:

1. $f(x) = \lambda e^{-\lambda x}$ (5.40)

2. $f(x)$ (5.41)

3. $g(x) = \beta e^{-\beta x}$ (5.42)

where:
- λ is the specific threshing/segregation rate for events a and b, respectively
- β is the specific separation rate for event c

At this point it is necessary to emphasize that these probability density functions are defined on the threshing space length x, and λ and β are space increments between respective successive event changes, while Huynh et al. (1982) assumed similar rates expressed as waiting times. This is not quite correct since the speed of the material (implicit time rate) varies along the length x (Gasparetto et al., 1989; Wacker, 1985; Miu, 2002a).

The distribution frequency of unthreshed grain percentage into the threshing space is a continuous variable. Hence, the *percentage of unthreshed grain* s_n is given by integrating Equation 5.40 over the length x as follows:

$$s_n(x) = 1 - \int_0^x \lambda e^{-\lambda \xi} d\xi = e^{-\lambda x} \qquad (5.43)$$

At the end of the threshing space ($x = l_c$ for tangential unit or $x = L$ for axial unit), the unthreshed grain becomes threshing loss V_t, that is,

$$V_t = s_n(l_c) = e^{-\lambda l_c}, \text{ for a tangential unit} \qquad (5.44)$$

$$V_t = s_n(L) = e^{-\lambda L}, \text{ for an axial unit} \qquad (5.45)$$

Free grain kernels can separate; therefore, their passage through the straw mat and openings (random events b and c) is independent and successive. According to the probability theory, the *joint probability density* $s_d(x)$ of the sum of two independent and steady random variables with the densities $f(x)$ and $g(x)$ equals the convolution (Bosch, 1996) of their individual probability densities:

$$s_d(x) = f * g = \int_0^x f(z)g(x-z)dz \qquad (5.46)$$

By solving this equation, we get what actually is the *grain separation frequency*:

$$s_d(x) = \frac{\lambda\beta}{\lambda - \beta}\left(e^{-\beta x} - e^{-\lambda x}\right) \qquad (5.47)$$

The *cumulative distribution function* $s_s(x)$ of separated grain is found by integrating the grain separation frequency $s_d(x)$ to get

$$s_s(x) = \frac{1}{\lambda - \beta}\left[\lambda\left(1 - e^{-\beta x}\right) - \beta\left(1 - e^{-\lambda x}\right)\right] \qquad (5.48)$$

Since the material throughput is constant from the initial assumption, in the cross section of the threshing space at any current position x of separation length, the *mass balance* can be written using the *unthreshed grain* $s_n(x)$, *separated grain* $s_s(x)$, and *free grain fraction* $s_f(x)$, as follows:

$$s_n(x) + s_f(x) + s_s(x) = 1 \qquad (5.49)$$

From Equations 5.43, 5.48, and 5.49, we get the percentage of *free, separable grain* $s_f(x)$, as follows:

$$s_f(x) = \frac{\lambda}{\lambda - \beta}\left(e^{-\beta x} - e^{-\lambda x}\right) \qquad (5.50)$$

At the end of threshing space, the free grain becomes *segregation and separation loss* V_{gs}:

$$V_{gs} = s_f(l_c) = \frac{\lambda}{\lambda - \beta}\left(e^{-\beta l_c} - e^{-\lambda l_c}\right), \text{ for a tangential unit} \qquad (5.51)$$

$$V_{gs} = s_f(L) = \frac{\lambda}{\lambda - \beta}\left(e^{-\beta L} - e^{-\lambda L}\right), \text{ for an axial unit} \qquad (5.52)$$

Figures 5.17 and 5.18 show the graphs of unthreshed grain $s_n(x)$, free grain $s_f(x)$, and cumulative separated grain $s_s(x)$ for a tangential and an axial threshing unit, respectively. It can be noticed that the fraction of unthreshed grain decreases exponentially. The fraction of free grain rises to a peak value and then decreases due to grain separation.

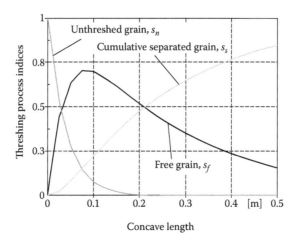

FIGURE 5.17
Variation of unthreshed grain, free grain, and cumulative separated grain in a tangential unit.

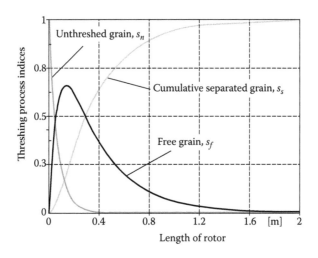

FIGURE 5.18
Variation of unthreshed grain, free grain, and cumulative separated grain in an axial unit.

Based on Equations 5.43, 5.48, 5.49, and 5.50, we can write the continuity equation of the process quality indices over the length of threshing space as (Miu, 2001b)

$$\frac{1}{\beta}\frac{\partial^2 s_s}{\partial x^2} + \frac{\partial s_s}{\partial x} = \lambda s_n \tag{5.53}$$

Grain separation frequency $s_d(x)$ depends on the *fraction of separable, segregated grain* $s_g(x)$, which reaches the concave/cage surface, by a coefficient that is the linear segregation rate λ. Thus, the fraction of separable, segregated grain $s_g(x)$ is given by the following equation:

$$s_g(x) = \frac{\beta}{\lambda - \beta}\left(e^{-\beta x} - e^{-\lambda x}\right) \tag{5.54}$$

At the end of the threshing space, the separable, segregated grain $s_g(x)$ becomes *pure separation loss* V_s as follows:

$$V_s(l_c) = \frac{\beta}{\lambda - \beta}\left(e^{-\beta l_c} - e^{-\lambda l_c}\right), \text{ for a tangential unit} \tag{5.55}$$

$$V_s(L) = \frac{\beta}{\lambda - \beta}\left(e^{-\beta L} - e^{-\lambda L}\right), \text{ for an axial unit} \tag{5.56}$$

Consequently, the *fraction of free, unsegregated grain* $s_{ng}(x)$ can be calculated with the following equation:

$$s_{ng}(x) = s_f(x) - s_g(x) = e^{-\beta x} - e^{-\lambda x} \tag{5.57}$$

At the end of the threshing space, the unsegregated grain becomes *segregation loss* V_{ng}:

$$V_{ng}(l_c) = e^{-\beta l_c} - e^{-\lambda l}, \text{ for a tangential unit} \tag{5.58}$$

$$V_{ng}(L) = e^{-\beta L} - e^{-\lambda L}, \text{ for an axial unit} \tag{5.59}$$

Separation efficiency describes the separation intensity along the length of the separation space (Wacker, 1985). According to Miu et al. (1997), the *continuous separation efficiency* $eff_s(x)$ at the current position x of the separation length is defined as the ratio between the separated grain mass on the differential interval dx and the available grain mass to be separated to the rear of the threshing unit. This can be mathematically expressed as follows:

$$eff_s = \frac{s_d}{1 - (V_t + s_s)} \tag{5.60}$$

The grain separation efficiency represents the probability of separation of the remaining available (separable) grain. Grain separation efficiency is a useful index for comparison of the design versus work of similar threshing units with different sizes.

In the following, a procedure of assessing the model parameter β is proposed. Let us consider the concave and cage geometry as shown in Figure 5.19a and b. According to the screening theory (Wessel, 1967; Huynh et al., 1982), it is assumed that a free kernel will pass through an opening of the concave (or cage) if the projection of the kernel on the opening surface is within that area. Thus, the theoretical probability for a kernel to pass through an opening concave is

$$p_{th} = \frac{(a_1 - a_2 - d_e)(b_1 - b_2 - d_e)}{a_1 b_1} \tag{5.61}$$

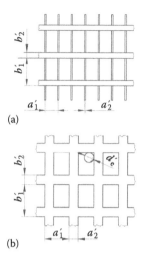

FIGURE 5.19
(a) Geometry of the concave openings. (b) Geometry of the rotor cage openings.

where:
 a_1 is the centerline distance between rods, mm
 a_2 is the rod diameter, mm
 b_1 is the centerline distance between bars, mm
 b_2 is the width of a bar, mm
 d_e is the equivalent diameter of a kernel, mm

Analogously, the theoretical probability of passage for a kernel through a cage opening is given by the following equation:

$$p'_{th} = \frac{(a'_1 - a'_2 - d_e)(b'_1 - b'_2 - d_e)}{a'_1 b'_1}$$

(5.62)

According to Huynh et al. (1982), the conditional probability of a grain passage through a given opening (Figure 5.20) can be written as follows:

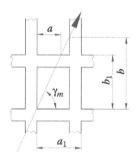

FIGURE 5.20
Scheme of local trajectory of a free (separable) kernel.

$$p = \beta a = \beta \frac{b_1}{tg\gamma_m} \tag{5.63}$$

where:
    a is the available average distance of passage for a kernel through an opening (measured in the x direction), mm
    γ_m is the angle of the material trajectory, degrees (Miu, 2002a)
    p is the overall probability

Since a kernel can separate through a concave opening or cage opening, this separation event can be partitioned into *two mutually exclusive events* (Miu et al., 1997). In this case the overall probability p represents a weighted value of the probabilities p_c and p_m, with the weighting relative to the concave wrap angle α and, respectively, cage wrap angle α_m:

$$p = \frac{p_c\alpha + p_m\alpha_m}{\alpha + \alpha_m} \tag{5.64}$$

The above calculations can be summarized using the following formula (Miu and Kutzbach, 2000):

$$p = \eta\left(1 - \frac{\pi d_e^2}{4s_o}\right) \tag{5.65}$$

The model parameters λ and β can be better expressed as functions of design parameters (concave, grates, and helical blade geometry), functional parameters (feeding velocity, rotor speed, feedrate, and concave clearance) of the threshing unit, and material properties (bulk density, moisture content, an equivalent diameter of kernels [Mohsenin, 1986]). This mathematical model has been validated with comprehensive experimental data obtained from testing of different threshing units (tangential and axial units).

Cumulative grain separation percentage, threshing losses, and separation losses were used for identification/validation of process rates β and λ with very good agreement, as follows (Miu, 1994, 1995; Miu et al., 1997; Miu and Kutzbach, 2000):

   $\beta = 2.9–6.0$ m^{-1} and $\lambda = 21.2–64.4$ m^{-1} ($R^2 = 0.998–0.9996$) for a tangential unit

   $\beta = 2.6–3.8$ m^{-1} and $\lambda = 3.0–24.0$ m^{-1} ($R^2 = 0.999–0.9998$) for two different axial units
     (with axial vs. tangential feeding)

Figure 5.21 shows the graphs of the indices of threshing and separating processes with different settings of functional parameters of an axial threshing unit.

Grain separation depends on the type, variety, and maturity of cereals, as well as the design and functional parameters of the threshing unit. In Chapter 2 we emphasized that the precision of evaluation of threshing and separation rates depends on the number of collection points (number of boxes), especially in the beginning of the cumulative separation curve due to its characteristic S-shape. As a consequence, it is completely inappropriate to compare these models with other models found in the literature when a reduced amount of experimental data (collection points, especially at the beginning of the separation zone) is available. As an example, 10 collection points of data make a bigger difference

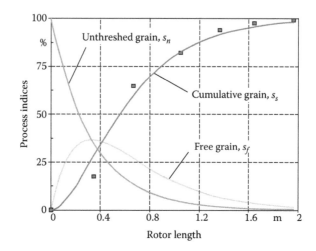

FIGURE 5.21
(See color insert.) Graphs of process indices of a different axial threshing unit.

than 6 points per trial. However, it is true that more collection points translate into a longer period and higher costs of experiments. If given no choice, the practitioner shall balance the costs with the accuracy of experimental results.

By deriving Equation 5.47 and equating it with zero, we get the position along the rotor length where maximum separation occurs. Therefore, this point can be found using the following equation:

$$x_{sd\,max} = -\frac{1}{\lambda - \beta}\ln\frac{\beta}{\lambda} \tag{5.66}$$

In the next sections, we will apply this mathematical model to a series of tangential units and to an axial unit with tangential feeding. Further development includes models of the influences of functional and constructional parameters of threshing units on the process indices β and λ.

5.7.3 Grain Processing in a Tangential Threshing Unit Series

In modeling the grain processing in a series of tangential threshing units, we take into account that the series input of all grains is 1 (or 100%), and then the output of the first tangential unit becomes the input for the next one in line. The rates β and λ are specific to each threshing unit if their design or functional parameters differ.

Thus, for the first threshing unit, Equations 5.44 and 5.51 become

$$V_{t1} = s_n\left(l_{c1}\right) = e^{-\lambda_1 l_{c1}} \tag{5.67}$$

$$V_{gs1} = s_f(l_{c1}) = \frac{\lambda_1}{\lambda_1 - \beta_1}\left(e^{-\beta_1 l_{c1}} - e^{-\lambda_1 l_{c1}}\right) \tag{5.68}$$

where l_{c1} is the length of the first unit of the series.

For the second tangential unit of the series, Equations 5.43 and 5.48 become

$$s_{n2}(x_2) = V_{t1}e^{-\lambda_2 x_2} \tag{5.69}$$

$$s_{s2}(x_2) = \left(V_{t1} + V_{gs1}\right)\frac{1}{\lambda_2 - \beta_2}\left[\lambda_2\left(1 - e^{-\beta_2 x_2}\right) - \beta_2\left(1 - e^{-\lambda_2 x_2}\right)\right] \tag{5.70}$$

where x_2 is the variable expressing the current position associated with the length of the second unit's concave.

Thus, the second unit's mission will be to thresh only the threshing loss of the first unit, and then to separate the resulting grain plus the grain separation loss of the preceding unit.

Other equations from Section 5.7.2 will be adapted accordingly for the second unit, as well as for all of the following units that compose the series of threshing units.

5.7.4 Grain Processing in an Axial Unit with Tangential Feeding

Axial units with tangential feeding are being used in Gleaner and Laverda combines. For further development of the model, we take into consideration the following additional assumptions:

- While feeding, no return of fed material should be possible.
- The threshing process begins immediately in the feeding section of the axial threshing unit; therefore, the grain starts separating through the cage openings in the feeding zone.
- Material is distributed uniformly over the width L_f of the feeding section.
- The material feedrate is constant (steady-state process).

We define the total length of threshing length L as the total length of the feeding, threshing, and separating sections of the axial unit (Figure 5.22).

Let x be the associated variable of the threshing space length, $x \in [0, L]$. Let q_s be the nominal grain feedrate. The specific material $\dot{q}_s(x)$ that is fed into the threshing unit on an infinitesimal width dx, for $0 \le x \le L_f$, is

$$\dot{q}_s(x) = \frac{dq_s}{dx} \tag{5.71}$$

Consequently,

$$q_s(x) = \dot{q}_s x \tag{5.72}$$

Let $q_s(L_f)$ represent 100% of the grain feedrate. Within any cross section of the feeding zone at current position $x \in [0, L_f]$, the grain feedrate is proportional to x as follows:

$$q_s(x) \approx \frac{1}{L_f}x \tag{5.73}$$

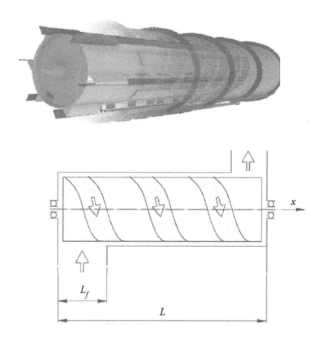

FIGURE 5.22
Representation of an axial unit with tangential feeding.

In the following, we will consider the same probabilistic laws (Equations 5.40 through 5.42) that describe the events into which the threshing and separating process can be divided. The rate parameters β and λ are constant over the lengths L_f and L, and are equal to the average number of corresponding events per unit length.

The frequency of unthreshed grain into the threshing space is a continuous variable. The fraction of unthreshed grain $s_{n1}(x)$ in the feeding zone, $0 \le x \le L_f$, is

$$s_{n1}(x) = \frac{1}{L_f} x \left(1 - \int_0^{x<L_f} \lambda e^{-\lambda s} ds \right) = \frac{1}{L_f} x e^{-\lambda x} \tag{5.74}$$

Making the first derivative

$$\frac{ds_{n1}(x)}{dx} = 0$$

we find that $s_{n1}(x)$ has a maximum at $x = 1/\lambda$.

The fraction of unthreshed grain $s_{n2}(x)$ over the length $L_f \le x \le L$ is

$$s_{n2}(x) = 1 - s_{n1}(L_f) \int_0^x \lambda e^{-\lambda(s-L_f)} ds = e^{-\lambda x} \tag{5.75}$$

At the end of the threshing space ($x = L$), the unthreshed grain becomes threshing loss V_t:

$$V_t = s_{n2}(L) = e^{-\lambda L} \tag{5.76}$$

Separation of free grain kernels through the straw mat and openings of the cage and concave occurs due to the independent and successive random events b and c. According to the probability theory, the joint density $s_d(x)$ of the sum of independent steady random variables equals the convolution of their densities. Therefore, we get the joint density that is the separation frequency:

$$s_d(x) = f(x) * g(x) = \int_0^x f(z)g(x-z)dz = \frac{\lambda\beta}{(\lambda-\beta)}\left(e^{-\beta x} - e^{-\lambda x}\right) \tag{5.77}$$

The cumulative distribution function $s_{s1}(x)$ of separated grain over the length $[0, L_f]$ is found by integrating the density function $s_d(x)$, and then multiplying by proportionality factor x/L_f. Hence,

$$s_{s1}(x) = \frac{1}{(\lambda-\beta)}\frac{x}{L_f}\left[\lambda\left(1-e^{-\beta x}\right) - \beta\left(1-e^{-\lambda x}\right)\right] \tag{5.78}$$

For $L_f \le x \le L$, the cumulative distribution function $s_{s2}(x)$ of separated grain is found by integrating Equation 5.77 only (i.e., the proportionality factor equals 1), as follows:

$$s_{s2}(x) = \frac{1}{(\lambda-\beta)}\left[\lambda\left(1-e^{-\beta x}\right) - \beta\left(1-e^{-\lambda x}\right)\right] \tag{5.79}$$

It can be noted that

$$s_{s2}(L_f) = s_{s1}(L_f) \tag{5.80}$$

Therefore, the cumulative distribution function $s_s(x)$ over the length $[0, x]$ is

$$s_s(x) = \begin{cases} s_{s1}(x)\big|_{0 \le x \le L_f} \\ s_{s2}(x)\big|_{L_f \le x \le L} \end{cases} \tag{5.81}$$

In the cross section of threshing space, at position x, the balance equations can be written as

$$s_{n1}(x) + s_{f1}(x) + s_{s1}(x) = \frac{x}{L_f}, \quad x \le L_f \tag{5.82}$$

$$s_{n2}(x) + s_{f2}(x) + s_{s2}(x) = 1, \quad L_f \le x \le L \tag{5.83}$$

where $s_{f1}(x)$ and $s_{f2}(x)$ represent free, separable grain proportions corresponding to different positions x over the rotor length.

Consequently, we get

$$s_{f1}(x) = \frac{\lambda}{(\lambda - \beta)} \frac{x}{L_f} \left(e^{-\beta x} - e^{-\lambda x} \right)$$

(5.84)

$$s_{f2}(x) = \frac{\lambda}{(\lambda - \beta)} \left(e^{-\beta x} - e^{-\lambda x} \right)$$

(5.85)

It can be noted that

$$s_{f2}(L_f) = s_{f1}(L_f)$$

(5.86)

At the end of the threshing space, the free grain becomes segregation and separation loss V_{gs}:

$$V_{gs} = s_{f2}(L) = \frac{\lambda}{\lambda - \beta} \left(e^{-\beta L} - e^{-\lambda L} \right)$$

(5.87)

Figure 5.23 shows the predicted unthreshed grain $s_n(x)$, free grain $s_f(x)$, and cumulative separated grain $s_s(x)$ for an axial threshing unit with tangential feeding.

Notice that the fraction of unthreshed grain decreases with rotor length, but reaches a maximum within the feeding zone, and then decreases exponentially. The graph of unthreshed grain shows the primary difference in functioning between this type of axial unit and all the others with axial feeding. In those cases, the fraction of free grain increases with rotor length until it reaches a maximum at the end of the feeding section. Later, it decreases due to grain separation.

The continuity equation of the process quality indices over the length of threshing space is

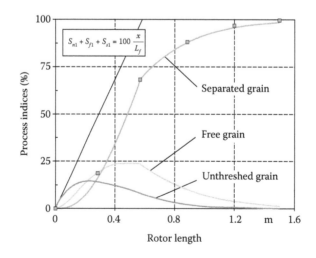

FIGURE 5.23
Variation of unthreshed, free, and separated grain in an axial unit with tangential feeding.

$$\frac{1}{\beta}\frac{\partial^2 s_s}{\partial x^2} + \frac{\partial s_s}{\partial x} = \lambda s_n \tag{5.88}$$

Grain separation frequency $s_d(x)$ depends on the *fraction of separable, segregated grain* $s_g(x)$ that reaches the concave/cage surface, by a coefficient that is the linear segregation rate λ. Thus, the fraction of separable, segregated grain $s_g(x)$ is given by the following equations:

$$s_g(x) = \frac{\beta}{(\lambda - \beta)}\frac{x}{L_f}\left(e^{-\beta x} - e^{-\lambda x}\right) \text{ for } x \le L_f \tag{5.89}$$

$$s_g(x) = \frac{\beta}{\lambda - \beta}\left(e^{-\beta x} - e^{-\lambda x}\right) \text{ for } L_f \le x \le L \tag{5.90}$$

At the end of the threshing space, the separable, segregated grain $s_g(x)$ becomes *pure separation loss* V_s as follows:

$$V_s(L) = \frac{\beta}{\lambda - \beta}\left(e^{-\beta L} - e^{-\lambda L}\right) \tag{5.91}$$

Consequently, the *fraction of free, unsegregated grain* $s_{ng}(x)$ can be calculated with the following equation:

$$s_{ng}(x) = s_f(x) - s_g(x) \tag{5.92}$$

That leads to

$$s_{ng1}(x) = s_{f1}(x) - s_{g1}(x) = \frac{x}{L_f}\left(e^{-\beta x} - e^{-\lambda x}\right) \tag{5.93}$$

$$s_{ng2}(x) = s_{f2}(x) - s_{g2}(x) = \left(e^{-\beta x} - e^{-\lambda x}\right) \tag{5.94}$$

At the end of the threshing space, the unsegregated grain becomes *segregation loss* V_{ng}:

$$V_{ng}(L_f) = \left(e^{-\beta L_f} - e^{-\lambda L_f}\right) \tag{5.95}$$

The continuous separation efficiency $eff_s(x)$ at the current position x of the separation length is defined as the ratio between the separated grain mass on the differential interval dx and the available grain mass that could be separated at the rear of the threshing unit. The separation efficiency on the rotor length is composed of two formulae corresponding to the feeding and threshing zones, respectively:

$$eff_s(x) = \begin{cases} eff_{s1}(x)\big|_{0 \le x \le L_f} \\ eff_{s2}(x)\big|_{L_f \le x \le L} \end{cases} \tag{5.96}$$

Consequently, for feeding and threshing zones, these formulae can be mathematically expressed as follows:

$$eff_{s1}(x) = \frac{s_{d1}(x)}{1 - (V_t + s_{s1}(x))} \tag{5.97}$$

$$eff_{s2}(x) = \frac{s_{d2}(x)}{1 - (V_t + s_{s2}(x))} \tag{5.98}$$

Test runs with an axial unit (Figure 5.24) were conducted in the laboratory with dry winter wheat. The specific rates β and λ have been determined using multiple nonlinear regression analysis of the cumulative separated grain fraction, threshing and separation losses, and the generalized functions that describe the influence of crop properties and the functional and design parameters of the axial unit (Miu et al., 1998b). The overall quality of the regression analysis of the cumulative grain separation over the length of the rotor was assessed based on the value of the coefficient of determination R^2.

According to the experimental results, the identified values of the coefficients range as follows:

$\beta = 3.03\text{--}3.95 \text{ m}^{-1}$

$\lambda = 3.95\text{--}5.06 \text{ m}^{-1}$

$R^2 = 0.981\text{--}0.996$

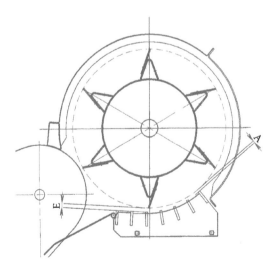

FIGURE 5.24
Cross section of axial threshing unit in tangential feeding zone.

In terms of losses, at $x = 1.8$ m, the model shows the following predicted values over the entire range of testing conditions:

- Threshing loss $V_t = s_{n2}(1.8) = 0.011\%\text{--}0.082\%$
- Separation loss $V_s = s_{f2}(1.8) = 0.32\%\text{--}1.48\%$

These results show an excellent agreement between theoretical predictions and the experimental data.

5.7.5 Models of Threshing and Separating Process Rates

The mathematical model of threshing and separating processes in threshing units, described in the previous section, uses two coefficients: λ, the linear rate of threshing, and β, the linear rate of separation. Using physical process analysis, dimensional analysis, and nonlinear multiple regression techniques based on the principles described in Chapter 2, the linear rate λ and β models are influenced by the following parameters:

- Crop properties: Crop type and variety, and moisture and bulk density of the MOG.
- Functional parameters: MOG feedrate, rotor speed, and concave clearance.
- Design parameters: Concave wrap angle, dimension of concave and cage openings, and length of the rotor.

In developing the theory, the following steps were taken:

- Analysis of the influence of the above-mentioned parameters on grain separation and separation losses
- Analysis of the dependence of process indices on the values of coefficients β and λ
- Comparison of the variation of both coefficients
- Identification of the functions that describe the influence of the parameters on the coefficient values
- Combination and compression of these functions into a single function for every coefficient β and λ
- Generalization of the functions for prediction of the values of coefficients β and λ

The analysis of the values of coefficients β and λ revealed the following conclusions:

- Coefficients β and λ are in a certain reverse proportional ratio.
- Grain separation is proportional to coefficient β, which represents the linear rate of separation.
- Losses are proportional to coefficient λ.
- Variation of coefficients β and λ in wheat and barley (summer and winter varieties) displays similar shapes of the respective curves when a certain functional or design parameter varies.

Because of the large number of variables, we were faced with the following challenges:

- Finding out the proper forms of the generalized functions of coefficients β and λ
- Minimizing the number of coefficients from equations

Using a nonlinear multiple regression technique, we found two compact generalized functions of coefficients β and λ, as follows:

$$\beta = k_\beta p \sqrt{\frac{v u_p \sqrt{q_p}}{\sqrt{\rho_p} e^{\left(\frac{q_p}{q_{po}} + \frac{u_p}{u_{pm}}\right)}}} \tag{5.99}$$

where:

p is the probability of grain passage through the openings
v is the rotor speed, m/s
q_p is the MOG throughput, kg/s
q_{po} is the optimum working MOG throughput, kg/s
u_p is the MOG moisture content (wet basis), %
u_{pm} is the minimum MOG moisture content (wet basis), %
ρ_p is the MOG bulk density (free density), kg/m³
k_β is the coefficient that depends on crop type and variety

The optimum working MOG throughput q_{po} represents the MOG throughput corresponding to the maximum value of the cumulated fraction of separated grain for a given axial threshing unit. The value of u_{pm} represents the MOG moisture content (wet basis) corresponding to the minimum grain damage when MOG throughput has a value of q_{po}. That means the MOG moisture content u_{pm} corresponds to the grain moisture content when the grain is more resistant to dynamic forces of impact (Wacker, 1988).

Dimensional analysis shows that coefficient β implicitly depends on the rotor angular velocity and on the separation surface (i.e., rotor diameter and length).

The generalized function of coefficient λ is

$$\lambda = k_\lambda \sqrt{\frac{\rho_p v \delta_e}{q_p \sqrt{u_p}}} e^{\left(\frac{q_p}{q_{po}} + \frac{u_p}{u_{pM}} - \frac{v}{v_0}\right)} \tag{5.100}$$

where:

k_λ is the coefficient that depends on the crop type and variety
δ_e is the exit concave clearance, mm
u_{pM} is the maximum MOG moisture content (wet basis), %
v_o is the optimum working rotor speed, m/s

The value of v_o represents the rotor speed corresponding to the maximum value of the cumulated fraction of separated grain (i.e., to the feedrate q_{po}).

In wheat and barley, the value of maximum MOG moisture content falls within the range

$$u_{pM} = (1.5\ldots2)u_{pm} \tag{5.101}$$

The following subsections are based on the author's work in partnership (Miu et al., 1998) with German colleagues at Hohenheim University, Stuttgart, Germany.

5.7.5.1 Influence of MOG Feedrate

In axial threshing units, at low material throughput, the relatively thin layer of material is driven quickly by the active elements of the rotor, the crop is not sufficiently processed, and therefore the threshing and separation occur later. That means coefficient λ increases and β decreases.

The grain separation improves with increasing the MOG feedrate because of the increasing material thickness. At a higher MOG feedrate, the straw layer in the narrow concave clearance is thicker. This has the effect of damping the beating energy via the active elements of the rotor, toward the depth of the material layer; consequently, the grain separation decreases. The influence of the MOG feedrate q_p on coefficients β and λ establishes an exponential dependence, which mathematically can be written as

$$\beta = \frac{aq_p^b}{e^{cq_p}} \tag{5.102}$$

and

$$\lambda = \frac{ke^{mq_p}}{q_p^b} \tag{5.103}$$

where:
q_p is the MOG feedrate
e is the base of natural logarithms
a, b, c, k, m, n are the coefficients

Above, the following was taken into account:
q_p^b, q_p^n are the positive influence of the MOG feedrate on grain separation
e^{cq_p}, e^{mq_p} are the negative influence of the decreasing of straw permeability and the transmissibility of beating energy across a greater thickness of the crop mat

The influence of the MOG feedrate and rotor speed on coefficient λ, threshing loss, and grain separation in barley, when MOG moisture content varies between 10% and 50% (wet basis), is shown in Figures 5.25 through 5.27.

5.7.5.2 Influence of Rotor Speed

The detachment of the grain from ears and grain separation mainly take place due to the beating energy of the rotor. The greater the rotor speed, the stronger is the action of the active elements (rasp bars, straight bars), and the greater are the centrifugal force and grain separation. And yet, with increasing rotor speed, the grain separation reaches a maximum

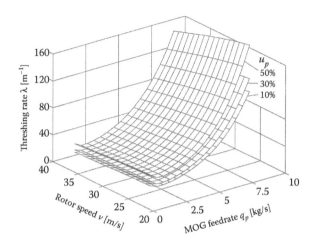

FIGURE 5.25
Influence of MOG feedrate and rotor speed on barley threshing rate λ.

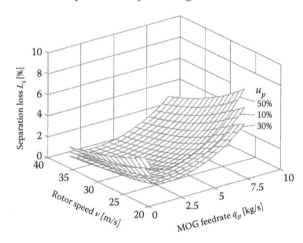

FIGURE 5.26
Influence of MOG feedrate and rotor speed on barley grain separation loss.

(Sonnenberg, 1971; Lo, 1978; Wacker, 1985). At higher rotor speed, the stationary time of the material in the axial threshing unit becomes shorter and shorter, and the straw permeability decreases because of a stronger fragmentation of the MOG. Following the abovementioned statements and on the basis of my own experimental results, we assumed that coefficients β and λ bear exponential relationships with rotor speed, with the forms

$$\beta = \frac{av}{e^{bv}} \tag{5.104}$$

and

$$\lambda = \frac{v^2}{ke^{mv}} \tag{5.105}$$

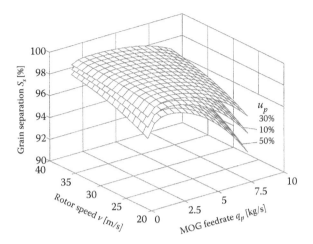

FIGURE 5.27
Influence of MOG feedrate and rotor speed on barley grain separation.

where:

v is the rotor speed

a, b, k, m are the coefficients (other than in Equations 5.102 and 5.103)

Above, the following was taken into account:

v, e^{mv} are the positive influence of increasing the centrifugal force and threshing action

v^2, e^{bv} are the negative influence of decreasing crop stationary time and straw permeability

The influence of the rotor speed and MOG feedrate on coefficient λ, threshing loss, and grain separation in wheat, when MOG moisture content varies between 10% and 50% (wet basis), is shown in Figures 5.28 through 5.30.

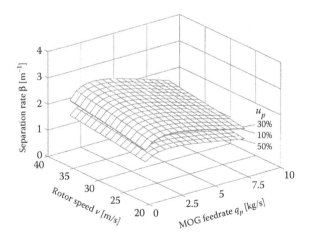

FIGURE 5.28
Influence of rotor speed and MOG feedrate on wheat threshing rate β.

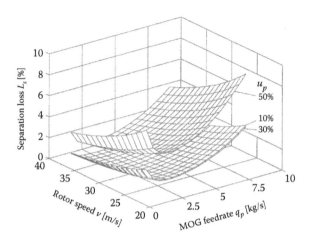

FIGURE 5.29
Influence of rotor speed and MOG feedrate on wheat grain separation loss.

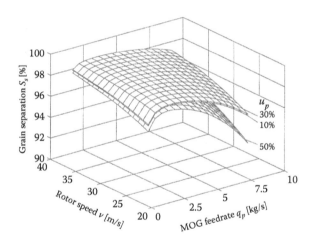

FIGURE 5.30
Influence of rotor speed and MOG feedrate on wheat grain separation.

5.7.5.3 Influence of Concave Clearance

A reduced number of researchers have already reported experimental data concerning the influence of concave clearance on grain separation in axial threshing units (Bjork, 1988a; Wacker, 1990; Miu et al., 1997).

Thus, the influence of concave clearance on grain separation s_s can be described by the equation

$$s_s = \frac{\xi \delta_i{}^\eta}{e^{\zeta \delta_e}} \tag{5.106}$$

where:

δ_i is the concave clearance at entrance
δ_e is the concave clearance at exit
ξ, η, ζ are the coefficients

At the exit concave clearance $\delta_e = 0$, the rotor is blocked, and consequently, the grain separation $s_s = 0$. With an increase in concave clearance, grain separation increases until it reaches a maximum, and then slowly decreases. The same observations have been reported in tangential units by Caspers (1973). The influence of a concave clearance on coefficient β can be neglected. Starting from Equation 5.103, we found for coefficient λ a close form of relationship:

$$\lambda = \frac{k\delta_e^{\ m}}{e^{n\delta_e}}$$ (5.107)

where k, m, n are the coefficients.

Equation 5.104 was built taking into account the following:

$\delta_e^{\ m}$ is the negative influence of the concave clearance due to decreasing the beating intensity

$e^{n\delta e}$ is the positive influence of the concave clearance due to increasing the material compression

k is the coefficient that depends on the rotor length (concave length)

5.7.5.4 Influence of MOG Moisture

In an axial threshing unit, due to a long trajectory of the material into the threshing space, the dryer the straw, the more broken it will be. Thus, at very low MOG moisture content, due to a stronger MOG separation, the grain separation loss increases. The straw mass acts as a viscoelastic layer (Baader et al., 1969), because of the behavior of individual straw pieces and their kinetic friction (Huisman, 1983). At high MOG moisture content, the coefficient of sliding friction between grain and straw increases, and the elasticity and size of the MOG decrease; therefore, the separation loss also increases (Figure 5.31). The graphics from Figures 5.32 and 5.33 show the predicted grain separation and separation losses in summer barley and wheat, respectively, when the MOG feedrate, MOG moisture content, and rotor speed vary. It can be observed that separation loss rises with an increase in MOG moisture content.

5.7.5.5 Influence of MOG Bulk Density

Spatially variable field operations depend on various parameters, such as in-field variation of crop yield (Reitz and Kutzbach, 1996) and MOG yield. McGechan and Glasbey (1982) reported the short-term variability in apparent MOG yield within a field or within 1 day. The MOG yield strongly depends on the MOG moisture content and MOG bulk density.

The research of Voß (1970) shows a nonlinear dependence between straw bulk density and moisture content, which is reflected in the equation of the rates β and λ. According to Pickett and West (1988), the straw bulk density falls within the range of $14 \div 34$ kg/m³ in

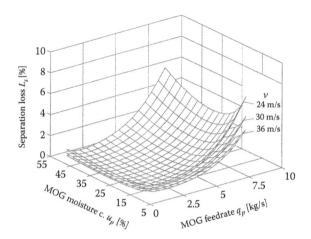

FIGURE 5.31
Influence of MOG moisture and feedrate on separation loss in barley.

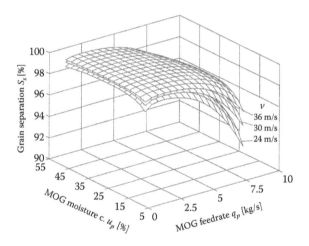

FIGURE 5.32
Influence of MOG moisture and feedrate on grain separation in barley.

wheat and $20 \div 22$ kg/m³ in barley. Figure 5.34 shows the predicted grain separation in winter wheat as functions of MOG moisture content and MOG (straw) bulk density for different values of rotor speed.

5.7.5.6 Influence of Concave/Cage Opening Design

Grain can pass through the concave and cage openings. As the opening surfaces differ from concave to cage, it is necessary to compute the mean surface of an opening as a ponderal value of concave and cage opening surfaces (Miu et al., 1998a,b).

Let us define the opening ratio η of the separation surface as

$$\eta = \frac{S_o}{S} \tag{5.108}$$

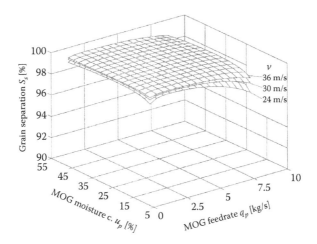

FIGURE 5.33
Influence of MOG moisture and feedrate on grain separation in wheat.

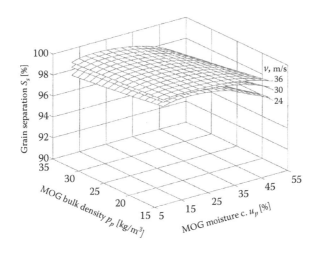

FIGURE 5.34
Influence of MOG bulk density and moisture on grain separation in wheat.

where:

s_o is the mean surface of the concave and cage openings (ponderal value)
s is the mean surface that contains an opening with the surface s_o

Thus, we can use Equation 5.65 to calculate the probability p for a grain to pass through an opening of the concave/cage.

Let us consider the opening is a square with the side a, so that $s_o = a^2$. In Figure 5.35, the graphics show the predicted data in winter barley as functions of the opening ratio η, and dimension a of a square opening for different values of the MOG feedrate. The designer can then choose the proper dimensions of the openings corresponding to the surface s_o.

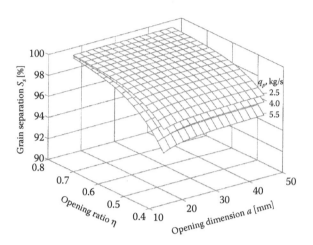

FIGURE 5.35
Predicted influences of opening dimension and opening ratio in wheat.

5.8 Modeling of MOG Fragmentation and Separation Processes

5.8.1 Conjoint Mathematical Model of MOG Processing in Tangential/Axial Threshing Units

Miu (1999, 2000) developed the conjoint mathematical model of MOG fragmentation and separation for both tangential and axial threshing units. In developing this mathematical model for both types of threshing units, the following assumption has been considered: the MOG fragmentation and separation begins effectively

- At the entrance in the tangential unit
- In the threshing section of the axial threshing unit

Such an assumption could be eliminated, though, if a precise correction of the initial fraction of unfragmented MOG (e.g., 95% instead of 100%) is known.

Due to the forced passage of vegetal material through the threshing space, the bonds between the grain and their flowery cover are destroyed, and simultaneously, leaf and stalk portions of the plant mass undergo deformation and fragmentation processes.

The seizure of stalks by the cylinder or rotor depends on several factors, such as the type and construction of the threshing unit, gaps between their various elements, peripheral speed, MOG throughput, and physical conditions of the plant mass.

In the threshing space, the MOG becomes a mixture of long stalks, chaff, and fragments of the spikes, stalks, and leaves. The chaff and other small components of MOG are potentially able to pass through the openings of the concave and grates.

Applying *mutatis mutandis*, for data analysis, the methods of determining and expressing particle size described by ASAE S319.3 and ANSI/ASAE S424.1, we can assert that

- The analysis of the mass percentage distribution of fragmented MOG in the threshing space is based on the assumption that these percentages are logarithmic normally distributed.
- The size of MOG particles should be reported in terms of geometric mean length l_{gm} and standard deviation S_{gml} by mass percentage.
- Calculated values are obtained as follows:

$$l_{gm} = \log^{-1} \frac{\sum \left(m_i \log \overline{l_i} \right)}{\sum m_i} \tag{5.109}$$

$$S_{gml} = \log^{-1} \left[\frac{\sum m_i \left(\log \overline{l_i} - \log l_{gm} \right)^2}{\sum m_i} \right]^{1/2} \tag{5.110}$$

where:
- l_i is the nominal length of the MOG particles of the ith interval of screening (sorting)
- $\overline{l_i}$ is the geometric mean length of the articles on the ith interval of screening
- m_i is the mass percentage on the ith interval at the conditions of screening
- l_{gm} is the geometric mean length
- S_{gml} is the standard deviation

MOG fragments passing through a concave or grate openings are considered to have a geometric mean length of half the opening's diagonal. That is what we define as being the *separable MOG fraction*. That means $l_{gm} = l_{50} =$ MOG fragment length at 50% probability, and $S_{gml} = l_{84}/l_{50}$, where l_{84} is the MOG fragment length at 84% cumulative probability.

Development of the model is based on several judicious assumptions:

- The MOG fragmentation is performed in the threshing space.
- The MOG fragmentation and separation process can be divided into the next three random events:
  - The deformation and fragmentation of the ears, leaves, and stalks
  - The movement of separable MOG particles through the straw mat to the concave and grates, due to the airstream and centrifugal force
  - The passage of separable MOG particles through the openings of the concave grates and cage

In the previous section, we defined the *length of threshing space* where l is the length of the concave (for a tangential unit), and L is the length of the threshing and separation sections of the rotor (for an axial threshing unit).

The general probabilistic laws that describe the above random events are respectively defined by the following exponential probability density functions:

1. $f(x) = ae^{-bx}$ (5.111)

2. $f(x) = ae^{-bx}$

3. $g(x) = ae^{-ax}$ (5.112)

where:
  x is the current position along the length of the threshing space
  b is the linear rate of MOG fragmentation and dispersion
  a is the linear rate of MOG separation

The above-mentioned probability density functions are defined on the threshing space length x; a and b are space increments between respective successive event changes over the length x.

The fraction of fragmented MOG (small enough to become separable) is a continuous variable and can be computed by integrating Equation 5.111 over the length x. Hence, the *fraction $pp_n(x)$ of unfragmented (unseparable) MOG* is

$$pp_n(x) = 1 - \int_0^x ae^{-bs}ds = 1 - \frac{a}{b}\left(1 - e^{-bx}\right) \tag{5.113}$$

Fragmented and separable MOG can separate; therefore, their passage through the straw mat and openings (random events 2 and 3) is independent and successive. According to the probability theory, their joint probability density $s_d(x)$ of the sum of two independent and steady random variables with the densities $f(x)$ and $g(x)$ equals the convolution (Equation 5.46) of their individual probability densities.

Consequently, the *MOG frequency distribution* or *probability density function $pp_d(x)$ of separated MOG particles* results in

$$pp_d(x) = f * g = \frac{a^2}{b-a}\left(e^{-ax} - e^{-bx}\right) \tag{5.114}$$

The *cumulative distribution function $pp_s(x)$ of separated MOG* is found by integrating the density function (5.114). We get

$$pp_s(x) = \frac{a}{b(b-a)}\left[b\left(1 - e^{-ax}\right) - a\left(1 - e^{-bx}\right)\right] \tag{5.115}$$

The *total cumulated percentage of separated MOG* over the length of the threshing space is $pp_s(l)$ for a tangential threshing unit and $pp_s(L)$ for an axial unit.

Since the material throughput is constant (as per initial assumption), in the cross section of threshing space at any current position x of separation length, the mass balance of MOG fractions can be written as a summation of the unfragmented MOG $pp_n(x)$, cumulative separated MOG $pp_s(x)$, and separable, fragmented MOG $pp_f(x)$ fractions as follows:

$$pp_n(x) + pp_f(x) + pp_s(x) = 1 \qquad (5.116)$$

According to the initial condition, $pp_f(0) = 0$, we get

$$pp_f = \frac{a}{b-a}\left(e^{-ax} - e^{-bx}\right) \qquad (5.117)$$

According to the probabilistic theory, the MOG separation is proportional to the amount of fragmented, separable MOG:

$$\frac{d}{dx}(pp_s) = a \cdot pp_f \qquad (5.118)$$

The values of model parameters a and b were found through nonlinear regression analysis with very good agreement, as follows: $a = 1.8 \div 4.0$ m^{-1} and $b = 10.1 \div 20.4$ ($R^2 = 0.992$–0.9994) for a tangential threshing unit (Miu, 1999) and $a = 0.4 \div 2.2$ m^{-1} and $b = 2.7 \div 17$ for two different axial units.

Figures 5.36 and 5.37 show the graphs of MOG processing in a tangential and an axial threshing unit, respectively.

The above-described models help to simulate and optimize the threshing process and unit design and reduce time and costs in carrying out laborious experiments.

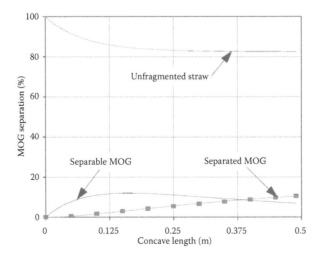

FIGURE 5.36
(See color insert.) MOG processing in a tangential unit (feedrate 6.2 kg/s, cylinder speed 29.6 m/s).

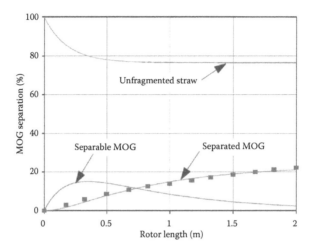

FIGURE 5.37
MOG processing in an axial unit (feedrate 7.5 kg/s, cylinder speed 30.6 m/s).

5.9 Modeling of Threshing Unit Power Requirement

5.9.1 Literature Review of Mathematical Modeling of Threshing Unit Power Requirement

The threshing unit consumes the major part of combine harvester power used in grain harvesting (Kepner et al., 1982; Bjork, 1988b; Kim and Gregory, 1989a). That becomes more evident as, with the axial-flow design, a high percentage (up to 75%) of the total power could be consumed by the rotor. Thus, modeling the power requirement of the threshing unit represents a key in computing the total power of the combine engine. Concurrently, a model of the cylinder/rotor power requirement helps in the simulation and optimization of the threshing and separating processes, as well as in the threshing unit design using evolutionary computation (Miu and Perhinschi, 2001).

Research carried out by Dolling (1957), Segler and Wienecke (1961), Arnold (1964), Eimer (1977), and Kanafojski (1973) has shown that the specific power requirement of a tangential unit varies within the 1–2.5 kWh/t$_{MOG}$ range. In a pull-type combine with a tangential threshing unit and two separating rotors, the power requirement has increased linearly with MOG feedrate (PAMI, 1987).

Huynh et al. (1982) assumed that the threshing process is determined by friction and compression phenomena. Hence, the processing power can be calculated using the following equation:

$$P = \frac{D}{2}\mu BLk_1\left(\frac{\delta_e}{\delta_i}\right)^{k_2}$$

(5.119)

where:

μ   is the coefficient of friction between material and concave
B is the cylinder width
L is the concave length

Kim and Gregory (1989a) developed a power requirement model for tangential threshing units as follows:

$$P = P_0 + kz \frac{h_a}{\delta} nL(q_p + q_s)$$ (5.120)

where:

P_0 is the idle power of the threshing unit cylinder
h_a is the height of the material fed into the threshing unit
n is the cylinder speed, rpm

Research carried out by Wacker (1985) on an axial threshing unit shows that the idle power is proportional to the cube of the rotor speed v, as follows:

$$P_0 = 0.14 \cdot 10^{-3} v^3$$ (5.121)

Specific energy requirements for wheat threshing and separation increase linearly by increasing the rotor speed and MOG throughput. Wacker also investigated the influence of spiral position and MOG moisture content on the specific power requirement of the tested axial threshing unit.

Harrison (1991) described the variation of the rotor power requirement at different MOG throughputs, rotor speeds, and concave clearances. The power required for the axial-flow rotor was extremely sensitive to any change in the throughput, doubling for a 50% increase of it.

Outside of the author's paper (Miu, 2001a), there are no published mathematical functions that describe the influence of design and functional parameters of the axial threshing unit on the power consumption.

5.9.2 Idle Power of an Axial Threshing Unit

The main components of an axial threshing unit are the rotor (feeding impeller, and threshing and separating zones) and a cage with concave and grates. The grates are located in the separating zone of an axial threshing unit. From the design point of view, they are basically very similar to a concave.

When functioning, a rotor is considered to be both

- An impeller, which forces the air to move through the threshing space in the axial and radial directions
- A rigid body, which processes vegetal material through the threshing space

When idling (no throughput material), the rotor requires an *idle power* P_o. When threshing, the rotor requires a *process power* P_{pr}. The process power is directly proportional to the material MOG throughput q_p by a specific process power p_{sp}, that is,

$$P_{pr} = p_{sp}q_p \tag{5.122}$$

Consequently, the total power P_t required by a rotor of an axial unit is

$$P_t = P_o + P_{pr} \tag{5.123}$$

Applying the principle of kinetic energy of a rotor with rotary motion with respect to time t, we get

$$P_o = \frac{E_{co}}{t} = \frac{1}{2}J\omega^2 \frac{1}{t} = k_o R L v^3 \tag{5.124}$$

where:
  J is the moment of inertia of the rotor with respect to the axis of rotation (kg m^2)
  ω is the angular velocity of the rotor (s^{-1})
  R is the rotor radius (m)
  L is the rotor length (m)
  v is the rotor speed (m/s)
  k_o is the coefficient, which includes carbon steel density (kg/m^3)

For an impeller, with negligible friction losses and no pneumatic losses, the idle power can be calculated with the following formula:

$$P_o = Q\rho_a W g \tag{5.125}$$

where:
  Q is the volumetric airflow rate (m^3/s)
  ρ_a is the density of air (kg/m^3)
  W is the total theoretical head (m)
  g is the acceleration due to gravity (m/s^2)

Equation 5.125 is for an ideal situation in which the airflow through the rotor's active elements (rasp bars) is frictionless, and the fluid velocity vectors can be accurately determined from the bars' angles.

The total head is the sum of a velocity head and a pressure head (Henderson et al., 1997). The total head is proportional to the square of the rotor speed and to the difference in the squares of the inner and outer radii of active elements of the rotor. Since the rotor moves a quantity of air against a certain resistance, which may be attributable to the friction of cage spirals and separation openings, a part of the velocity head should be converted into a pressure head. Following the above-mentioned statements, the following equation for idle power P_o (kW) was found:

$$P_o = 10^{-3} c_o \rho_a R L \delta_R v^2 \tag{5.126}$$

where:
  c_o is the coefficient (s^{-1})
  δ_R is the difference of the inner and outer radii of active elements of the rotor

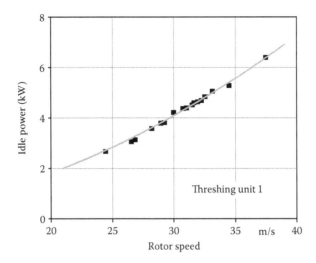

FIGURE 5.38
Rotor idling power dependence on the peripheral speed.

An example of variation of idle power versus rotor speed is shown in Figure 5.38. The determination coefficient achieved for the nonlinear regressions was $R^2 = 0.98$, and the standard error of the estimates was $e_s = 0.11$ kW.

5.9.3 Models of Input Variable Influence on Axial Threshing Unit Power

In this section we analyze and model mathematically the dependence of threshing unit specific power p_{sp} (kW/(kg/s)) on MOG throughput q_p, rotor speed v, concave clearance δ_e, and MOG moisture content u_p.

5.9.3.1 Effect of MOG Feedrate

In axial threshing units the specific process power increases linearly with MOG feedrate (Wacker, 1985) or with MOG feedrate to the power of $1.6 \div 2$ (Bjork, 1988b; Kim and Gregory, 1989a). The power required for the axial-flow rotor is extremely sensitive for a change in the feedrate, doubling for a 50% increase (Harrison, 1991).

The influence of the MOG feedrate on specific process power is higher in barley than in wheat, as shown in Figures 5.38 and 5.39.

At low MOG throughput, the relatively thin layer of material is driven quickly by the active elements of the rotor. At higher MOG feedrate, the straw layer is thicker. This has the effect of damping the beating energy via the active elements of the rotor and the specific process power increases.

The influence of MOG feedrate q_p on the specific process power p_{sp} (kW/(kg/s)) establishes an exponential dependence, which mathematically can be written as

$$p_{sp} = a \frac{e^{bq_p}}{q_p^b} \tag{5.127}$$

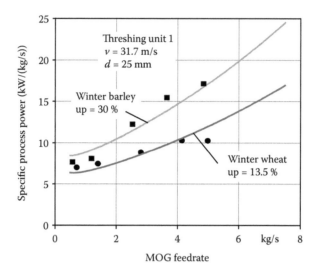

FIGURE 5.39
Specific processing power dependence on the MOG feedrate.

where:
  q_p is the MOG feedrate (kg/s)
  e is the base of natural logarithms
  a is the coefficient
  b is the exponent

Above, the following was taken into account:

e^{bq_p} is the negative influence of decreasing the transmissibility of the beating energy across a greater thickness of the crop mat

q_p^b is the positive influence of decreasing the friction of the crop when the layer is thin

5.9.3.2 Effect of Rotor Speed

The rotor speed has no significant effect on the process power (Bjork, 1988b; Harrison, 1991). Wacker (1985) showed that the specific energy requirement increases linearly by increasing the rotor speed. The greater the rotor speed, the stronger is the action of the active elements (rasp bars, straight bars), and the greater is the straw fragmentation and, consequently, the specific process power. At the same time, increasing the rotor speed does not mean a proportional increase of axial velocity of the material. Therefore, it can be assumed that the specific process power bears with the rotor speed the following relationship:

$$p_{sp} = \frac{e^{av}}{bv} \tag{5.128}$$

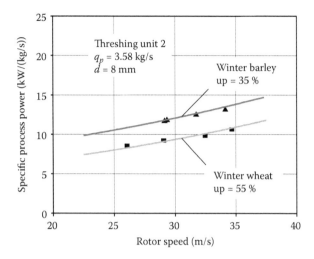

FIGURE 5.40
Specific processing power dependence on the rotor peripheral speed.

where:

v is the rotor speed (m/s)

a, b are the coefficients (other than in previous equations)

In Equation 5.128, the term e^{av} expresses the negative influence of intensifying the straw fragmentation when the rotor speed increases; therefore, the specific power increases. In accordance with Equation 5.128, Figure 5.40 displays the dependence of the specific processing power on the rotor peripheral speed in wheat and barley.

The graph shown in Figure 5.41 displays the influence of the rotor speed and MOG throughput at two MOG moisture contents of barley on a specific process power of an axial threshing unit.

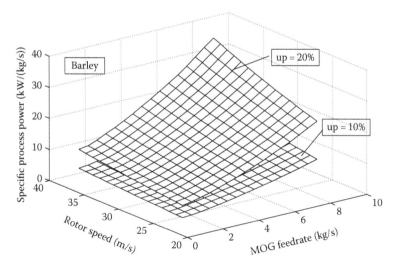

FIGURE 5.41
Specific processing power dependence on the rotor speed and MOG feedrate.

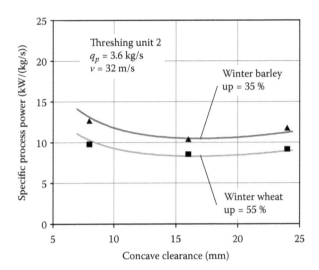

FIGURE 5.42
Specific process power dependence on the concave clearance.

5.9.3.3 Effect of Concave Clearance

An example of how the concave clearance influences the variation of specific power of an axial threshing unit is shown in Figure 5.42.

On one hand, when the material MOG feedrate is constant, decreasing the concave clearance determines an increase of the pressure the material encounters within the threshing space, and thus an increase of specific power consumption. On the other hand, the layer of material should be quickly driven toward the exit of the threshing unit. This has a slight effect on decreasing the power consumption, though. Both effects can be described by the following equation:

$$p_{sp} = \frac{e^{a\delta_e}}{b\delta_e^c}$$

(5.129)

where:
δ_e is the concave clearance at the outlet (mm)
a, b, c are the regression coefficients/exponents

The above-mentioned coefficients depend on crop type, variety, and maturity stage.

5.9.3.4 Effect of MOG Moisture Content

At low moisture content of MOG, the straw size decreases; therefore, the specific power requirement decreases. At very high MOG moisture content, the coefficient of the sliding friction between the straw and the rotor cage decreases.

Consequently, the specific power requirement was found to be best described by the equation

$$p_{sp} = \frac{au_p}{e^{cu_p}}$$

(5.130)

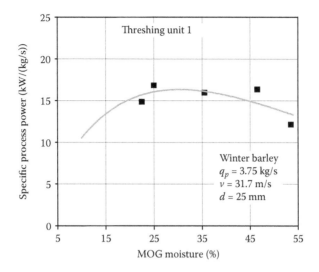

FIGURE 5.43
Specific process power dependence on the MOG moisture content (w.b.).

where:

u_p is the MOG moisture content (wet basis, %)

a, b, c are the coefficients

Figure 5.43 shows how Equation 5.130 can be fitted to experimental data obtained when threshing barley with an axial unit. Figure 5.44 shows the cumulated effects of concave

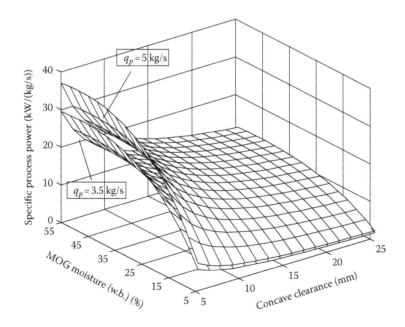

FIGURE 5.44
Specific process power dependence on the concave clearance and MOG moisture content.

clearance and MOG moisture content on specific power consumption, in wheat, for two different MOG feedrates.

5.9.3.5 Generalized Model of Input Variable Influence on Specific Process Power

Using Equations 5.127 through 5.130 and a multiple regression technique, we developed a compact generalized function of specific process power as follows:

$$p_{sp} = au_p \frac{e^{\left(\sqrt{q_p} + c\left(v + \delta_e - u_p\right)\right)}}{q_p^b v^c \delta_e} \tag{5.131}$$

where the coefficients (exponents) take values within the following ranges: $a = 0.6\text{–}1.4$, $b = 0.2\text{–}0.55$, and $c = 0.02\text{–}0.055$ (in wheat and barley).

The trends predicted by the above-mentioned model are similar to the observed data with a global coefficient of determination $R^2 > 0.8$. This model is further used in the next section as part of the outlined GA for optimization of the threshing unit process and design.

5.10 GA-Based Optimization of the Threshing Unit Process and Design (MATLAB® Application)

5.10.1 Introduction

Increasing the field performance of the combine harvester depends on the improvement of the threshing and separating processes, as well as the threshing unit design. The contemporary design procedure of threshing units is mainly based on general and manufacturer expertise that relies heavily on expensive, time-consuming, and laborious experiments performed on different crops on full-scale prototypes, aimed at tuning and improving functional and design parameters.

Following the author's work (Miu, 2001a,b; Miu and Perhinschi, 2001), in this section we will integrate the mathematical models outlined in this chapter within a GA with application in axial threshing unit optimization.

GAs are parameter iterative search techniques, which rely on analogies to natural biological processes. They simulate the evolution of species and individual selection based on Darwin's principle of the *survival of the fittest*. GAs work simultaneously with a set (*population*) of potential solutions (called *individuals*) of the investigated problem. The degree to which such solutions meet certain performance requirements is evaluated and used to select "surviving" individuals that will "reproduce" and generate a new population. Then individuals will undergo alterations similar to those of natural *genetic mutation* and *crossover*.

Search is governed by two major contradictory elements: use of already acquired information about good solutions and thorough investigation of the entire solution space. The first is enhanced by biased selection toward the fittest individuals, while the second is enhanced by alterations of the genetic material in the new generation. Emphasis on the first element decreases the convergence time with the risk of reaching a local extreme. The emphasis on the second element increases the convergence time, but will also improve the chances of reaching a global extreme.

The selection scheme is biased toward high-performance solutions. A careful selection of GA structure and parameters will ensure a good chance of reaching the global optimum solution after a reasonable number of iterations.

Therefore, readers are advised to first read Section 2.5.2, and then continue on with the current section. However, in the following, we will synthesize the major aspects of developing the GA for evolutionary optimization of threshing unit design and functional parameters. A good understanding of the method, structure, and searching parameters may help in the development of similar algorithms for other processes and working units, not necessarily connected to the combine harvester's processes and design. A ready-to-use MATLAB script of the program is given in the Appendix.

5.10.2 Statement of the Optimization Problem and Object

Crop properties are components of the property vector $g = [g_1, ..., g_n]^T$.

The specifications of a threshing unit that processes the considered crop are integrated within the following vectors:

- Vector x of m design parameters, $x = [x_1, ..., x_m]^T$
- Vector ξ of r functional parameters, $\xi = [\xi_1, ..., \xi_r]^T$
- Vector y of p performance criteria, $y = [y_1, ..., y_p]^T$

Some functional parameters that rely on other combine subassemblies and influence threshing unit operation are also included in the vector ξ.

The problem is finding the optimum parameter configuration $\{x_i, \xi_i\}_0$ that has to be applied to the crop property configuration $\{g_i\}_0$ for a maximum, possibly nonlinear, evaluation function F_E that is developed based on threshing unit performance criteria. The solution of the problem will be constrained by a number of restrictions expressed in terms of function inequalities.

5.10.3 Process Model Integration

The evolutionary computation and decision process in searching for the optimal solution of the problem is mainly based on the mathematical models developed by the author as described in this chapter.

Due to the nature of the threshing and separation process and the goals of the threshing unit operation, seven required objectives define the vector of performance criteria, as given below:

$$y = \left[s_s, V_s, pp_s, p_{sp}, s_v, V_t, \varepsilon_1 \right]^T$$

where:

s_s is the cumulative fraction of separated grain (Equation 5.48)
V_s is the grain separation loss fraction (Equation 5.56)
pp_s is the cumulative fraction of separated MOG (Equation 5.115)
p_{sp} is the specific power requirement (Equation 5.131)
s_v is the grain damage fraction (not yet published)

V_t is the grain threshing loss fraction (Equation 5.45)
ε_1 is the material sliding coefficient (cinematic condition) (Equation 5.48)

Two continuity equations of the process quality indices on the threshing length can be written:

$$\frac{1}{\beta}\frac{\partial^2 s_s}{\partial x^2} + \frac{\partial s_s}{\partial x} = \lambda s_n \qquad\qquad \text{(see 5.53)}$$

$$\frac{1}{a}\frac{\partial^2 pp_s}{\partial x^2} + \frac{\partial pp_s}{\partial x} = abpl_n \qquad\qquad (5.132)$$

where:
   x is the current position on the threshing length (concave or rotor length)
   s_n is the fraction of unthreshed grain (Equation 5.43)
   pl_n is the percentage of undetached chaff, that is, $pp_n(L)$ (Equation 5.113)

The rate b of MOG fragmentation and rate a of MOG separation are functions of the following parameters:

$$a = a(q_p, v, \delta_e, s_o, \eta, d_p, u_p) \qquad\qquad (5.133)$$

$$b = b(q_p, \delta_e, u_p) \qquad\qquad (5.134)$$

where d_p is the average diameter of the plant (straw).

The model also takes into account external perturbations that apply to the system, such as weed percentage, mean length of the weeds, crop cutting height, feeding MOG density, and variability of MOG moisture content in the field during the day (Heger, 1974).

The decision process is based on *a priori* articulation of preferences; the required individual objectives are combined into a single utility function (evaluation function) prior to optimization. Although such an evaluation function is difficult to formalize in every detail, we used approaches based on weighting coefficients and goal values (performance measure parameters). Weighting coefficients are real values, which express the relative importance of the objectives and control their involvement in the overall evaluation function. Goal values indicate desired levels of performance in each objective dimension.

5.10.4 Genetic Algorithm

The GA starts with an initial randomly chosen population of potential solutions (individuals). To each individual a genetic representation (*chromosome*) is assigned. This means that every set of parameters is coded in such a way that a tractable form results and allows an easy application of genetic operators (*crossover* and *mutation*). Evaluation of each chromosome in each generation gives the measure of its "fitness." The fittest individuals are selected to *reproduce*, that is, to form the new generation (the next iteration). Genetic operators are further applied to alter the genetic information, and hence introduce new solutions into the search scheme. The cycle is repeated a prescribed number of times or until some convergence criterion is met (Davis, 1991; Michalewicz, 1996).

In terms of genetic representation, a floating-point representation is used to code real values of x_k ($k = m + r$) variables (parameters that have to be determined) as a floating-point vector (chromosome) with k components in the range [0, 1].

The GA begins with an initial population to avoid erroneous results due to the stochastic nature of the search algorithm. Population is constant through the process, and it is randomly initiated. Therefore, k random uniformly distributed numbers over the interval [0, 1] are generated to form a chromosome (individual), which represents the initial solution to the problem.

5.10.4.1 Evaluation Function

The parameters associated with the seven operation requirements listed in the performance vector y are computed for each individual, and a performance measure is assigned to each of the requirements except the first. If a solution is not viable (requirement number 1 not satisfied), then the computation of all other parameters is no longer necessary, and each of the seven performance measures is set to −0.5. Otherwise, each performance measure will take a value in the range [0, 1] according to fuzzy-logic schemes in Figures 5.45 and 5.46, depending on the specific requirement.

In the domain [a_j, b_j] the performance measure curves are third-order polynomials. If any of the performance measure exceeds limit c_j, then the accepted index i_a is set to zero and the configuration is penalized by a factor f_p. In this way, the complete elimination of unviable configurations is avoided because they can contain useful genetic information for the next generation. "Bad" parents may have "good" children though.

The performance vector y is multiplied by the weighting vector W. The evaluation function F_E has the expression

$$F_E = \begin{cases} W \cdot y & \text{for } i_a \neq 0 \\ W \cdot y / f_p & \text{for } i_a = 0 \end{cases} \tag{5.135}$$

As far as the seven required objectives are concerned, it has been assumed that they are equally important; therefore, the weighting vector W has all components equal to 1/7 in the base version. It is likely the algorithm with equal weights to converge to a local

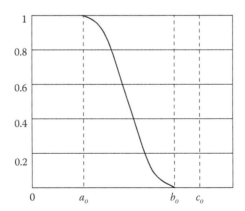

FIGURE 5.45
Fuzzy-logic scheme of performance measure for minimized requirements.

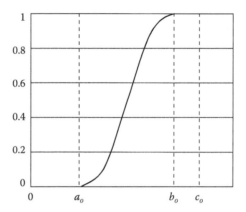

FIGURE 5.46
Fuzzy-logic scheme of performance measure for maximized requirements.

optimum when it does not meet one of the requirements. That could be due to the fact that there are large, easily reachable regions of the solution space where all the requirements are met except one. To avoid this problem, we used selective weights in the vector **W**.

Using selective weights does not mean emphasizing unequal importance of different performance elements, but directing the search toward narrow regions of the solution space (Perhinschi, 1997).

5.10.4.2 Individual Selection

A roulette wheel selection technique is used (Michalewicz, 1996). Each individual is evaluated by computing F_{Ei} ($i = 1$–pop_size). The total fitness F_T of a population is given by

$$F_T = \sum_{i=1}^{pop\_size} F_{Ei} \tag{5.136}$$

The probability p_i of each individual selection is

$$p_i = \frac{F_{Ei}}{F_T} \tag{5.137}$$

The cumulative probability q_i for each individual to be selected is

$$q_i = \sum_{j=1}^{i} p_j \tag{5.138}$$

Each time an individual for a new generation is selected, a random number n_r is generated in the range [0, 1]. If $n_r < q_1$, then select the first individual; otherwise, select the ith individual such that $q_{i-1} < n_r \le q_i$. The best individual gets more copies in the new generation, and the worst are eliminated.

However, due to the probabilistic nature of the process, many times the best individual does not survive to the next generation, especially in an algorithm setup that allows for high variability. Elitist selection strategy will ensure that the best individual survives unaltered into the new generation. When the elitist strategy is applied, we check whether the best individual made it through the above selection process. If it did not, an arbitrary individual is simply replaced to generate the new population. A second check is necessary to see if an unaltered copy of the best individual is still present in the new population after genetic operators are applied. If it is not, then again, an arbitrary individual is simply replaced with the best one.

5.10.4.3 Genetic Alteration

Once a new population is selected, crossover and mutation are applied randomly. Single-point (gene) crossover and uniform mutation have been used in this study. The single-point crossover operation consists of splitting two chromosomes in two parts at the same randomly determined crossover point, and then building up two new chromosomes by combining the first part of one chromosome with the second part of the other. The probability that an individual undergoes crossover is the crossover rate.

Mutation alters a gene, a position in the binary string, with a probability equal to the mutation rate m_{bin}. The alteration consists of changing 0 to 1, or vice versa. Uniform operators keep the mutation rate constant for all bits throughout the entire iterative process. Let the number of bits that would represent an algorithm parameter be n_{bit}. For the floating-point representation, a gene is a component of the floating-point vector. The mutation rate m_{float} is

$$m_{float} = 1 - \left(1 - m_{bin}\right)^{n_{bit}} \tag{5.139}$$

It introduces more variability of populations.

5.10.4.4 Application

Considered was the threshing of wheat, whose properties are integrated into the following explicit vector:

$$g = [l_p, \lambda_r, \varepsilon_b, u_p, u_s, d_e, l_s, \rho_p] \tag{5.140}$$

where:
 l_p is the mean length of the plants
 λ_r is the grain-to-MOG ratio
 ε_b is the weeds-to-MOG ratio
 u_p is the MOG moisture content (wet basis)
 u_s is the grain moisture content (wet basis)
 d_e is the geometric mean diameter of the grain kernel
 l_s is the mean of the grain kernel maximum diameter
 ρ_p is the MOG bulk density

The values of the following parameters have to be optimized:

$$x_k = [q_p, v, \delta_e, s_o, \eta, L, R, u_p, h]^T \tag{5.141}$$

where:

q_p is the MOG feedrate
v is the rotor speed
δ_e is the exit concave clearance
s_o is the mean surface of the concave and cage openings
η is the opening ratio of the separation surface
L is the rotor length
R is the rotor radius
h is the plant cutting height

These parameters have to be found within the limits indicated in Table 5.3.

The values a_o, b_o, and c_o of the desired levels of performance parameters are listed in Table 5.4.

Two different weighting vectors W have been considered:

- Equal components of the weighting vector
- Selected components of the vector (some larger than others) (Table 5.5)

TABLE 5.3

Limits of Wanted Design and Process Parameters of Axial Threshing Unit

Parameter	q_p	v	δ_e	s_o	η	L	R	u_p	h
Limits	kg/s	m/s	mm	mm²		m	m	%	m
Minimum	2	25	5	200	0.1	1.2	0.2	15	0.04
Maximum	7	35	25	900	0.6	2	0.45	50	0.4

TABLE 5.4

Values of Performance Measure Parameters

Parameter/Performance Measure	a_o	b_o	c_o
Specific power requirement, p_{ts} (kW/kg$_{MOG}$/s)	6.5	9.8	11.5
Separation loss, L_s (%)	0.5	2.5	4
Grain damage, s_v (%)	0.8	2	3
Grain separation, s_s (%)	97.2	99.4	95
MOG separation, pp_s (%)	18	23	28
Material slip, σ	0.58	0.65	0.75

TABLE 5.5

Versions of Weighting Vector

Parameter/Version	GA1	GA2
Specific power requirement, p_{ts}	1/6	3/11
Separation loss, L_s	1/6	1/11
Grain damage, s_v	1/6	2/11
Grain separation, s_s	1/6	3/11
MOG separation, pp_s	1/6	1/11
Material slip, σ	1/6	1/11

The purpose of selective weighting coefficients is not especially to emphasize the unequal importance of different good elements, but to control their involvement in the overall utility measure and to direct the search toward narrow regions of the solution space.

The initial population has 100 individuals. The crossover rate is 0.5, and the mutation rate is 0.03.

5.10.4.5 Results

For each version of GA, five runs of a hundred generations each have been performed. The averages over these runs of the best individual evaluation function and generation average are plotted versus generation number in Figures 5.47 and 5.48.

All five runs for each version yield best individuals that satisfy the performance criteria within the range 91 ÷ 100%. The stronger restrictions come from specific power

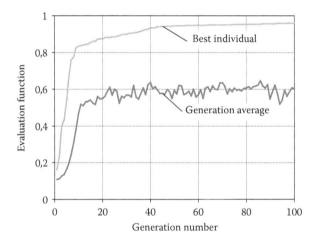

FIGURE 5.47
Best individual evaluation function and generation average for equal weights (GA1).

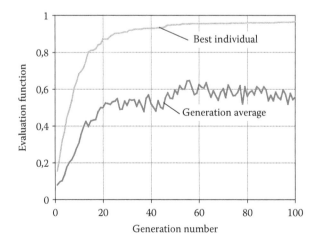

FIGURE 5.48
Best individual evaluation function and generation average for selective weights (GA2).

TABLE 5.6

Best Individual Components

Parameter	q_p	v	δ_e	s_o	η	L	R	u_p	h
	kg/s	m/s	mm	mm²		m	m	%	m
Value (GA1)	4	30.7	24.3	295	0.58	1.994	0.26	15.61	0.14
Value (GA2)	3.8	31.2	24.8	307	0.56	1.990	0.24	15.13	0.24

requirement and separation loss performance criteria. Although both of them restrict the throughput increase, an essential improvement in performance is emphasized in comparison with the experimental data. It is implicitly assumed that there is at least one solution, which satisfies all constraints, although in practice that cannot always be guaranteed. The best individual was found using selective weights (GA2) (Table 5.6).

The set of best performance parameters that were obtained are

$$p_{ts} = 7.09 \text{ kW/kg}_{MOG}/s$$
$$L_s = 0.65\%$$
$$s_v = 0.78\%$$
$$s_s = 99.26\%$$
$$pp_s = 18.36\%$$
$$\sigma = 0.67$$

and the evaluation function has the mean value 0.971.

It should be noted that general algorithm uses a model for grain damage, which was validated for other axial threshing units. Since not all goals can be simultaneously 100% met, a *family* of points (individuals) can be closest to the solution for the problem. Note that optimal solutions obtained with both GAs are very closed. That shows the developed GAs are robust in the beginning of the formulated problem.

5.10.4.6 Conclusions

After a reasonably limited number of runs, the following conclusions may be drawn:

- Although it is implicitly assumed that there is at least one solution, which satisfies simultaneously all goals and constraints, optimal solutions obtained with two GAs are approximately identical, and the evaluation function has the greatest value: 0.971.
- Selective weights affecting the elements composing the evaluation function can be successfully used to direct the search toward narrow regions of the solution space. In this case, the evolution of the best individual evaluation function seems to be very smooth and precise, and higher values are obtained (Figure 5.48).
- Using the elitist strategy is recommended because it reduces the convergence time and allows for parameter fine-tuning.
- Experimental results indicate that the GA is a promising design and optimization tool for solving agricultural machinery-related problems for the best approximation solution and for minimizing the execution time.

5.10.4.7 GA Program Installation

The MATLAB script of the GA program is given in the Appendix. It contains nine files. Here are the installation instructions:

1. Edit and save the files from the Appendix into a folder, for example, D:matlgenp; make sure this path is indicated in applicable files.
2. Create two empty text files in the same directory: FUZrez.txt and graf.txt.

In MATLAB, in the corresponding directory, write *genpet*, and then press ENTER.

References

Alferov, S.A. and V.S. Braginec. 1972. Grain threshing and separation in threshers as a random process [in Russian]. *Traktori i Sel'chozmashini* 42(4):23–26.

ANSI/ASAE S424.1. 1997. Method of determining and expressing particle size of chopped forage materials by screening. In *ASAE Standards 1997.* 44th ed. St. Joseph, MI: American Society of Agricultural Engineers.

Arnold, R.E. 1964. Die Bedeutung einiger Einflussgrössen auf die Arbeit der Schlagleisten-trommel [The significance of influence parameters on the work of a rasp bars cylinder]. *Grundlagen der Landtechnik* 14(21):22–28.

Arnold, R.E. 1979. Limited field trials of an international 1460 axial-flow combine. National Institute of Agricultural Engineering Report MI/1/8177/79. Silsoe, England: National Institute of Agricultural Engineering.

ASAE S319.3. 1996. Method of determining and expressing fineness of feed materials by sieving. In *ASAE Standards 1997.* 44th ed. St. Joseph, MI: American Society of Agricultural Engineers.

Baader, W., H. Sonnenberg, and H. Peters. 1969. Die Entmischung eines Korngut - Fasergut - Haufwerkes an einer vertikal schwingenden horizontalen Unterlage [The segregation of grain and straw on a vertically vibrating horizontal surface]. *Grundlagen der Landtechnik* 19(5):149–157.

Bjork, A. 1988a. Computer modelling of grain separation and grain separation losses for a rotary combine. ASAE Paper 88-102. St. Joseph, MI: American Society of Agricultural Engineers.

Bjork, A. 1988b. Power requirement for threshing and separation in a rotary combine. ASAE Paper 88-204. St. Joseph, MI: American Society of Agricultural Engineers.

Bosch, K. 1996. *Grosses Lehrbuch der Statistik [Statistics Handbook].* Vienna: R. Oldenbourg Verlag.

Brohnstein, I.N. and K.A. Semendyayev. 1985. *Handbook of Mathematics.* 3rd ed. Berlin: Springer.

Caspers, L. 1973. Die Abscheidefunktion als Beitrag zur Theorie des Schlagleistendresch-werk [Separation function as a contribution to the theory of rasp bars threshing unit]. PhD dissertation, Institut für Landmaschinenforschung der Forschungsanstalt für Landwirtschaft, Braunschweig-Völkenrode.

Davis, L. 1991. *Handbook of Genetic Algorithms.* New York: Van Nostrand Reinhold.

Dolling, C. 1957. Der Leistungsbedarf von Mähdreschern [Power requirement of combine harvesters]. *Landtechnische Forschung* 7(2):33–40.

Eimer, M. 1977. Optimierung der Arbeitsqualität des Schlagleistendreschwerks [Optimization of the quality of rasp-bars thresher]. *Grundlagen der Landtechnik* 27(1):12–17.

Eimer, M. 1980. Axialdrusch-geignetes Dreschprinzip für Mittleeuropa? [Axial threshing: A principle for Middle Europe?]. *Landtechnik* 35(6).

Filatov, N.V. and P.G. Chabrat. 1969. Verstärkung der Kornabscheidung auf Hordenstrohschüttlern [Strengthening of grain separation on straw walkers]. *Landtechnische Forschung* 17(6):196–200.

Gasparetto, E., P. Febo, D. Pessina, and E. Rizatto. 1989. High speed movie observation of an axial flow combine harvester cylinder. In *Land and Water Use*. Rotterdam: Balkema.

Harrison, H.P. 1991. Rotor power and losses of an axial-flow combine. *Transactions of the ASAE* 34(1):60–64.

Heger, K. 1974. A model for determining the moisture content of wheat grain at harvest time. In *Agrometeorology of the Wheat Crop*. WMO 396. Offenbach.

Henderson, S.M., R.L. Perry, and J.H. Young. 1997. *Principles of Process Engineering*, 4th edn, ASAE Textbook 801 M0297, St. Joseph, MI: ASAE.

Huisman, W. 1983. Optimum cereal combine harvester operation by means of automatic machine and threshing speed control. PhD dissertation, Agricultural University Mansholtlaan, Wageningen.

Huynh, V.M., T. Powel, and J.N. Siddall. 1982. Threshing and separating process: A mathematical model. *Transactions of the ASAE* 25(1):65–73.

Jalnin, V.E. 1982. Nekotorie tendentii zarubejnogo kombainostroenie [Some trends in the foreign construction of combine harvesters]. *Mehanizatiia i Elektrifikatiia selskogo-hoziaistva* 12.

Kanafojski, C. 1973. *Grundlagen erntetechnischer Baugruppen [Fundamentals of Construction of Harvesting Modules]*. Berlin: VEB Verlag.

Kepner, R.A., R. Bainer, and E.L. Barger. 1982. *Principles of Farm Machinery*. 3rd ed. Avi Publishing.

Kim, S.H. and J.M. Gregory. 1989a. Power requirement model for combine cylinders. ASAE Paper 89-1592. St. Joseph, MI: American Society of Agricultural Engineers.

Kim, S.H. and J.M. Gregory. 1989b. Straw to grain ratio for combine simulation. ASAE Paper 89-1593. St. Joseph, MI: American Society of Agricultural Engineers.

Klenin, N.I. and S.G. Lomakin, 1972. Körnerabscheidung durch den Dreschkorb [Grain separation through the concave]. *Deutsche Agrartechnik* 22(7):321–323.

Kutzbach, H.D. 2003. Approaches for mathematical modelling of grain separation. Presented at ICCHP Conference, Louisville, KY.

Lo, A. 1978. Untersuchungen zum Druschverhalten von Körnermais in einem Axialdresch-werk [Researches on threshing processes of corn in an axial thresher]. PhD dissertation, Hohenheim University, Forschungsbericht Agrartechnik der max-Eyth-Gesellschaft (MEG) 27.

Maertens, K. and J. De Baerdemaeker. 2003. Flow rate based prediction of threshing process in combine harvesters. *Applied Engineering in Agriculture* 19(4):383–388.

Mailander, M. 1984. Development of a dynamic model of a combine harvester in corn. ASAE Paper 84-1588. St. Joseph, MI: American Society of Agricultural Engineers.

McGechan, M.B. and C.A. Glasbey. 1982. An assessment of the relative benefits of different forward speed control for combine harvesters. Departmental Note SIN/352. Penicuik: Scottish Institute of Agricultural Engineering.

Michalewicz, Z. 1996. *Genetic Algorithms + Data Structure = Evolution Programs*. Berlin: Springer-Verlag.

Miu, P. 1993. Axial threshing unit: Contributions to the study of material kinematics. *Scientific Bulletin: Series D: Mechanical Engineering* 55(1–2):95–105.

Miu, P. 1994. Concave separation in a tangential threshing unit. ASAE Paper 94-1544. St. Joseph, MI: American Society of Agricultural Engineers.

Miu, P. 1995. Modelarea procesului de treier la combinele de recoltat cereale [Mathematical modeling of threshing process in cereal combine harvesters]. PhD dissertation, Politehnica University of Bucharest.

Miu, P. 1999. Mathematical modeling of material other-than-grain separation in threshing units. ASAE Paper 99-3208. St. Joseph, MI: American Society of Agricultural Engineers.

Miu, P. 2000. Modeling of the separation process of MOG components in threshing units. In XXVIII International Symposium, Opatija, Croatia, February 2000, pp. 183–190.

Miu, P. 2001a. Modeling of power requirement in axial threshing units. ASAE Paper 01-3045. St. Joseph, MI: American Society of Agricultural Engineers.

Miu, P. 2001b. Optimal design and process of threshing units based on a genetic algorithm. I. Algorithm. ASAE Paper 01-3124. St. Joseph, MI: American Society of Agricultural Engineers.

Miu, P. 2002a. Kinematic model of material movement through an axial threshing unit. ASAE Paper 02-3052. St. Joseph, MI: American Society of Agricultural Engineers.

Miu, P. 2002b. Mathematical model of threshing process in an axial unit with tangential feeding. CSAE Paper 02-219. Saskatoon, Saskatchewan: Canadian Society of Association Executives.

Miu, P. 2004. Combine harvester processor: Theory of crop movement and mathematical modeling of the process. Report for Professional Services Contract Agreement I4-13150-00 with Iowa State University.

Miu, P., F. Beck, and H.-D. Kutzbach. 1997. Mathematical modeling of threshing and separating process in axial threshing units. ASAE Paper 97-1063. St. Joseph, MI: American Society of Agricultural Engineers.

Miu, P. and H.-D. Kutzbach. 2000. Simulation der Dresch- und Trennprozesse in Dreschwerken [Simulation of threshing and separation processes in threshing units]. *Agrartechnische Forschung* 6:1–7.

Miu, P. and H.-D. Kutzbach. 2007. Mathematical model of material kinematics in an axial threshing unit. *Computers and Electronics in Agriculture* 58(2):93–99.

Miu, P. and H.-D. Kutzbach. 2008a. Modeling and simulation of grain threshing and separation in threshing units. Part I. *Computers and Electronics in Agriculture* 60(1):96–104.

Miu, P. and H.-D. Kutzbach. 2008b. Modeling and simulation of grain threshing and separation in threshing units. Part II. Application to tangential feeding. *Computers and Electronics in Agriculture* 60(1):105–109.

Miu, P. and M.G. Perhinschi. 2001. Optimal design and process of threshing units based on a genetic algorithm. II. Application. ASAE Paper 01-3125. St. Joseph, MI: American Society of Agricultural Engineers.

Miu, P., P. Wacker, and H.-D. Kutzbach. 1998a. A comprehensive simulation model of threshing and separating process in axial units. Part I. Further model development. In *Int. Conf. Agr. Eng Proceedings*, Norway, vol. 2, pp. 765–766, AgEng Paper 98-A-115.

Miu, P., P. Wacker, and H.-D. Kutzbach. 1998b. A comprehensive simulation model of threshing and separating process in axial units. Part II. Model validation. In *Int. Conf. Agr. Eng Proceedings*, Norway, vol. 2, pp. 767–768, AgEng Paper 98-A-116.

Mohsenin, N.N. 1978. *Physical Properties of Plant and Animal Materials*. New York: Gordon and Breach Science Publishers.

PAMI. 1987. Evaluation Report 532. Versatile trans-axial 2000 pull-type combine. Humboldt, Saskatchewan: Prairie Agricultural Machinery Institute.

Perhinschi, M.G. (1997). Controller design for autonomous helicopter model using a genetic algorithm. Presented at American Helicopter Society 53rd Annual Forum, Virginia Beach, VA.

Pickett, L.K. and N.L. West, 1988. Agricultural machinery: Functional elements: Threshing, separating and cleaning. In *Handbook of Engineering in Agriculture*, 65–84. Vol I. Boca Raton, FL: CRC Press.

Reitz, P. and H.-D. Kutzbach. 1996. Investigations on a particular yield mapping system for combine harvesters. *Computers and Electronics in Agriculture* 14:137–150.

Rusanov, A.J. 1971. Zavisimost raboty molotil'no—separirujescego ustroistva ot diametra barabana i dliny podbaran'ja [Dependence of threshing work on cylinder diameter and concave length]. *Mechanizacija i Elektrifikacija sel'skogohoziastva* 29(8):16–18.

Rusanov, A.I. 2001. Concerning the article "Grain combine harvesters: Returning to the past" [in Russian]. *Traktori i sel'chozmasiny* 71(4):45–48.

Rusanov, A.I. 2002. Additional notes to the article of E.V. Zhalnin "Great combine harvesters: From the past to the future" [in Russian]. *Traktori i sel'chozmasiny* 72(7):40–43.

Segler, G. and F. Wienecke. 1961. Dreschverluste und Leistungsbedarf des Mähdreschers beim Verarbeiten von Getreide mit Grüngutbesatz [Threshing loss and power consumption at green cereal processing]. *Landtechnische Forschung* 11(5):141–144.

Serii, F.G. and I.N. Kosilov. 1986. *Zernouborociniie combaini* [*Cereal Combine Harvesters*]. Moskow: Agropromizdat.

Sonnenberg, H. 1971. Korn-Stroh-Trennung mit einer Umlenktrommel [Grain and straw separation in a threshing unit]. *Grundlagen der Landtechnik* 21(6):169–172.

Trollope, J.R. 1982. A mathematical model of the threshing process in a conventional combine-thresher. *Journal of Agricultural Engineering Research* 27(2):119–130.

Vasilenko, I.F. 1954. *Kompendium der sowjetischen Landmaschinentechnik* [*Compendium of Soviet Agricultural Machinery*], 198–199. Berlin: VEB Verlag Technik.

Vetrov, Ye.F. 2002. In reference to the article of E.V. Zhalnin "Great combine harvesters: From the past to the future" [in Russian]. *Traktori i sel'chozmasiny* 72(5):45–48.

Vogt, C. 1982. Entwicklungstendenzen im Getreidebau [Trends in development of construction for cereals]. *Landtechnik* 37(5):226–228.

Voß, H. 1970. Ermittlung von Stoffgesetzen für Halmgut [Determination of stems break laws]. PhD thesis, Technical University Carolo-Wilhelmina, Braunschweig.

Wacker, P. 1985. Untersuchungen zum Dresch- und Trennvorgang von Getreide in einem Axialdreschwerk [Researches on threshing and separation process in an axial threshing unit]. PhD dissertation, Hohenheim University, Forschungsbericht Agrartechnik der MEG 117.

Wacker, P. 1988. Vergleich von Axial- und Tangentialdreschsystemen in Getreide [Comparison of axial and tangential threshing units in cereal harvesting]. *Landtechnik* 43(6):264–266.

Wacker, P. 1990. Einflüßgrossen auf die Arbeitsqualität von Axial- und Tangentialdresch-werken [Influence parameters on the quality of work of axial and tangential threshing units]. *Agrartechnik* 40(3):102–104.

Wessel, J. 1967. Grundlagen des Siebens und Sichtens. Teil II. [Fundamentals of screening and winnowing. Part II]. *Aufbereitungstechnik* 13(4):167–180.

Bibliography

Audsley, E. 1991. A method for minimizing the total cost of combine harvesting. Note 51. Silsoe, England: British Society for Research in Agricultural Engineering.

Bäck, T., U. Hammel, and H.P. Schwefel. 1997. Evolutionary computation: Comments on the history and current state. *IEEE Transactions on Evolutionary Computation* 1(1):3–16.

Beck, F. 1999. Simulation der Trennprozesse im Mähdrescher [Simulation of separation processes in combine harvesters]. PhD dissertation, Stuttgart University, Fortschritt-Berichte, VDI Reihe 14(66).

Böttinger, S. 1993. Die Abscheidungsfunktion von Hordenschüttler und Reinigungsanlage in Mähdrescher [The separation function on straw walkers and cleaning unit of combines]. PhD dissertation, Stuttgart University, VDI Verlag GmbH Düsseldorf.

Campbell, D.W. 1980. Modelling the combine harvesting. PhD dissertation, University of Saskatchewan, Saskatoon.

Casandroiu, T. 1985. Studiul procesului de treier al stiuletilor de porumb in aparate de treier cu raza variabila [Study of corn threshing process in threshers with variable radius]. PhD dissertation, Polytechnic Institute of Bucharest.

Cooper, G.F. 1972. Concave separation from laboratory tests. *Transactions of the ASAE* 15(5):865–869.

DIN 11390. 1991. Prüfverfahren für Mähdrescher [Testing procedures for combine harvesters]. Berlin: Beuth Verlag.

Gasparetto, E., M. Zen, and A. Guadagnin. 1988. Ultra high speed movie observation of a conventional threshing mechanism working on wheat. In *Grain and Forage Harvesting*. St. Joseph, MI: American Society of Agricultural Engineers.

Gieroba, J. 1972. Investigations on threshing cereal crops with the use of a threshing cone unit [in Polish]. *Wydzial Nauk Rolniczych i Lesniych Roczniki nauk Rolniczych* D:142, Polska Akademia Nauk.

Glasbey, C.A. and M.B. McGechan. 1983. Threshing loss stochastic variability on combine harvesters. *Journal of Agricultural Engineering Research* 28(2):163–174.

Goldberg, D.E. 1989. *Genetic Algorithms in Search, Optimisation and Machine Learning*. Reading, MA: Addison-Wesley.

Gregory, J.M. 1988. Modeling applications in agricultural engineering: Combine model for grain threshing. *Mathematical Computer Modelling* 11:506–509.

Herbsthoffer, F.J. 1988. Mähdrescherkonzept mit konischem Axialdreschwerk [Concept of a combine with a conical axial thresher]. Düsseldorf: VDI/MEG Koloquim Landtechnik, Verein Deutscher Ingenieure, Heft 6.

ISO 8210. 1989. Equipment for harvesting—Combine harvesters—Test procedure. Geneva: International Organization for Standardization.

Jbail, F.A.M. 1994. Studiul rezistentei la desprindere a semintelor din spicele cerealelor paioase [Study of the grain resistance at detachment from cereal ears]. PhD dissertation, Politehnica University of Bucharest.

Kutzbach, H.-D., P. Wacker, and P. Reitz. 1996. Developments in European combine harvesters. AgEng Paper 96A-069.

Lalor, W. and W.F. Buchele. 1963a. Designing and testing of a threshing cone. *Transactions of the ASAE* 6(2):73–76.

Lalor, W. and W.F. Buchele. 1963b. The theory of threshing cone design. *Journal of Agricultural Engineering Research* 19(8):35–40.

Letosnev, M.N. 1955. *Sel'skohoziaistvennye masini* [*Agricultural Machinery*]. Sel'chozgiz.

Miles, G.E. and Y.J. Tsai. 1987. Combine systems engineering by simulation. *Transactions of the ASAE* 30(5):1277–1281.

Nath, S., W.H. Johnson, and G.A. Milliken. 1982. Combine loss model and optimization of the machine system. *Transactions of the ASAE* 25(2):308–312.

Neculaiasa, V. and I. Danila. 1995. *Working Processes and Harvesting Machinery* [in Romanian]. Iasi, Romania: A92 Publisher.

Nyborg, E.O., H.F. McColly, and R.T. Hinkle. 1969. Grain combine loss characteristics. *Transactions of the ASAE* 12(6):727–732.

Quick, G.R. 2002. Setting combines for best seed and field corn quality at harvest. ISU Extension AE-3112. Ames: Iowa State University.

Segarceanu, M.I., I. Paunescu, et al. 1983. Experimentarea modelelor de combine cu aparate de treier cu flux axial [Testing of axial flow combine harvesters report]. Romania: Polytechnic Institute of Bucharest.

Wacker, P. 1984a. Gutgeschwindigkeitsmesseinrichtung für ein Axialdreschwerk [Material speed measurement in an axial threshing unit]. Presented at the 10th International Congress for Agricultural Engineering, Budapest.

Wacker, P. 1984b. Die Korn- und Strohabscheidung in axial Dreschwerken [Grain and material-other-than-grain separation in axial threshing units]. Presented at the 10th International Congress for Agricultural Engineering, Budapest.

6

Separation Process and Operation of Straw Walkers

6.1 Introduction

Most of the material other than grain (MOG) and threshing unit grain losses are delivered to the straw walkers. Depending on material type and properties, and on the threshing unit design and operation, the material mixture discharged by a threshing unit onto the straw walkers (Figure 6.1) consists of long straw (20%–50%), short straw (10%–35%), chaff (10%–15%), weed components, and grain losses of the threshing unit, as well as other components of vegetal origin, insects, and small mineral components.

The *straw walkers* of a conventional combine harvester execute the following functions:

- Recover the grain losses of the threshing unit through separation
- Transport the recovered grains to the cleaning unit
- Separate the straw from the MOG mixture
- Transport the straw toward the rear of the combine for discharging on the field

The most difficult task of straw walkers is separating residual grains (free or unthreshed) from straw. In fact, the grain-separating capacity of straw walkers determines the total material (MOG) feedrate of conventional combine harvesters.

Straw walker performance is mainly sensitive to combine inclinations (flow uniformity), MOG federate, and moisture content. To increase the efficiency of grain separation, on top of the straw walkers is mounted a straw-mat agitator (ruffling drum) to aerate the straw as it speeds along and to allow the free fall of the grain for efficient separation.

A great advantage of using straw walkers is maintaining grain quality and straw size and condition when compared with the process performed by an axial threshing unit, which eliminates the straw walkers, but damages the grain to a certain extent; however, an axial threshing unit offers the advantage of a threefold feedrate increase.

At their rear, most modern combines are equipped with a chopper–spreader that grabs the straw while in motion from straw walkers, chops the straw, and spreads it evenly over the field (Figure 6.2). The debris from the cleaning sieve reaches the spreader as well, mixes with the chopped straw, and follows the same path before reaching the soil. We discuss such working units in Chapter 9.

6.2 Construction of Straw Walkers

The straw walkers of conventional combine harvesters can be classified based on the following criteria:

FIGURE 6.1
Discharged straw on straw walkers by threshing unit. (Courtesy of CLAAS, Harsewinkel, Germany.)

FIGURE 6.2
Lateral view of a straw chopper/chaff spreader. (Courtesy of CLAAS, Harsewinkel, Germany.)

- Construction shape: Straw walkers with shakers, oscillating platform, sliding screens, and grain separation-intensifying devices (retractable-finger drum, oscillating stellate discs, or oscillatory forks [used in the past])
- Motion type: Straw walkers with uniform motion or oscillatory motion
- Motion phase: Synchronous or independent

A modern conventional combine harvester may have four to seven independent straw walkers (shakers or shoes), which are mounted on a crank axle (front side) and motion-driving crankshaft (rear).

Each straw walker (Figure 6.3a and b) is formed of four to six shaking sections whose screens are inclined to form a series of steps (cascades) that intensify the grain separation. Moreover, the wall of each cascade is a screen as well (Figure 6.4). Each straw walker may have 4–11 cascades and 5–12 screens (sieves) for grain recovery from the straw mat.

The sidewalls of each straw walker have a linear sawtooth profile forming retaining elements, which, together with the section cascades, push the straw rearward to be discharged directly on the field, or to be chopped and then spread over the field.

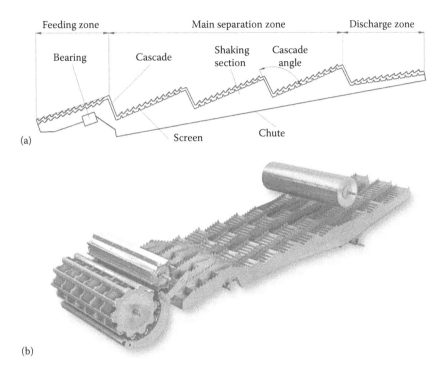

(a)

(b)

FIGURE 6.3
(a,b) Functional zones and elements of a straw walker endowed with four cascades.

FIGURE 6.4
Shape of cascade slotted screens. (Courtesy of Sampo Rosenlew, Pori, Finland.)

The straw walkers have three functional zones: feeding, separation, and discharge. Each straw walker performs a plane-parallel motion within a vertical plane, in which the trajectory of every point results from a translational motion with an arbitrarily selected pole, around which rotational motion occurs due to the driving rotation of the crankshaft.

For the straw walkers whose shakers have an independent motion, the position of each crank of the crankshaft and axle has an angular offset whose value depends on the number of shakers, as shown in Figure 6.5 for independent straw walkers with four, three, and six shakers at 90°, 120°, and 60°, respectively. For straw walkers with five shakers, the angular offset of the cranks is 72°. The independent straw walkers are very efficient when the MOG contains a relatively big fraction of short straws and fine chaff.

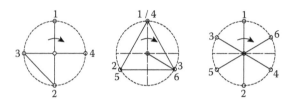

FIGURE 6.5
Schematic of crank angular offset for straw walkers with four, three, and six shakers.

The straw walkers with synchronous motion are divided into two groups with opposed motion direction (e.g., a group of shakers 1, 3, and 5, and another group of shakers 2, 4, and 6). In such a case, the cranks of the crankshaft and crank axle have a 180° angular offset. These straw walkers are efficient when the MOG contains a relatively big fraction of long straws.

The overall dimensions of a shaker are length of 2.75–4.5 m (108–177 in.), width of 230–355 mm (9–14 in.), and length of cranks of 40–60 mm (1.58–2.375 in.). The rotation of the crankshaft can be set between 180 and 270 rpm. Data about the straw walker separation area is in Table 1.1. Depending on the type and design of the threshing unit, as well as on the crop, the load of the straw walkers varies between 1.3 and 2 kg/s/m MOG feedrate.

6.3 Theory and Modeling of Grain Separation of Straw Walkers

6.3.1 Literature Review of Mathematical Models

Due to the shaking motion the MOG and threshing unit grain losses are subjected to, the material jumps from one position to another; thus, the grains separate through the screens (mounted at the top of each section) and cascade screens, and then they are collected on the chute of each straw walker and directed forward to the feeding zone of the cleaning unit (oscillating grain pan). At the same time, the straws are pushed rearward and eventually are thrown on the field in the form of a swath or chopped and spread evenly on the field to serve as an organic fertilizer after their incorporation into the soil.

The assessment of straw walker performance when working at a relatively high material throughput can be quantified based on process indices: fraction of separated residual grain (free grains and unthreshed grains), grain separation loss, fraction of separated MOG (mostly chaff and small straws), and specific energy consumption. Mechanically, the straw walkers are evaluated based on the vibration spectra, magnitude, and resonance frequency induced through the combine chassis.

Many researchers developed mathematical representations of the grain separation process on straw walkers. According to Vasilenko (1954), Filatov and Chabrat (1967), Reed et al. (1987), and Wang et al. (1987), the grain g_s separated over the length of straw walkers can be described by the following equation:

$$g_s = g_0\left(1 - e^{-bx}\right) \tag{6.1}$$

where:

g_0 is the residual grains coming onto the straw walkers
x is the current position along the straw walker chute
e is the base of natural logarithms
b is the exponent that is determined experimentally

Based on the rules of mathematical modeling outlined in Chapter 2, the derivative of Equation 6.1, representing the density function of grain separation, has no maximum value; thus, this kind of equation implies a certain error.

Gregory and Fedler (1987) derived a mathematical relationship to predict grain separation in combines (including straw walkers) based on Fick's law of gas diffusion.

Huynh and Powell (1978) presented a quantitative description of the separation process in wheat. The process was characterized by three parameters: mean time of grain passage through the straw mat ($1/\lambda_m$), mean time of grain passage through the screen ($1/\lambda_c$), and crop dwell time t to predict the grain loss g_ς for a give shoe design:

$$g_\varsigma = \frac{\lambda_m e^{-\lambda_c t} - \lambda_c e^{-\lambda_m t}}{\lambda_m - \lambda_c} \tag{6.2}$$

Böttinger (1993) developed a mathematical model to describe the separation process of grain on both straw walkers and the cleaning unit, as follows:

$$g_\varsigma = -d\frac{ab}{b-a}x^c \left(e^{\frac{-b}{c+1}x^{c+1}} - e^{\frac{-a}{c+1}x^{c+1}} \right) \tag{6.3}$$

where the coefficients a, b, c, and d are determined experimentally and depend on the MOG feedrate, grain/MOG ratio, and crop that is harvested.

Ivan and Popescu (2008) developed a theory that considers the material displacement on the shaker, crankshaft speed, crankshaft length, and Froude number—thus the angle of the crankshaft at which the material starts getting out from contact with the sawtooth-like retaining elements of the chute walls.

Steponavicius and Butkus (2012) carried out research to determine the inclination angle of the straw walker sections (22°, 10°, and 18°); the grain separation reached a maximum when the rotation of the crankshaft was 205–225 rpm. The intensity of grain separation varies with the rotational speed of the crankshaft or the inclination angle of straw walker sections.

6.3.2 Mathematical Model of Grain Separation on Straw Walkers

The author's model that follows quantifies the grain separation and defines the grain separation efficiency along the straw walkers' length for comparison of the straw walkers' working processes.

Straw walkers are fed with a mixture of fragmented MOG and grains, which can be divided into two groups:

- Threshing loss (fragments of unthreshed ears) V_t (Equation 5.44)
- Threshing unit segregation and separation loss V_{gs} (Equation 5.51 or 5.52)

The segregation and separation loss are the biggest fractions of threshing unit losses; these are free grains (detached from the ears). Separation of these grains as well as the unthreshed grains occurs when a kernel or fragment of ear bounces through an opening of the straw walker screen.

Developing the theory implies consideration of the following commonsense assumptions (Miu and Kutzbach, 1997):

- The material feedrate is constant; this means, besides regular interactions between the straw and grains, no additional feeding turbulences or transient state can influence the separation process.
- At the very beginning of the feeding area, the residual grains from the threshing unit are uniformly distributed in the straw mass across the working width of the straw walkers. This translates to equal probabilities for each grain to migrate through the straw mat.

Let us consider a shaker of length L_s expressed in meters or feet (Figure 6.6).

The physical separation process of grains (free or unthreshed) on the straw walkers can be divided into two successive and separate events:

1. Migration of the grains through the MOG mass (segregation phase)
2. Passage of segregated grains through the openings of the straw walker screen (separation phase)

Applying the theory of statistics in regard to assumption 2, any grain is equally likely to reach the screen surface at any distance x (current position) associated with the length of a straw walker. Then, any grain segregated on the top surface of the screen is equally likely to pass through the screen at any distance x.

The probabilistic laws that respectively describe events 1 and 2 have been identified as follows:

1. $f_1(x) = ae^{-ax}$ (6.4)

2. $f_2(x) = (a-b)e^{-bx}$ (6.5)

where:
    a is the specific rate of grain segregation (m^{-1})
    b is the specific rate of grain separation through the screen (m^{-1})

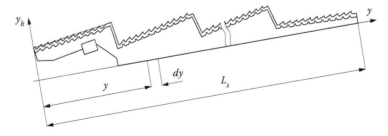

FIGURE 6.6
Schematic diagram of straw walker.

Rates a and b are considered constant; they respectively represent space increments between successive event occurrences. Since the material speed over the length of straw walkers decreases, the time interval each increment a and b is covered varies as well. To establish the time rates corresponding to space increments a and b, extensive experiments and very accurate measurements are required.

Events 1 and 2 are independent; therefore, according to the probability theory, the *joint probability density* $g(x)$ of the sum of two independent and steady random variables with densities $f_1(x)$ and $f_2(x)$ equals the convolution of their individual probability densities, as follows:

$$g(x) = f_1 * f_2 = \int_0^x f_1(z) f_2(x - z) dz \tag{6.6}$$

By solving this equation, we get what actually is *the probability density function of grain separation*, starting from the initial position within the MOG mixture and ending with grain passage through the screen:

$$g(x) = a\left(e^{-bx} - e^{-ax}\right) \tag{6.7}$$

Since the maximum fraction of the total grain that can be separated by straw walkers is $V_t + V_{gs}$, the *grain separation density function* or *separation frequency* $g_d(x)$ is

$$g_d(x) = a\left(V_t + V_{gs}\right)\left(e^{-bx} - e^{-ax}\right) \tag{6.8}$$

Equation 6.8 is expressed in terms of the grain fraction out of the total residual grains being fed onto the straw walkers. If one needs to express it in terms of percentage, then Equation 6.8 should be multiplied by 100 (%); the same rule should be followed for the next applicable equations.

By integrating Equation 6.8 over the length of the straw walkers, we obtain the *cumulative distribution function* $g_s(x)$ of separated grain:

$$g_s(x) = \frac{V_t + V_{gs}}{b}\left[a\left(1 - e^{-bx}\right) - b\left(1 - e^{-ax}\right)\right] \tag{6.9}$$

The cumulated fraction of separated grain over the length L_s is given by Equation 6.9 for $x = L_s$ as follows:

$$g_s(L_s) = \frac{V_t + V_{gs}}{b}\left[a\left(1 - e^{-bL_s}\right) - b\left(1 - e^{-aL_s}\right)\right] \tag{6.10}$$

Since the material throughput is constant (according to the first assumption), in every cross section of the straw walkers' operating space that is perpendicular to the direction of the straw movement, the grain mass balance can be written as

$$g_s(x) + g_r(x) = V_t + V_{gs} \tag{6.11}$$

where $g_r(x)$ is the *fraction of residual grains* that still exists within the straw mat. That means

$$g_r(x) = \left(V_t + V_{gs}\right)\left\{1 - \frac{1}{b}\left[a\left(1 - e^{-bx}\right) - b\left(1 - e^{-ax}\right)\right]\right\}$$ (6.12)

At the end of the straw walkers, the fraction of residual grains becomes *grain separation loss* g_ς of the straw walkers:

$$g_\varsigma = \left(V_t + V_{gs}\right)\left\{1 - \frac{1}{b}\left[a\left(1 - e^{-bL_s}\right) - b\left(1 - e^{-aL_s}\right)\right]\right\}$$ (6.13)

Figure 6.7 shows the graphs of the density function $g_d(x)$ and cumulative separated grains $g_s(x)$ when the threshing unit losses $V_t + V_{gs} = 0.2$ (i.e., 20%). Notice that the fraction of the grains residing within the straw decreases exponentially, while the cumulative separated grain fraction increases and asymptotically tends to the $V_t + V_{gs}$ ordinate.

As x approaches $+\infty$, Equation 6.10 has the true limit

$$\lim_{x \to \infty} g_s(x) = V_t + V_{gs}$$ (6.14)

if, and only if, the rates satisfy the condition $a = 2b$, that is, the specific separation rate is half of the segregation rate.

Consequently, Equation 6.9 becomes

$$g_s(x) = \left(V_t + V_{gs}\right)\left(1 - e^{-bx}\right)^2$$ (6.15)

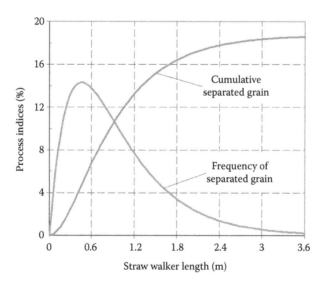

FIGURE 6.7
(See color insert.) Straw walker process indices.

Equation 6.15 was developed based on a comprehensive theory; it is very compact and simpler than other models known from the literature. The exponent b, which represents the specific separation rate, is proportional to the conditional probability of a kernel passage through an opening of a straw walker's screen.

Straw walker separation efficiency quantifies the separation intensity along the length of straw walkers. We define the *continuous separation efficiency* $s_{ef}(x)$ *of straw walkers* at the current position x of separation length as the ratio between separated grain mass on the differential interval dx and available grain mass to be separated to the rear of the threshing unit. This can be mathematically expressed as follows:

$$s_{ef}(x) = \frac{g_d(x)}{\left(V_t + V_{gs}\right) - \left(g_\varsigma + g_s(x)\right)} \tag{6.16}$$

Equation 6.16 is equivalent to the following equation:

$$s_{ef}(x) = \frac{g_d(x)}{g_r(x) - g_\varsigma} \tag{6.17}$$

Using Equation 6.9, the nonlinear regression analysis of available experimental data showed that $a = 1.9–4.54$ and $b = 0.9–2.26$. It can be noted that $a \approx 2b$, that is, Equation 6.15 is correct.

Using the experimental data published by Beck (1999, Figure 63) in his PhD dissertation, by regression analysis made with Equation 6.12, we found the values of coefficients a and b as shown in Table 6.1. The agreement of experimental data (wheat MOG federate = 4.1 kg/ (s·m) with model predicted data is shown in Figure 6.8, and the variation of corresponding process indices is illustrated in Figure 6.9.

The mathematical model described above is mainly intended as a simulation and design optimization tool when a minimum of experimental data was previously acquired.

Another simulation approach of grain separation through the straw was described by Lenaerts et al. (2010) considering discrete element modeling of bendable straw particles by segmentation of cylindrical elements. The results have been compared with the experimental data obtained by Beck (1992). The grain particles were assigned a spherical shape

TABLE 6.1

Results of Regression Analysis of Experimental Data Published by Beck (1999)

Wheat MOG Feedrate, kg/(s·m)	a	b	Coefficient of Determination, R^2
4.1	1.9334	1.0313	0.995
2.7	2.5097	1.2794	0.999
1.4	2.9306	1.4561	0.998

Source: From Beck, F., Simulation der Trennprozesse im Mähdrescher (Simulation of separation processes in combine harvesters), PhD dissertation, Hohenheim University, Forschritt-Berichte VDI, Reihe 14(92), 1999, figure 63.

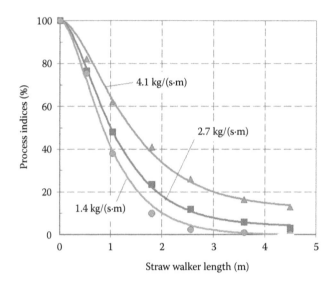

FIGURE 6.8
(See color insert.) Agreement of model-predicted data with experimental data published by Beck (1999). (Data from Beck, F., Simulation der Trennprozesse im Mähdrescher [Simulation of separation processes in combine harvesters], PhD dissertation, Hohenheim University, Forschritt-Berichte VDI, Reihe 14(92), 1999, figure 63.)

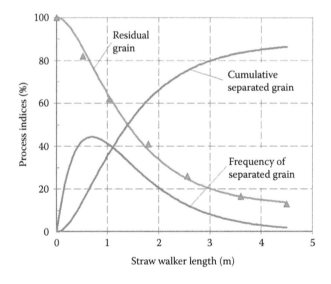

FIGURE 6.9
(See color insert.) Indices of grain separation process on straw walkers (experimental data published by Beck, 1999). (Data from Beck, F., Simulation der Trennprozesse im Mähdrescher [Simulation of separation processes in combine harvesters], PhD dissertation, Hohenheim University, Forschritt-Berichte VDI, Reihe 14(92), 1999, figure 63.)

even though in reality they are spherical. Although the method looks good, a series of trade-offs had to be made with simulation time, and only a qualitative assessment was intended. To quantitatively validate this method, larger numbers of simulation times and correct particle properties are required. We hope that the data from this book will help in such a matter.

6.4 MOG Motion on Straw Walkers

The motion of MOG on straw walkers is complex, being determined by the crankshaft angular speed, design parameters such as angles of the cascade walls, sawtooth profile dimensions and angles, number of walkers and their motion phase, and the properties and condition of the crop. On this matter, the Russian literature offers extensive theoretical studies.

On straw walkers, the MOG behaves as a body with a combination of elastic–plastic properties that allow the material during its movement to repetitively undergo a series of compression and elongation deformations. This kind of behavior is even more amplified when an agitating system is used on top of the straw walkers. One of the most used of these systems is the drum with retractable fingers. Readers are advised to consider its design and the theory of finger kinematic parameters, as explained in Section 4.2.3.

6.5 Straw Walker Design Considerations

The design of straw walkers must initially take into consideration the design specifications of the combine harvester (MOG throughput in different crops q_p, fraction of grain losses of the threshing unit $V_t + V_{gs}$, allowable grain loss of the straw walkers g_ς, etc.).

In the following, we describe some considerations and formulae for calculating the constructional and functional design parameters of straw walkers with shakers. The main design parameters are the length L_s, the total width W_s, the cascade angles c_a, the cascade height c_h, and the design parameters of the screen openings.

The length of straw walkers L_s can be calculated with Equation 6.13 based on the maximum allowable straw walker grain loss g_ς (<0.5%) and threshing unit losses $V_t + V_{gs}$ using computational numerical methods.

Alternatively, Equation 6.15 can be used for $x = L_s$; thus, the length of straw walkers can be calculated as follows:

$$L_s = \frac{1}{b} \ln \frac{1}{1 - \sqrt{1 - \dfrac{g_\varsigma}{V_t + V_{gs}}}}$$

(6.18)

The width W_s of straw walkers depends on the length of the tangential threshing unit; thus,

$$W_s = (1.1...1.2)L$$

(6.19)

where L is the length of the cylinder(s) of the tangential threshing unit.

Considering a number n_s of shakers, the width w_s of a shaker can be calculated with the following formula:

$$w_s = \frac{W_s - (n_s - 1)c}{n_s}$$

(6.20)

where $c = 5$–10 mm (0.2–0.4 in.) is the clearance between every two adjacent shakers. A tighter clearance helps to avoid the damaging of the grains that escape between the shakers, but it is technologically more difficult and expensive to achieve.

As specified in the beginning of this chapter, the number of the shakers $n_c = 3$–7.

Table 1.1 indicates the technical specifications of the straw walkers of several combine harvesters made by prestigious manufacturers.

References

Beck, F. 1992. Meßverfahren zur Beurteilung des Stoffeigenschaftseinflusses auf die Leistung der Trennprozesse im Mähdrescher [Measurement methods to assess the influence of material properties on the power for separation processes in combine harvesters]. PhD dissertation, Stuttgart University, Forschritt-Berichte VDI, Reihe 14(54).

Beck, F. 1999. Simulation der Trennprozesse im Mähdrescher [Simulation of separation processes in combine harvesters]. PhD dissertation, Hohenheim University, Forschritt-Berichte VDI, Reihe 14(92).

Böttinger, S. 1993. Die Abscheidungsfunktion von Hordenschüttler und Reinigungsanlage in Mähdrescher [The separation function on straw walkers and cleaning unit of combines]. PhD dissertation, Stuttgart University, VDI Verlag Düsseldorf.

Filatov, N.V. and P.G. Chabrat. 1967. Verstärkung der Körnerabscheidung auf Hordenschüttlern [Strenghtening of grain separation on the straw walkers]. *Landtechnische Forschung* (6):196–200.

Gregory, J.M. and C.B. Fedler. 1987. Mathematical relationship predicting grain separation in combines. *Transactions of the ASAE* 30(6):1600–1604.

Huynh, V.M. and T. Powell. 1978. Cleaning shoe performance prediction. ASAE Paper 78-1565. St. Joseph, MI: American Society of Agricultural Engineers.

Ivan, G. and S. Popescu. 2008. A new theory regarding the vegetal matter displacement on shaker straw walkers disposed on two axes at conventional cereal harvesting combines. Report 818-1616-1-SM. Bucharest: National Institute of Research and Development of Machines and Equipment for Agriculture and Food Industry.

Lenaerts, B., E. Tijskens, J. De Baerdemaeker, and W. Saeys. 2010. Simulation of grain-straw separation by a discrete element approach with bendable straw particles. Belgium: KU Leuven University.

Miu, P. and H.D. Kutzbach. 1997. Mathematical modeling of grain separation process over the length of straw walkers. ASAE Paper 97-1062. St. Joseph, MI: American Society of Agricultural Engineers.

Reed, W.B., G.C. Zoerb, and F.W. Bigsby. 1987. A laboratory study of grain separation. *Transactions of the ASAE* (3):452–460.

Steponavicius, D. and V. Butkus. 2012. The empirical model of grain separation on straw walker. *Journal of Food, Agriculture, and Environment* 10(3–4):296–302.

Vasilenko, I.F. 1954. *Kompendium der sowietischen Landmaschinentechnik.* [*Compendium of Soviet Agricultural Machinery Technique*]. Verlag Technik, Berlin.

Wang, G., G.C. Zoerb, and F.W. Bigsby. 1987. A new concept in combine separation analysis. *Transactions of the ASAE* 30(4):899–903.

Bibliography

Boyce, H.B., R.T. Pringle, and B.M.D. Willis. 1974. The separation characteristics of a combine harvester and a comparison of straw walker performance. *Journal of Agricultural Engineering Research* (19):77–84.

Kutzbach, H.D. 1996. Körnerfrüchternte [Grain harvesting]. *Jahrbuch Agrartechnik* 163–142.

Neculaiasa, V. and I. Danila. 1995. *Working Processes and Harvesting Machinery* [in Romanian]. Iasi, Romania: A92 Publisher.

Wessel, J. 1967. Grundlagen des Siebens and Sichtens [Fundamentals of sieves and screens]. *Aufbereitungstechnik* (13):167–180.

7

Cleaning Unit Process and Operation

7.1 Introduction

The grain–material-other-than-grain (MOG) mixtures separated through the concave/cage of the threshing unit and through the straw walker screens are delivered to the cleaning unit of the combine harvester by means of the *oscillating gain pan*, which moves the material rearward due to its oscillating movement.

The duties of the *cleaning unit* (*cleaning shoe*) of a combine harvester are as follows:

- Separate the free grains from the grain–chaff–small straw mixture and forward them to the grain auger.
- Separate small fragments of unthreshed fragments of ears, panicles, and so forth, called *tailings*, from the chaff–small straw mixture and advance them to the auger of the tailing return system to rethresh them.
- Discharge the MOG mixture on the soil or to the straw–chaff chopper/spreader.

The cleaning unit (Figure 7.1) generally consists of a grain pan, two sieves mounted one above the other, and a *fan* that blows air upward through the bottom of the sieves toward the rear of the combine harvester. Between the grain pan and the top sieve (also called *chaffer*), there is a winnowing step. The top sieve has openings whose size is adjustable, while the *bottom sieve* is changeable for different crops. The air blows the chaff and straw fragments off, while the grains separate through the openings of the sieves, and then an inclined wall moves them to the clean grain auger. At the end of the top sieve, there is a *chaffer extension* that favors the tailings through which to separate.

After *preliminary grain segregation* through the MOG mixture on the top of the grain pan, the *grain separation process* on the cleaning unit occurs due to the combination of the forces generated by the oscillatory motion in opposite phases of the sieves and the gravitational force. The *grain cleaning* occurs due to differences in terminal velocities, shapes, and dimensions of grain, ear fragments, and chaff. Table 3.5 indicates the dimensions of selected crop grains, while Tables 3.8 and 3.9 show the values of drag coefficients and terminal velocities for grains and MOG components of different crops.

The next section describes the key differences in the design and function of different cleaning units and their main components.

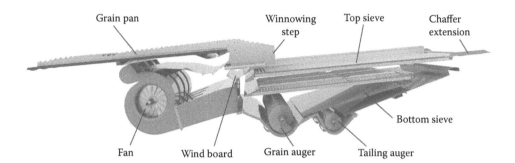

FIGURE 7.1
Main components of a cleaning unit. (Courtesy of CLAAS, Harsewinkel, Germany.)

7.2 Construction of Cleaning Units

7.2.1 General

In terms of design, there are cleaning units with different grain pans, with two or three sieves, and with fans of various shapes and drives. Then, the cleaning system may have a fixed mounting position or may be *tilted* to compensate the combine inclination when working on hillside fields.

Most combine harvesters have a regular grain pan or a smaller one, as shown in Figure 7.2. In the second case, the main transportation function of grain and the MOG mixture from the axial threshing unit is accomplished by a set of parallel augers due to the length of the axial threshing unit. This design is also used on John Deere combines, but the grain pan is inclined at a higher angle.

When a conventional combine has a series of tangential threshing units, *two grain pans* may collect and transport the separated material from the threshing units and straw walkers, respectively. Such a design (Figure 7.3) includes a set of three sieves (chaffer, middle

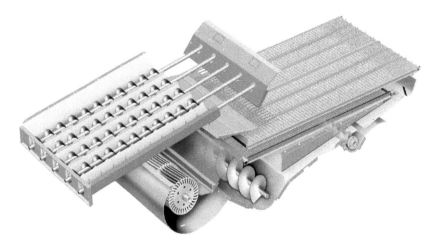

FIGURE 7.2
Cleaning unit with a short grain pan fed by a system of parallel augers. (Courtesy of Case IH, Racine, WI.)

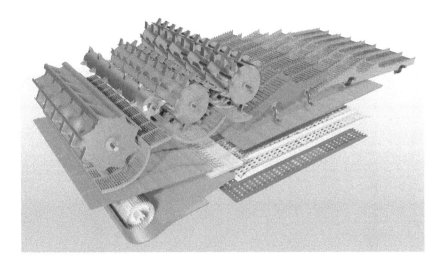

FIGURE 7.3
Cleaning unit with two grain pans and three sieves. (Courtesy of SAME Deutz-Fahr, Treviglio, Italy.)

sieve, and bottom sieve) and is used in the Topliner-3 combine series made by SAME Deutz-Fahr. A similar solution (Figure 7.4) is used in the axial-flow RTS-1 combine harvester of the same manufacturer.

The separation and cleaning processes performed by a cleaning unit are very sensitive to the tilting angle of the combine when harvesting on side slopes. This is due to the fact that the material slides to the downhill side of the sieves, and the airflow escapes on the upper-hill side. The gathering of material on the downhill side of the sieves has a negative effect on material transportation and grain segregation and separation, thus resulting in grain losses on the field. A higher flow of the air passes through the upper-hill side of the sieves,

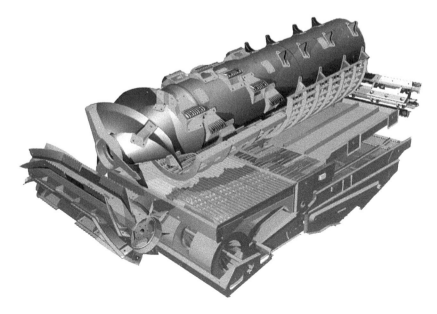

FIGURE 7.4
Cleaning unit with two grain pans under an axial threshing unit. (Courtesy of SAME Deutz-Fahr, Treviglio, Italy.)

FIGURE 7.5
Laterally tilted cleaning unit. (Courtesy of Case IH, Racine, WI.)

throwing out the grains, while separated grains are accompanied by a greater fraction of MOG. The sieves of a *tilted cleaning unit* rotate in the transversal plane on a combine harvester, as shown in Figure 7.5. This cleaning unit has a synchronous movement, contrasted with the combine frame rotation when the combine harvester works on side slopes.

The CLAAS combine design includes additional lateral oscillation of the sieves with respect to the cleaning unit frame that conveys the material to the upper side of the inclined sieve. Figure 7.6 synthesizes the kinematic schematics of cleaning units implemented on combine harvesters (including hillside combines) of different manufacturers worldwide

Passive crop-guiding systems		Slope-independent systems	
Slope guide	Deflection guide	Slope leveling	Rotary sieve
Several manufacturers		New Holland	Laboratory
Additional pneumatical forces		Additional mechanical forces	
From the side	From the bottom	Lateral oscillations	Conveyors
Allis chalmers	Laboratory	CLAAS	MDW

FIGURE 7.6
Synthesis of different kinematics designs of cleaning unit sieves when working on side slopes.

FIGURE 7.7
Grain pan general construction. (Courtesy of CLAAS.)

(adaptation of Figure 1.289, Kutzbach and Quick, 1999). On the hillside or sidehill combines, since the combine body is leveled for the side slopes, there is no need to separately adjust the inclination of the sieves.

7.2.2 Grain Pan

The grain pan of a cleaning unit has two tasks: *transportation* of the separated grain–MOG mixture from the threshing unit/straw walkers to the chaffer and *segregation* of the grain through this mixture on the pan's top surface. When segregated grains fall through the pan extension on the chaffer, their separation is accelerated, while the airflow is very efficient over the area of this winnowing step. The widespread design of the grain pan among combine harvester manufacturers is shown in Figure 7.7. The surface of the grain pan is usually divided by equidistant walls that are perpendicular to its surface. The wall roll is to prevent the lateral movement or agglomeration of the MOG and grain mixture, knowing that the airflow distribution in the transversal cross section is not uniform.

7.2.3 Chaffer and Bottom Sieve

The chaffer (top sieve) performs the separation of small fragments of straw from the grain–MOG mixture. The chaffer consists of a Graepel sieve (Figure 7.8) or a sieve with

FIGURE 7.8
Graepel sieve design.

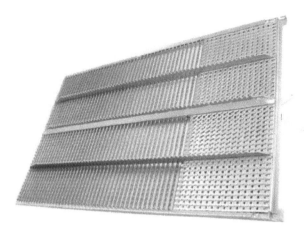

FIGURE 7.9
Common design of a chaffer with adjustable louvers.

adjustable louvers (Figures 7.9 and 7.10); the latter type of sieve is used by most combine manufacturers.

The adjustable louvers are mounted on rods whose ends can rotate with an angle between 0° and 45° into the sieve frame, thus varying the size of the sieve openings. The profile of adjustable louvers can be curved (Petersen type) or of a spline shape; they can be made of galvanized steel, stainless steel, or plastic. The length of the chaffer is 1.1–1.5 m (43.3–59 in.).

The bottom sieve of the cleaning shoe consists of a perforated sheet metal whose openings may have a round or rectangular shape. In some cases, a sieve with adjustable louvers can be used as well.

The sieves of the cleaning shoe have an inclination angle of $\alpha = 1°–7°$, with the lifted end of the sieve located at the exit where fragmented straw and chaff are discharged onto the field. In order to recover most unthreshed, fragmented ears, the chaffer extension can be inclined at an angle of $\alpha_1 = 7°–30°$.

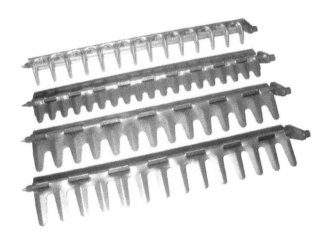

FIGURE 7.10
Different sizes of adjustable louvers. (Courtesy of A&I Products, Rock Valley, IA.)

One of the most important functional parameter of the sieves is the *Froude* number, which expresses the relationship between the forces that act on a particle lying on the sieve, the acceleration of sieve oscillations, and gravity. This number can be expressed as follows:

$$Fr_v = \frac{a\omega^2 \sin(\beta - \alpha)}{g \cos \alpha} \tag{7.1}$$

where:
- a is the amplitude of sieve vibration, m
- ω is the angular velocity of the driving shaft, s^{-1}
- β is the angle of oscillation direction
- α is the angle of sieve inclination
- g is the gravitational acceleration, m/s^2

The amplitude of oscillations is usually within the range of 0.015–0.040 m (0.59–1.575 in.), and angle $\beta = 23°–33°$ and the rotation of driving shaft $n = 200–340$ rpm. Thus, the value of the Froude number is $Fr_v = 1.4–2.7$.

The load of a chaffer with adjustable louvers is $q_c = 1.75–2.75$ kg/m^2/s, while for the bottom sieve $q_{bs} = 1.0–1.5$ kg/m^2/s. Uniform feeding of the cleaning shoe over the entire width is also very important for efficient grain separation and cleaning.

7.2.4 Fans

7.2.4.1 Fan Construction

The grain-cleaning process occurs on the cleaning unit due to pneumatic forces created by the airflow generated by the fan. The role of the airflow is to move the light components of the MOG mixture and evacuate them out of the combine. The pneumatic forces of the air must be higher than gravitational forces acting on MOG particles (chaff, fragments of straw, and leaves) and lower than the gravitational force acting on grain. Thus, a certain range of air velocity is required to simultaneously satisfy both conditions. Besides, the airflow should have an even distribution across the width of the sieves and decrease significantly from the front side to the rear side of the sieves to avoid throwing some grain on the field. Values of the air velocity should be 6–8 m/s (19.7–26.25 ft/s) in the winnowing steps, 5 m/s (16.4 ft/s) at the front of the chaffer, and less than 3 m/s (9.8 ft/s) at the end. The airflow direction in the winnowing steps should be approximately 30° to the horizontal, while on the sieve, it should be within the 20°–30° range.

The airflow required for grain cleaning is generated by a fan or several fans mounted on the same shaft driven through a driving transmission by the combine engine.

The fan types mostly used in a combine harvester cleaning unit are the centrifugal radial-flow fan and cross-flow fan. The axial-flow blowers may also be used; they could be oriented either in a *longitudinal-axial* position or, rarely, in a *cross-axial position* (Figure 7.11). The fan types are characterized by the path of the airflow through the fan.

Centrifugal fans are commonly used in applications with low to medium airflow rates at high pressures, being able to handle high-particulate airstreams, which includes the case of combine cleaning units.

The *radial-flow fan* (Figure 7.12) uses a rotating impeller to increase the velocity of the airstream. The impeller consists of a shaft mounted on two bearings, and four to six radial

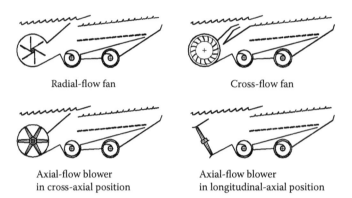

Radial-flow fan   Cross-flow fan

Axial-flow blower   Axial-flow blower
in cross-axial position   in longitudinal-axial position

FIGURE 7.11
Fan types of cleaning units.

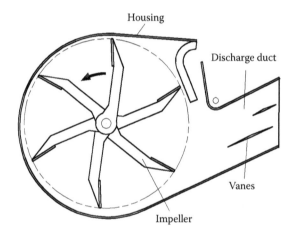

FIGURE 7.12
Centrifugal radial-flow fan.

blades mounted on radial arms. The impeller rotates into a sheet metal housing of cylindrical shape whose longitudinal axis does not coincide with the rotation axis of the impeller. On the housing there are air intake circular openings, but most of the airflow enters through the lateral sides. In some radial-flow fans, the airflow can be controlled by using a sliding throttle plate that changes the effective width of the impeller that is exposed to the airstream. The airstream then radially flows through the blades to the air discharge divergent duct, where one or two stabilizers (vanes) are mounted to modify the airflow direction. The straight-blade position can be radial or backward inclined; this position is characterized by the orientation angles α_1 and α_2 (Figure 7.13). By convention, the values of these angles are considered positive when the measuring direction from the blade to the radial direction coincides with the impeller rotation sense.

The design parameters of centrifugal radial-flow fans used in combine cleaning units have the following general values: outside diameter $D_o = 0.4–0.65$ m (15.75–25.75 in.), fan housing width $B_f = (1–2)D_o$, inside diameter $D_i = (0.35–0.5)D_o$, air intake opening diameter $D_{ai} = (0.62–0.83)D_o$, discharge duct height $h_d = (0.4–0.65)D_o$, $B_f/h_d = 2–5$, and impeller width $B_i = (0.98–0.99)B_f$. Regarding the straight-blade orientation, the angle values are $\alpha_1 = 0°–30°$

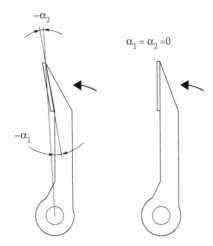

FIGURE 7.13
Orientation parameters of radial-flow fan blades.

and $\alpha_2 = 0°–10°$. The impeller rotation n_i must be continuously adjustable within the 500–1550 rpm range.

The *cross-flow fans* are highly efficient, being capable of generating high pressures at relatively low rotating speeds, in harsh operating conditions. The housing of a cross-flow fan (Figure 7.14) consists of a front guide and a rear guide divided by an open region (air intake) that extends over the entire length of the fan; then, in the air discharge duct, one or two stabilizers can adjust the direction of the airflow. The cross-flow fans have forward-curved blades relative to the rotation sense.

When the impeller rotates, the air is sucked through the air intake opening of the housing, then it transversally passes the inner side of the impeller, and then it is discharged through the duct. A cross-flow fan having forward-curved blades produces a relatively high dynamic pressure at low rotating speed because the air passes through the impeller blade twice. Low rotating speed translates to a low level of noise. Moreover, forward-curved blade fans do not require high-strength attributes due to relatively low rotating speeds.

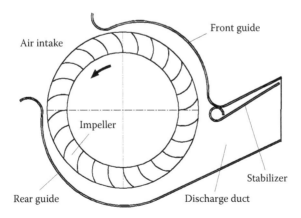

FIGURE 7.14
Model of a cross-flow fan.

The influence of the guides and the stabilizer design on fan performance is stronger than that of the impeller (Kim et al., 2008). Thus, their shape, as well as the air intake opening position, should be well considered during the fan design stage.

The design parameters of cross-flow fans used in combine cleaning units have the following general values: outside diameter D_o, inside diameter $D_i = (0.7–0.85)D_o$, and impeller width $B_i = (0.98–0.99)B_f$, where B_f is the fan housing width. Regarding blade orientation, the angle values are $\alpha_1 = 0°–20°$ and $\alpha_2 = 50°–65°$.

7.2.4.2 Background of the Fan Theory

Fan performance is typically indicated by a diagram of developed air pressure and power required over the range of generated airflow (Figure 7.15). Most fans have an operating region in which their curve slopes in the same direction as the system resistance curve. In this region, the fan operation may become unstable due to interaction between the fan and the system for which the airflow is generated. As a consequence, the fan attempts to generate more airflow, which leads to an increase of the system pressure, reducing the generated airflow. As airflow decreases, the system pressure also decreases, and the fan generates more airflow. Thus, the fan operation is characterized by a cyclic behavior that leads to poor fan efficiency and increases wear on the fan components.

In fan selection or design, there are several key pressure terms of interest, as follows: static pressure, dynamic pressure (also known as velocity pressure), and the total pressure that is the sum of the static and dynamic pressures. The static pressure p_s is the resistance opposed to the airflow (aided by the fan) that is independent of the air velocity. This is measured in inches of water gauge (wg) or pascals (1 in. wg = 249.089 Pa, 1 Pa = 1 N/m²).

The dynamic pressure p_d developed by the fan can be calculated with the following formula:

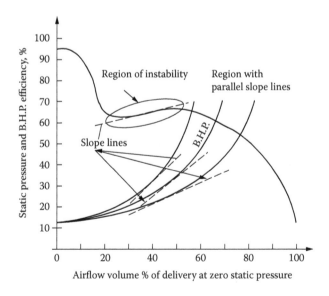

FIGURE 7.15
Characteristic curve of a forward-curved blade fan.

$$p_d = 9.807 v_a^2 \frac{\rho_a}{2g} \qquad (7.2)$$

where:

p_d is the dynamic pressure, Pa
v_a is the air velocity, m/s
ρ_a is the air density, $\rho_a = 1.226$ kg/m³ (0.075 lb/ft³ at 15°C)
g is the gravitational acceleration, $g = 9.807$ m/s²

Then, the total pressure is

$$p_t = p_s + p_d \qquad (7.3)$$

The air velocity in a duct can be calculated with the following formula:

$$v_a = 0.764 \sqrt{\frac{T p_d}{p_0}} \qquad (7.4)$$

where:

T is the temperature (K, i.e., °C + 273.15)
p_d is the dynamic pressure, Pa
p_0 is the barometric pressure, 101.325 KPa (absolute value)

Equation 7.4 can be simplified for standard conditions at 20°C and 101.325 barometric pressure, as follows:

$$v_a = 1.3 \sqrt{p_d} \qquad (7.5)$$

The volume of airflow rate V (L/s) can then be calculated with the following equation:

$$V = 1000 v_a A_d \qquad (7.6)$$

where A_d is the cross-sectional area of the discharge duct (m²).

The fan laws relate the performance variables, and they can also be written as similarity equations, as expressed below:

- The volume V of the airflow variation is proportional to the fan speed n (rpm):

$$\frac{V_2}{V_1} = \frac{n_2}{n_1} \qquad (7.7)$$

- The variation of the total differential pressure Dp_t is proportional to the square of the fan speed:

$$\frac{Dp_{t2}}{Dp_{t1}} = \left(\frac{n_2}{n_1}\right)^2 \qquad (7.8)$$

The total differential pressure can be calculated as follows:

$$Dp_t = p_{so} + p_{do} - p_{si} - p_{di} \tag{7.9}$$

where:

p_{so} is the static pressure at outlet, Pa (gauge)
p_{do} is the dynamic (velocity) pressure at outlet, Pa (gauge)
p_{si} is the static pressure at inlet, Pa (gauge)
p_{di} is the dynamic (velocity) pressure at inlet, Pa (gauge)

- The power required by a fan varies in proportion to the cube of the fan speed:

$$\frac{P_{f2}}{P_{f1}} = \left(\frac{n_2}{n_1} \right)^3 \tag{7.10}$$

The total fan efficiency η_t may be expressed as follows:

$$\eta_t = \frac{VDp_t}{1000P_{fi}} \tag{7.11}$$

where P_{fi} is the power input to the fan shaft (kW).

7.2.5 Fan Airflow Parameters and Position Relative to the Sieve

There are certain design and functioning requirements for a fan to handle the MOG feedrate on the sieves in connection with its position relative to the sieve length. The main requirements refer to the airflow rate, speed, discharge angles, duct shape design, dimensions, and position.

The magnitude of the airflow rate V (m³/s) generated by the fan should be correlated with the feedrate q_{pc} of light fragments of MOG that flow onto the cleaning unit sieves. This is a fraction of the entire MOG feedrate and can be calculated as follows:

$$q_{pc} = q_c + q_{ss} \tag{7.12}$$

where:

q_c is the chaff feedrate, kg/s
q_{ss} is the small straw and threshed ear feedrate, kg/s

The required airflow rate to be generated by the fan can be calculated with the following formula:

$$V = \frac{q_{pc}}{\gamma \rho_a} \tag{7.13}$$

where:

γ is the coefficient of gravimetric concentration of MOG particles in the airflow (nondimensional)
ρ_a is the air density, kg/m³

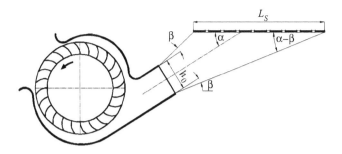

FIGURE 7.16
Fan position and airflow pathway relative to the sieve.

The coefficient of gravimetric concentration of MOG particles in the airflow can be quantified experimentally by the method of physical gravimetry. Its value is $\gamma = 0.2$–0.3.

The airflow speed should be uniformly distributed over the cross section of the fan discharge duct, and its vertical component must be higher than the terminal velocity of the chaff, and fragments of straw and threshed ears in the floating zone above the chaffer. Tables 3.8 and 3.9 show the values of drag coefficients and terminal velocities for grains and MOG components of different crops. Thus, the terminal velocity of the chaff is 1–5.15 m/s, and of straw of various lengths, 2.2–6.8 m/s. Moreover, at the exit from the discharge duct, the airflow spreads laterally within the angle $\beta = 12°$–$15°$ (Figure 7.16). As a consequence, depending on the fan type and its position relative to the sieve, the airflow speed at the exit from the duct should be 1.5–1.75 times higher than the terminal velocity of MOG particles in the floating zone over the chaffer. The angle α of the main direction of the airflow takes values within the 25°–32° range. The value of this angle should be correlated with the angle of adjustable louvers that forms the chaffer.

7.3 Modeling of Grain Separation on Cleaning Unit

7.3.1 Literature Review

Modeling of grain separation on the cleaning unit refers to the mathematical description and quantification of the frequency distribution and cumulative distribution of cleaned, separated grain through the sieves of the cleaning unit, which is subjected to the vibratory action of the sieves, forces generated by the pressure of the fan airflow, and gravitational forces that act on the grains. Related literature mentions different approaches that expand from physical measurements and empirical equations to analytical equation systems and, lately, stochastic models. Other different models describe the influence of main process variables on grain separation and grain loss of the cleaning unit (Damm, 1972; Zehme, 1972; Huynh and Powell, 1978; Freye, 1980; Hübner and Bernhardt, 1998).

It is very important to consider the grain movement within the MOG layer over the cleaning shoe. Regarding this subject, Freye (1980) performed comprehensive research on the influence of mechanical and pneumatic forces that determine the movement of the grain–MOG mixture on the chaffer. He determined an analytical solution for the grain motion on the chaffer, as well as for the influence of several factors, such as material velocity and material motion phases. The influence of the flight phase, fluidization phase, and

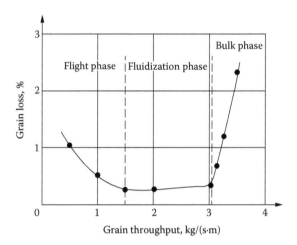

FIGURE 7.17
Influence of motion phases on grain loss in combine cleaning unit. (From Freye, Th., Untesuchungen zur Trennung von Korn-Spreu-Gemischen durch die Reiningungsanlage des Mähdreschers [Research on separation of grain-chaff mixtures in the cleaning unit of combine harvesters], dissertation, Universität Hohenheim, Forschungsbericht Agrartechnik MEG H. 47, 1980.)

bulk phase on the cleaning unit grain loss is shown in Figure 7.17. During the fluidization phase, the airflow disrupts the MOG layer, allowing an improvement of the layer penetration by grains. One of the most important factors that influences grain separation through a sieve is the opening ratio of the sieve (Bilanski and Dongre, 1968) surface, defined as

$$\eta = \frac{s_0}{s} \tag{7.14}$$

where:
s_0 is the mean surface of sieve openings
s is the mean surface of the sieve that contains an opening with the surface s_0

Beck (1999) and Kutzbach (2003) did a comprehensive inventory of mathematical models of grain separation in combine working units ranging from a simple exponential function with a constant separation rate to several exponential functions. Table 7.1 synthesizes such mathematical models that refer to the grain separation on a cleaning unit.

According to the screening theory (Wessel, 1967; Huynh and Powell, 1978, it is assumed that a free kernel will pass through an opening of the sieve if the projection of the kernel on the opening surface is within that area. Thus, the theoretical probability for a kernel to pass through an opening concave (sieve) is given by Equation 5.61, which relates to Figure 5.19. When successive contacts of a kernel with the sieve are possible, the resulting probability for the kernel to separate through an opening corresponding to the nth contact is

$$p_n = 1 - (1 - p)^n \tag{7.15}$$

The equations developed by Chrolikow (1974), Huynh and Powell (1978), and Böttinger (1993) (Table 7.1) for the remaining grain to be separated are quite similar. Our experience

TABLE 7.1

Review of Mathematical Models of Grain Separation (u_s) or Remaining Grain (r_s) on Combine Cleaning Unit

Contributor, Year	Grain Separation Model	Notes
Chrolikow, 1974	$r_s = \dfrac{1}{k_2 - k_1}\left(k_1 e^{-k_2 x} - k_2 e^{-k_1 x}\right)$	k_1, k_2: exponents
Huynh and Powell, 1978	$r_s = \dfrac{\tau_1 \tau_2}{\tau_2 - \tau_1}\left(\dfrac{1}{\tau_1} e^{-\frac{1}{\tau_2 v}} - \dfrac{1}{\tau_2} e^{-\frac{1}{\tau_1 v}}\right)$	v: speed, constant; τ: time constants
Kim and Gregory, 1991	$r_s = e^{-k_1 x\left(1 - e^{-k_2 x}\right)e^{-k_3}}$	k_1, k_2, k_3: exponents
Böttinger, 1993	$r_s = \dfrac{1}{b-a}\left(b e^{-\frac{a}{c+1}x^{c+1}} - a e^{-\frac{b}{c+1}x^{c+1}}\right)$	a, b, c: exponents
Beck, 1999	$u(x,t) = \dfrac{m_g}{\sqrt{4\pi D_x t}}\, e^{-\frac{x^2}{4D_x t}}$	m_g: grain mass; D_x: dispersion coefficient; t: time
Schreiber and Kutzbach, 2003	$r_s = \left(\dfrac{a}{c}x^c + 1\right)e^{-\frac{a}{c}x^c}$	a, c: exponents
Miu, 2003	$u_s = \left(1 - e^{-\lambda x^c}\right)$	c, λ: exponents

Note: x, current position along the sieve.

suggests that such functions might be better suitable for describing the separation frequency distribution underneath the considered sieve. The relationship developed by Kim and Gregory (1993) is hyperexponential, which suggests a rapid decrease of the remaining grain from the top of the sieve to be separated.

The need for at least two exponents can be explained by the nature of the grain separation process that results from the combination of grain segregation through the MOG–grain mixture flowing above the sieve and the grain separation of already segregated grain that reached the sieve surface and eventually finds an opening through the sieve.

Schreiber and Kutzbach (2003) conducted limit considerations for the equations developed by Böttinger and developed an equation of the $y = ax\,e^{-bx}$ type. Zhao (2002) used such equations for processing the experimental data. Figure 7.22 shows an example of modeling of experimental data with this model. However, the step gradient corresponding to the separation peak is not well predicted (Kutzbach, 2003).

Voicu et al. (2008) published their investigations on the adequacy of describing experimental data of grain separation by different, modified, statistical distributions as follows: normal, gamma, Weibull, and beta. According to this study, the best results were obtained with gamma and beta distribution since the correlation coefficient values were close to 1.

7.3.2 Beck's Model of Grain Separation on the Cleaning Shoe

The convection–diffusion model, published by Meinel and Schubert (1971), was further developed by Beck (1999). The stochastic grain movement through the MOG layer on the top of the sieve is the result of convection (by constant sinking) and diffusion (due to random scattering). The model of this process is well represented in Figure 7.18. In developing

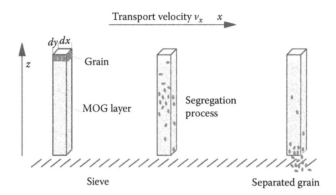

FIGURE 7.18
Model of grain separation through the MOG layer and chaffer of a cleaning unit. (From Beck, F., Simulation der Trennprozesse im Mähdrescher [Simulation of separation processes in combine harvesters], PhD dissertation, Hohenheim University, Forschritt-Berichte VDI, Reihe 14(92), 1999.)

the model, it was assumed that the grain is concentrated on top of the volume element of MOG layer. While the material moves over the sieve with a constant velocity v_x, the grain penetrates the MOG layer, and thus segregates until it reaches the top surface of the sieve; then the grain separates through the sieve openings. It was also assumed that the grain distribution within the MOG layer always follows a normal distribution whose shape evolves as shown in Figure 7.19.

Figure 7.20 shows the association of mathematical model functions to the physical model represented through a volume element of a grain–MOG mixture. The distribution function $u(z, t)$ (Table 7.1) is determined by the combination of a diffusion process characterized by a constant parameter D_z and a convection process characterized by an average sinking velocity v_z that takes place over the time t.

The grain distribution $u(z, t)$ follows the law of Fokker, Planck, and Kolmogorov (cited by Meinel, 1974; Beck, 1999) as follows:

$$\frac{\partial}{\partial t}u(z, t) = -v_z\frac{\partial u}{\partial z} + Dz\frac{\partial^2 u}{\partial z^2} \tag{7.16}$$

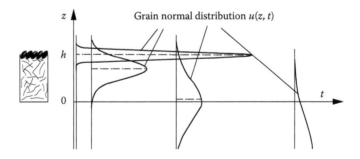

FIGURE 7.19
Qualitative representation of grain distribution transformation through convection and dispersion. (From Beck, F., Simulation der Trennprozesse im Mähdrescher [Simulation of separation processes in combine harvesters], PhD dissertation, Hohenheim University, Forschritt-Berichte VDI, Reihe 14(92), 1999.)

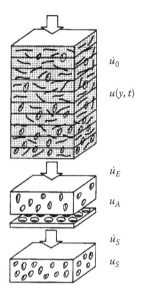

FIGURE 7.20
Association of mathematical model with physical model of grain segregation and separation. (From Beck, F., *Simulation der Trennprozesse im Mähdrescher* [Simulation of separation processes in combine harvesters], PhD dissertation, Hohenheim University, Forschritt-Berichte VDI, Reihe 14(92), 1999.)

The solution of Equation 7.16 leads to the normal distribution curve and has to satisfy the limit conditions as follows:

$$u(z, t = 0) = m_g \delta(z - h) \qquad \text{where} \quad \delta(z) = \begin{cases} 0 & \text{for } z \neq 0 \\ \infty & \text{for } z = 0 \end{cases}$$

$$u(z = \pm\infty, t) = 0 \tag{7.17}$$

where m_g is the total mass of the grain.

This means that the distribution function of the grain within the MOG layer is

$$u(z, t) = \frac{m_g}{\sqrt{4\pi D_z t}} e^{-\frac{(z - h - v_z t)^2}{4 D_z t}} \tag{7.18}$$

Separation of segregated grain on the top of the sieve is proportional to the mass of grain $u_A(t)$ that makes contact with the sieve, by a separation rate $1/t_A$. This means that the frequency of separated grain through the sieve is

$$\dot{u}_s(t) = \frac{1}{t_A} u_A(t) \tag{7.19}$$

The separation rate quantifies a time delay that depends on the sieve design, grain type and species, and opening ratio η of the sieve (see Equation 7.14).

FIGURE 7.21
Predicted grain distribution within the MOG layer (MOG mixture not displayed).

The model parameters for diffusion D_z, convection with the sinking speed v_z, and the separation rate were approximated by linear functions dependent on the parameters of the cleaning shoe.

Beck (1999) simulated the grain segregation through the MOG layer, as shown in Figure 7.21. At the feeding end of the sieve, the grain is considered to be concentrated in a thin layer on top of the MOG layer of height h, and then the grain penetrates the MOG layer with a random sinking velocity, which leads to various segregation times.

One of Beck's model deficiencies is the fact that, in reality, the grain is not distributed on top of the MOG layer, but within this layer. Readers should keep in mind that a random sinking velocity does not imply a constant speed v_x of the grain over the sieve length. The mathematical model described in the next section belongs to the author and attempts to eliminate the above-mentioned disadvantages.

7.3.3 Model of Grain Distribution and Separation on Cleaning Shoe

This section resumes the author's work (Miu, 2003) with additional explanations and further improvements.

The separated material from the threshing unit and straw walkers (if applicable) is composed of grain and MOG that consist of chaff, ear fragments (threshed or unthreshed), and straw and leaf fragments. This material is fed onto the cleaning shoe by an oscillatory grain pan whose movement is oppositely synchronized with the cleaning sieve motion. On the grain pan, grain presegregation may occur due to the following causes: vibratory motion of the grain pan and airflow blown by the fan in the winnowing step between the grain pan and the sieve (Figure 7.22). This is suggestively represented in Figure 7.23 (adaptation of Figure 29, Beck, 1999).

On a vibratory cleaning sieve, one can consider that within the MOG layer, a kernel (particle) jumps from one state to another with a particular transition rate. That is equivalent to the realization of a stochastic process of grain segregation with a specific segregation rate.

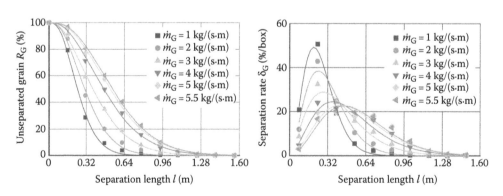

FIGURE 7.22
Prediction of remaining grain on the sieve and grain separation over the sieve length.

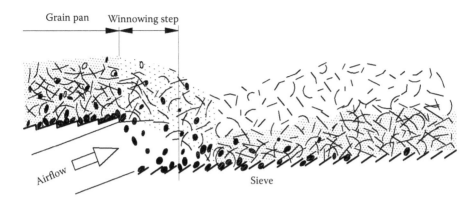

FIGURE 7.23
Grain presegregation in the winnowing step area.

When the grain reaches the top surface of the sieve, it may jump from a location to another, until it separates through an opening of the sieve. The overall process can be regarded as a continuous, bidimensional walk in the limit of small jumps. As a great number of such events are generated through the segregation/separation processes, the stochastic grain quantities of interest can be evaluated as ensemble averages.

Therefore, the overall movement of the grains within the MOG layer can be divided into the following sequential, at least two-by-two, components:

- Grain segregation to the top of the sieve (diffusion created by sieve vibrations)
- Grain transportation movement along the sieve
- Grain passing through the sieve openings

At the feeding end of the chaffer, the grains could be almost normally distributed within the MOG layer, over its height, displaying more or less a particular stage of segregation close to the chaffer top surface. Then, along the sieve, the *distribution $u(z, x)$ of the grains within the MOG layer* transforms to a distribution that is more and more skewed toward the MOG layer top. This is well represented in Figure 7.24. This transformation matches very well the *frequency (probability density function) of separated grain* through the sieve that

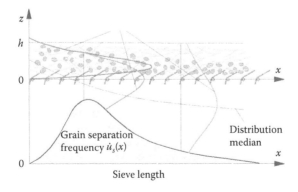

FIGURE 7.24
(See color insert.) Transformation of grain distribution $u(z, x)$ within the MOG layer, over the sieve length, to the grain separation frequency under the sieve.

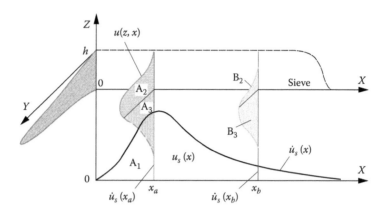

FIGURE 7.25
(See color insert.) Transformation of grain distribution $u(z, x)$ within the MOG layer to the grain separation frequency distribution $\dot{u}_s(x)$ under the sieve.

was determined from experimental data. Thus, the frequency of separated grain $\dot{u}_s(x)$ is perfectly described by *univariate Weibull distribution* (Equation 7.20) as a function over the length $0 \leq x \leq L_c$ of the chaffer, where x is the associated current position (Figure 7.25):

$$\dot{u}_s(x) = \lambda c x^{c-1} e^{-\lambda x^c} \tag{7.20}$$

where:
  λ is the scale parameter; $\lambda > 0$
  c is the shape parameter of the distribution; $c > 0$

The *fraction of the grain that is cumulatively separated* over an interval [0, x] can be calculated by integrating Equation 7.20:

$$u_s(x) = \int_0^x \lambda c \xi^{c-1} e^{-\lambda \xi^c} d\xi = 1 - e^{-\lambda x^c} \tag{7.21}$$

For instance, the fraction of separated grain $\dot{u}(x_a)$ at the location $x = x_a$ equals the area A_3 under the curve $u(z, x)$. That means

$$\dot{u}_s(x_a) = \int_0^h u(z, x_a) dz \tag{7.22}$$

If we integrate over the interval [0, x], we get the generalized form as follows:

$$1 - e^{-\lambda x^c} = 1 - \int_0^x \int_0^h u(z, x) \, dz dx \tag{7.23}$$

or, in a simplified form,

$$e^{-\lambda x^c} = \int_0^x \int_0^h u(z, x)\, dz\, dx \tag{7.24}$$

By deriving Equation 7.24 with respect to x, we get the following equivalent equation:

$$\lambda c x^{c-1} e^{-\lambda x^c} = -\int_0^h u(z, x)\, dz + k \tag{7.25}$$

where k is a constant from integrating the right-hand part of the equation.

Equation 7.25 describes the fraction of separated grain at any arbitrary location x ($0 \le x \le L_c$). It implies that the parameters of the function $u(z, x)$ are functions of x, λ, and c.

Grain separation frequency $\dot{u}_s(x)$ is proportional to the grain mass fraction $u_A(x)$, which is already segregated on the top surface of the sieve; we assume this proportionality is given by the decay rate λc. This leads to the following relationship (similar to Equation 7.19):

$$\dot{u}_s(x) = \lambda c u_A(x) \tag{7.26}$$

Consequently, the function that describes the *fraction of separable, segregated grain* $u_A(x)$ has the following form:

$$u_A(x) = x^{c-1} e^{-\lambda x^c} \tag{7.27}$$

At the end of the sieve, that is, where $x = L_c$, the separable, segregated grain becomes *sieve separation loss* $u_A(L_c)$, as expressed below:

$$u_A(L_c) = L_c^{c-1} e^{-\lambda L_c^c} \tag{7.28}$$

Within the cross section of the material layer on the sieve, at any position x along the sieve, the *equation of mass fraction balance* can be written as follows:

$$u_{ns}(x) + u_A(x) + u_s(x) = 1 \tag{7.29}$$

where $u_{ns}(x)$ is the *fraction of unsegregated grain* through the MOG layer, including the grain not segregated enough to reach the top surface of the sieve. Accordingly, this fraction of grain is

$$u_{ns}(x) = \left(1 - x^{c-1}\right) e^{-\lambda x^c} \tag{7.30}$$

It can be noted that although the exponent c must be higher than zero (per Weibull distribution definition), if $c = 1$, all separable grains are already segregated and reach the top of the sieve.

If $c \neq 1$, at the end of the sieve, the unsegregated grain becomes sieve *segregation loss* $u_{ns}(L_c)$:

$$u_{ns}\left(L_c\right) = \left(1 - L_c^{c-1}\right)e^{-\lambda L_c^c} \tag{7.31}$$

Segregation loss and separation loss represent the *total grain loss of the sieve* $u_L(L_c)$:

$$u_L\left(L_c\right) = u_A\left(L_c\right) + u_{ns}\left(L_c\right) = e^{-\lambda L_c} \tag{7.32}$$

Let us go back to Equation 7.25. The goal is to derive the expression of the function $u(z, x)$. Such a function must satisfy the following initial conditions:

$$u\left(z, x = 0\right) = \delta(h - z), \quad \text{where} \quad \delta(z) = \begin{cases} 0 & \text{for} & z = 0 \\ > 0 & \text{for} & 0 < z > h \\ 0 & \text{for} & z = h \end{cases} \tag{7.33}$$

$$\int_0^h h\left(z, 0\right)dz = 1$$

Then, this function should be defined for values $z < 0$ as well, to visualize the curve shape when sliding toward negative values of z (under the sieve). Instead of representing the function $u(z, x)$ in the coordinate system XOZ, it would be much more convenient to represent in the $\xi\Omega\zeta$ coordinate system the function $u(\zeta, \xi) = u(h - z, x)$, which results through a transformation series of reflection and sliding operations, as shown in Figure 7.26.

In order to derive the function $u(\zeta, \xi)$, a generalized bivariate Weibull distribution was considered by following the guidelines set up by Kundu (Dey and Kundu, 2012; Kundu and Gupta, 2014). The expression of the function $u(\zeta, \xi)$ is as follows:

$$u\left(\zeta, \xi\right) = \frac{a^2 b\left(c - b\right)\zeta^{b-1}e^{-a\zeta^{b-1}}}{1 - \exp^{-ah^b}}$$

$$a = 100e^{-\lambda\xi^c}, \qquad a > 0 \tag{7.34}$$

$$b = \xi^{c-1} + c - 1, \qquad b > 0$$

The denominator $\left(1 - \exp\left(^{-ah^b}\right)\right)$ is the complement of the survival function of the grain Weibull distribution along the ζ direction, at the top of the sieve ($\zeta = h$). Thus, the grain distribution within the MOG layer is inversely proportional to the complement of the survival function of the distribution. Physical interpretation of the denominator significance is as follows: if the segregated grain on the top of the sieve separates fast through the sieve, then the grain distribution within the MOG layer decreases fast and its curve displays a long tail.

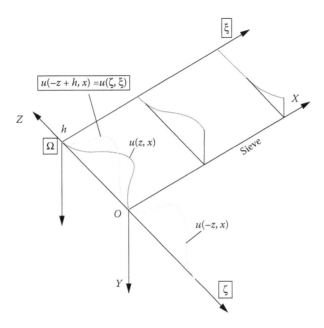

FIGURE 7.26
(See color insert.) Transformation of grain distribution $u(z, x)$ to the distribution $u(\zeta, \xi)$.

Figure 7.27 shows the graph of this function for a 1.6 m long sieve, $h = 0.1$ m, $\lambda = 1.1$, and $c = 2.45$. Note the Weibull distribution shape along and just above the sieve, as well as the grain flow in both directions within the layer of material.

Within the MOG layer, it is assumed that grain segregation follows a jump process as well as the physical laws of convection and diffusion (Meinel and Schubert, 1971; Beck, 1999; Beck and Kutzbach, 1995). Both above-mentioned assumptions are satisfied by the

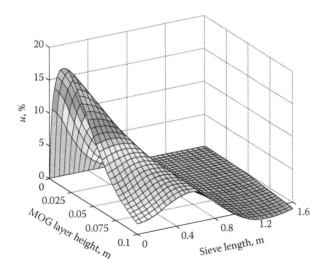

FIGURE 7.27
(See color insert.) Predicted grain distribution within MOG layer over the sieve length.

Fokker–Planck equation, which is a special type of master equation (van Kampen, 1992) in the limit of small jumps:

$$\frac{\partial}{\partial t} u(z, t) = -v_z \frac{\partial u}{\partial z} + D_z \frac{\partial^2 u}{\partial z^2} \tag{7.35}$$

where:
     v_z is the average sinking velocity (m/s)
     D_z is the constant of dispersion (m²/s)

Because we use the function $u(\zeta, \xi)$, Equation 7.35 can be applied as follows:

$$\frac{\partial}{\partial \xi} u(\zeta, \xi) = \frac{1}{v_\xi(\xi)} \left[-v_\zeta \frac{\partial u}{\partial \zeta} + D_\zeta \frac{\partial^2 u}{\partial \zeta^2} \right] \tag{7.36}$$

To find the variables of the process, we want to look at the stationary solution, which has to satisfy the following condition:

$$\lim_{x \to \infty} \frac{\partial}{\partial \xi} u(\zeta, \xi) = 0 \tag{7.37}$$

and hence,

$$\frac{d}{d\zeta} \left(-v_\zeta u + D_\zeta \frac{\partial u}{\partial \zeta} \right) = 0 \tag{7.38}$$

To satisfy Equation 7.38, the term

$$-v_\zeta u + D_\zeta \frac{\partial u}{\partial \zeta} = cst$$

must be constant. If we assume the stationary density $u_{st}(\zeta) \to 0$ for $\zeta \to 0$, then the constant in question must be zero, and we conclude that

$$v_\zeta = D_\zeta \frac{b - 1 - ab\zeta^b}{\zeta}$$

For comparison, nonlinear regressions of experimental data shown in Figure 7.22, using Equation 7.20, lead us to Figure 7.28. With coefficients of determination $R^2 > 0.99$, the regression results prove that the Weibull distribution describes perfectly the frequency of grain separation through the sieve.

This model represents an excellent simulation and design tool for saving time and costs during the testing and improvement of cleaning unit prototypes. For future work, the author recommends the analysis of the influence of constructional and functional parameters of the cleaning unit on the coefficients a and b, that is, λ and c.

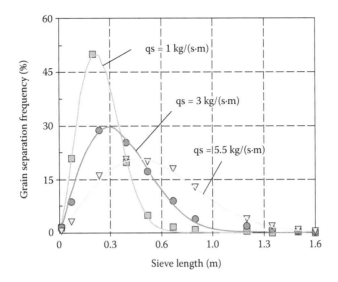

FIGURE 7.28
(See color insert.) Prediction of grain separation over the sieve (vs. Figure 7.22).

References

Beck, F. 1999. Simulation der Trennprozesse im Mähdrescher [Simulation of separation processes in combine harvesters]. PhD dissertation, Hohenheim University, Forschritt-Berichte VDI, Reihe 14(92).

Beck, F. and H.D. Kutzbach. 1995. Theoretische Untersuchungen zur Trennung von Korn/Stroh- und Spreu-Gemischen im Mähdrescher [Theoretical research on the separation of grain-straw-chaff mixtures in combine harvesters]. *Agrartechnische Forschung* 1(2):120–128.

Bilanski, W.K. and S.P. Dongre. 1968. Transporting wheat grain along the combine shoe. *Agricultural Engineering* 7:408–410.

Böttinger, S. 1993. Die Abscheidungsfunktion von Hordenschüttler und Reinigungsanlage in Mähdrescher [The separation function on straw walkers and cleaning unit of combines]. PhD dissertation, Stuttgart University, VDI Verlag GmbH Düsseldorf.

Chrolikow, W. 1974. Der technologische Vorgang bei der Trennung von Gemischen auf Schwingsieben. *Wissentschaftiches Zeitschrift* 23(2):431–433.

Damm, J. 1972. Der Sortiervorgang beim luftdurchströmten Schwingsieb. PhD dissertation. Stuttgart University, Fortschritt-Berichte VDI, Reihe 3(37).

Dey, A.K. and D. Kundu. 2012. Discriminating between the bivariate generalized exponential and bivariate Weibull distributions. *Chilean Journal of Statistics* 3(1):93–110.

Freye, Th. 1980. Untesuchungen zur Trennung von Korn-Spreu-Gemischen durch die Reiningungsanlage des Mähdreschers [Research on separation of grain-chaff mixtures in the cleaning unit of combine harvesters]. Dissertation, Universität Hohenheim, Forschungsbericht Agrartechnik MEG H. 47.

Hübner, R. and G. Bernhardt. 1998. Grundlagen zur rotierenden Reinigung im Mähdrescher. VDI/MEG-Tagung Landtechnik Garching. *VDI-Berichte* 1449:157–162.

Huynh, V.M. and T.E. Powell. 1978. Cleaning shoe performance prediction. ASAE Paper 78-1565. St. Joseph, MI: American Society of Agricultural Engineers.

Kim, S.H. and J.M. Gregory. 1991. Grain separation equations for combine chaffers and sieves. ASAE Paper 91-1603. St. Joseph, MI: American Society of Agricultural Engineers.

Kim, T.A., D.W. Kim, S.K. Park, and Y.J. Kim. 2008. Performance of a cross-flow fan with various shapes of a rear-guide and an exit duct. *Journal of Mechanical Science and Technology* 22:1876–1882.

Kundu, D. and A.K. Gupta. 2014. On bivariate Weibull-geometric distribution. *Journal of Multivariate Statistics* 123:19–29.

Kutzbach, H.D. 2003. Approaches for mathematical modelling of grain separation. Presented at the ASAE International Conference on Crop Harvesting and Processing, Louisville, KY, February 9–11.

Kutzbach, H.D. and G.R. Quick. 1999. Harvesters and threshers. In *CIGR Handbook of Agricultural Engineering*. Vol. III. *Plant Production Engineering*, ed. B.A. Stout and B. Cheze. St. Joseph, MI: American Society of Agricultural Engineers pp: 311–347.

Meinel, A. 1974. Klassierung auf Stößelschwingsiebmaschinen. VEB Deutcher für Grundstoffindustrie. PhD dissertation, Bergakademie Freiburg, Freiberger Forschungsheft A537.

Meinel, A. and H. Schubert. 1971. Zu den Grundlagen der Feinsiebung. *Aufbereitungstechnik* 12(3):128–133.

Miu, P. 2003. Stochastic modeling of separation process on combine cleaning shoe. Presented at the ASAE International Conference on Crop Harvesting and Processing, Louisville, KY, February 9–11.

Schreiber, M. and H.D. Kutzbach. 2003. Modellierung des Abscheideverhaltens von Mähdrescher-reinigungsanlagen. *Landtechnik* 58(4):236–237.

van Kampen, N.G. 1992. *Stochastic Processes in Physics and Chemistry*. Vol. 2. Amsterdam: North-Holland.

Voicu, G., T. Casandroiu, and C. Tarcolea. 2008. Testing stochastic models for simulating the seeds separation process on the sieves of a cleaning system and a comparison with experimental data. *Agriculturae Conspectus Scientificus* 73(2):95–101.

Wessel, J. 1967. Grundlagen des Siebens und Sichtens. Teil II, Das Siebklassiern. *Aufbereitungs-Technik* 13(4):167–180.

Zehme, C. 1972. Zur Entmischung einer homogenen Korn-Stroh-Spreu-Schüttung. *Deutsche Agrartechnik* 22(6):267–270.

Zhao, Y. 2002. Einfluss mechanischer und pneumatischer Parameter auf die Leistungsfähigkeit von Reinigungsanlagen im Mähdrescher. Forschungsbericht Agrartechnik VDI-MEG 387, PhD dissertation, Universität Hohenheim, Stuttgart.

Bibliography

Beck, F. 1992. Meßverfahren zur Beurteilung des Stoffeigenschaftseinflusses auf die Leistung der Trennprozesse im Mähdrescher [Measurement methods to assess the influence of material properties on the power for separation processes in combine harvesters]. Dissertation, Stuttgart University, Forschritt-Berichte VDI, Reihe 14(54).

DOE. 1989. *Improving Fan Systems Performance: A Sourcebook for Industry*. Prepared for U.S. Department for Energy by Lawrence Berkeley National Laboratory, Washington, D.C., and Resource Dynamics Corporation, Vienna, VA.

Jiang, S., W.K. Bilanski, and J.H.A. Lee. 1984. Analysis of the separation of straw and chaff from wheat by an air blast. *Canadian Agricultural Engineering* 26(2):181–187.

Neculaiasa, V. and I. Danila. 1995. *Working Processes and Harvesting Machinery* [in Romanian]. Iasi, Romania: A92 Publisher.

Witt & Son. 2013. Ventilator Grundlagen. IGW Ventilatoren Katalog. Pinnenberg, Germany: Witt & Son.

8

Grain Conveying Process and Equipment

8.1 Introduction

In most combine harvesters, conveying of the grains is needed in the following zones:

- Within the cleaning unit for directing the grain or unthreshed ears to the corresponding auger
- From the cleaning unit to the threshing unit for returning unthreshed grain in the ears
- From the cleaning unit to the grain tank for temporary storage of grains
- From the grain tank to discharging them into the transportation vehicle

Grain flow rate, conveying distance, incline, available space, efficiency, work quality, and economics influence the type, design, and operating parameters of the conveying equipment. This section provides an overview of grain conveying equipment design, operating parameters, and the process modeling approach in the design of combine harvesters.

Before designing any conveying system, the practitioner should consider the physical characteristics of the grains to be handled. In this regard, the most relevant properties are grain size, bulk density, moisture content, coefficient of friction with different materials, and other mechanical properties, such as compression stress, modulus of elasticity, and Reynolds number. These physical property values are given in Tables 3.1, 3.5, 3.6, and 3.10. Grains are relatively lightweight particles, free flowing, noncorrosive, and nonabrasive. According to the U.S. Conveyor Equipment Manufacturers Association, grains are classified as class I products.

Within the cleaning unit, conveying the grains and unthreshed ears is performed by the inclined walls of the cleaning shoe structure. The wall incline and harmonic motion determine the movement of the grains/unthreshed ears from the sieve/chaffer extension to the corresponding transversal auger, which collects and conveys them further. The oscillatory motion is induced by means of a crank–conrod mechanism; thus, the inclined walls and the sieves have similar kinematic indices characterized by the Froude number (see Equation 7.1).

For conveying the grains from the cleaning unit to the grain tank, and for discharging them from the grain tank, grain augers are used. For conveying the grains/unthreshed ears, scraper or bucket elevators are also used. We will refer to these conveyor types in the following subsections.

8.2 Auger Elevator

The auger elevator is also known as the screw elevator. In construction of agricultural harvesting machinery, it is used for the conveying of grains, ears, unthreshed ears, chaff, and chopped vegetal material. The auger conveyors have the advantages of compact design, low maintenance, and the ability of continuously conveying materials. Their disadvantages include their limited length and inability to transport sticky and lumpy materials (Shimizu and Cundall, 2001).

An auger elevator is a conveyor that moves the material from one elevation to a higher one, at a certain angle. It consists of a circular tube (barrel) with inside diameter D_b, in which a helical flight with external diameter D, rigidly mounted on a driving shaft with the diameter d, rotates with n rpm (Figure 8.1). The flight is made of a sheet metal of thickness e, and wraps up the shaft with the pitch p. The driving shaft is usually supported at its ends by two radial-axial bearings (not shown).

Practically, there is a radial clearance δ between the exterior side of the flight and the inside surface of the barrel, that is,

$$\delta = \frac{D_b - D}{2}, \text{ m} \tag{8.1}$$

The height h_o of the flight is

$$h_o = \frac{D - d}{2}, \text{ m} \tag{8.2}$$

The principle of conveying materials relies on the friction between the materials, the flight, and the barrel. An auger elevator is a volumetric conveying device. The volume of material that is conveyed in a given time depends on the material nature and properties, and on constructional and operational parameters, such as flight height h_o, flight pitch p, rotation n, incline α, and the degree ψ of filling of the flight channel with material.

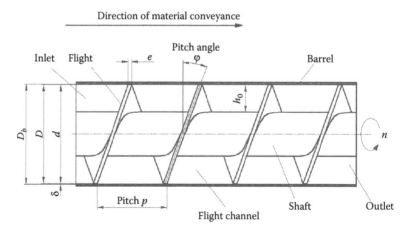

FIGURE 8.1
Design representation and parameters of a grain auger.

The performance and quality indices of an auger elevator are material feedrate q_m, conveying efficiency ε_c, and the power P required at the auger shaft.

The material feedrate q_m can be calculated with the following formula:

$$q_m = p\omega\gamma\psi \frac{D^2 - d^2}{8}, \text{ kg/s} \tag{8.3}$$

where:

ω is the angular speed of the auger, s^{-1}:

$$\omega = \frac{2\pi n}{60} \tag{8.4}$$

γ is the volumetric mass of the conveyed material, kg/m^3

The other design parameters are expressed in meters.

The filling coefficient ψ depends on the incline α of the auger conveyor as given by the following equation, which we determined using experimental data published by Verclyte (2013):

$$\psi = e^{-\frac{1}{\psi_0}\left(\frac{\alpha}{180}\right)^c} \tag{8.5}$$

where:

ψ_o = 0.4 is the filling coefficient of the flight channel of a horizontal auger conveyor
c = 1.53 is a coefficient

For grain conveying, here are the ranges of design parameter values: D = 100–300 mm (2–12 in.), d = 25–90 mm (1–3.5 in.), p = (0.6–1.25 D), and δ = 6–12 mm (0.25–0.5 in.). The length of the feeding section l_0 is equal to or bigger than the pitch p.

The *conveying efficiency* of an auger conveyor is defined as the ratio between the actual feedrate and the theoretical feedrate that results when the flight channel is completely full and the particles are conveyed with the axial speed of the auger (without slip or rotation). The efficiency decreases as the clearance between the barrel and screw flight increases.

Cyclic variations due to screw flight rotation, particle buildup on the auger, blockage in the auger feeding section, and an underpowered drive may lead to feedrate fluctuations and, consequently, to variable and lower efficiency (Dai et al., 2012).

The *power P* (kW), which is required for driving the auger, can be calculated as follows:

$$P = \frac{q_m \left(L\lambda + h\right)}{102} \tag{8.6}$$

where:

L is the total length of the conveyor, m
λ is the material resistance coefficient
h is the effective elevation at which the material is conveyed, m

The λ values are within the range 1.2–4.2.

The radial clearance δ is strongly influenced by the shape factor of the grains and the friction coefficient, but less so by the relative roundness of the flight edge, to commonly be within limits smaller than 0.4 mm (Rademacher, 1981).

8.2.1 Modeling of Conveying Process of Auger Elevator

Modeling of the grain conveying process performed by a grain auger is based on studying the behavior of the grain in the auger. Three methods have been used: discrete element method (DEM), computational fluid dynamics (CFD), and residence time distribution (RTD).

The *DEM* approach models each particle of a granular mixture as a separate entity; the particles interact among themselves and with conveyor surfaces through collisions or lasting contacts. DEM simulation keeps track of the position, velocity, and acceleration of the particles and reference surfaces, thus simulating the dynamic behavior of a granular flow. During simulation, the auger rotation, incline, and degree of flight channel filling with material are varied. Owen and Cleary (2009a,b) found two types of flow pattern: one pattern is exhibited at low angles and low filling degrees, while the other pattern occurs at higher angles and filling degrees. The research results displayed in Figure 8.2 come from simulation of superquadrics particles (SQBs) with the general form

$$x^N + \left(\frac{y}{B}\right)^N + \left(\frac{z}{C}\right)^N = s^N \tag{8.7}$$

that describes nonspherical particles. The power N is the shape factor, to smoothly change the shape from a sphere to a cube as N increases. B and C are the aspect ratios of the particle's first and second minor axes to the major axis. The authors found that increases in nonsphericity of the particles have a negligible effect on the particle flow patterns. The particle velocities and their axial and tangential components were invariant to changes of particle shape and friction between particles or with the auger wall.

The DEM method has two major disadvantages: (1) a large computational load at high particle concentrations and (2) difficulty in implementation of physical properties of particles, such as form and elasticity. In the next discussed method (CFD), all particles can be represented by one fluid.

The *CFD* method uses algorithms to analyze the fluid flow by simulating the interaction of the fluid with surfaces defined by boundary conditions. The laws of three fundamental principles are involved: conservation of mass (continuity), conservation of energy, and Newton's second law (dynamics). The obtained solution is as good as the initial/boundary

 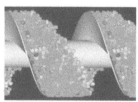 

| SQB super-quads; angle 0° | SQB super-quads; angle 30° | SQB super-quads; angle 60° |

FIGURE 8.2
(See color insert.) Influence of the auger inclination angle on the particle flow pattern. (From Owen and Cleary, 2009a and b.)

conditions considered for the mathematical model. The author recommends the CFD component of the ANSYS or COMSOL software package to perform CFD simulations.

When solving the CFD models, the software uses the finite element method with adaptive meshing (if selected) and error control. Predefined formulations of different types of flow make possible the use of physical quantities such as volumetric mass and viscosity of the material. Then, the software allows tracking the trajectories of particles based on Newton's second law.

In ANSYS, the rheology of a mixture of grains analyzes their complex behavior in transport or fluidized beds. The main factors governing the grain behavior are the volume fraction and shear rates. The Eulerian granular approach is based on a continuum, multi-dimensional model. Different particle models (DPMs) are available, such as a DPM for the dilute phase (steady and time dependent), the particle-tracking model with a probabilistic collision model, a dense-phase DPM for dense flows with large size distributions, and a macroscopic particle model for large particles.

The *RTD* approach deals with a probability distribution function that describes the time a particle (grain) spends in the conveyor, thus investigating the mixing and flow pattern. This method also facilitates the scale-up design of conveying equipment and optimization of process parameters.

For more reading on conveying process modeling, although not integrally applicable, practitioners are directed to Verclyte's thesis (2013), while becoming proficient in working with ANSYS or COMSOL software.

8.3 Scraper Elevator

In combine harvesters, one scraper elevator is used for lifting the grains from the cleaning unit grain auger to the bubble-up auger that, in turn, transfers the grains to the grain tank, while another scraper elevator is used to transfer the unthreshed ears from the cleaning unit tailing auger to the threshing unit for reprocessing them.

A scraper elevator is comprised of an endless metallic chain that is wrapped around a driving sprocket (at the lower end of the elevator) and around a tensioning (driven) sprocket at the upper end of the elevator. The chain supports a series of bulk material paddles (Figure 8.3) that lift the material from the lower opening and project it toward the discharge zone at the upper end of the elevator.

The paddles are usually made of plastic or rubber with fabric insertion layers. They can be flat or curved. The paddle shape is rectangular (Figure 8.4), with the following dimensions: $a = 32$–120 mm (1.25–4.75 in.), $b = 75$–300 mm (3–12 in.). The paddle pitch $p = 200$–500 mm (8–20 in.). The trunk is usually made of galvanized steel assembled as an integral staggered seam construction to provide a flush connection throughout the interior of the conveyor without use of welding and easy disassembly.

Usually, the paddle elevator conveys the material on the bottom side of the trunk. In this case, the material is easily discharged. If the grains are conveyed on the upper side, there is a flume to support the material during the conveying process. The inclination angle of the grain paddle elevator is $\alpha = 25$–$90°$ to the horizontal. The chain speed is within $v_c = 0.6$–2 m/s (2–6.5 ft/s).

Design and functional parameters are selected based on the nature and condition of the material, feedrate, length, and inclination of the conveyor.

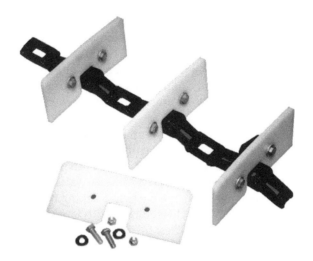

FIGURE 8.3
Paddles assembled on the chain and hardware. (Courtesy of May Wes, Hutchinson, MN.)

The material feedrate q_m of a paddle conveyor can be calculated with the following formula:

$$q_m = \frac{a^2 b v_c \gamma \psi}{2p \tan \alpha}, \text{kg/s}$$ (8.8)

where:
  γ is the volumetric mass of the conveyed material, kg/m³
  φ is the coefficient of filling the space between paddles with material

When the material is conveyed on the lower branch, the length of the discharge section depends on the maximum linear speed of the conveyor and the inclination angle. The

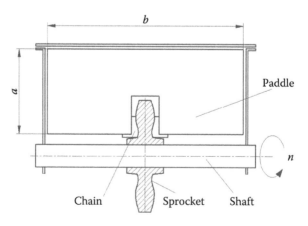

FIGURE 8.4
Shape and geometry of a paddle mounted on a chain.

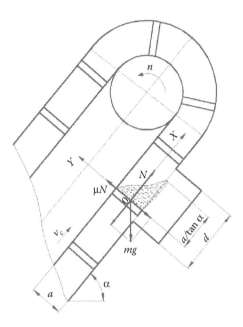

FIGURE 8.5
Calculation schematic for discharge section length.

design condition is that the length of the discharge section should allow the complete release of the material that is contained between each two consecutive paddles. The conveyor parameters to be considered in the design of the discharge section length are shown in Figure 8.5. The conveyor speed is v_c, and it is inclined with the angle α. Let us associate a Cartesian coordinate system *XOY* to the active face of a paddle.

We consider that the coefficient of the conveyor filling with material is 1; that means the material spreads on a length $a/\tan α$. During discharging, the following forces act on the material: gravitational force *mg*, conveying force *N*, and friction force μ*N* between the material and the paddle surface.

If we apply Newton's second law, we can write the following equation system:

$$N = mg \sin α$$

$$ma = mg \cos α - μN$$

$$a = \frac{d^2s}{dt^2} \tag{8.9}$$

where *a* is the material acceleration on its falling trajectory.

By integrating two times the second equation with respect to the time *t*, we get the equation of material speed v_s and the space *s* the material travels, as follows:

$$v_s = gt\left(\cos α - μ \sin α\right)$$

$$s = \frac{gt^2}{2}\left(\cos α - μ \sin α\right) \tag{8.10}$$

The length d of the discharge section must satisfy the following functional condition: the time t in which the material flows along the paddle height a must be less than or equal to the time in which the paddle travels the distance d; mathematically, this translates to the following equation:

$$d = v_c \left(\frac{2a \cos \phi}{g \cos(\alpha + \phi)} \right)^{1/2} \tag{8.11}$$

where the material repose angle

$$\phi = \arctan \mu \tag{8.12}$$

The *power P* (kW), which is required for driving the auger, can be calculated with Equation 8.6. The theoretical power should be increased by 10% to overcome the losses due to friction between conveyor parts, and power drive and transmission efficiency losses.

8.4 Bucket Elevator

Bucket elevators are the most used conveying systems for transporting the grains vertically or at high inclination angles. They require limited horizontal space, and because they are enclosed in the housing, they spread a low quantity of dust. A bucket elevator (Figure 8.6) consists of buckets attached to a belt transmission and transports the grain from the inlet hopper to the discharge spout. The driving pulley is always at the top. The buckets can be loaded directly from the inlet hopper, or from the back, or from both sides. The material is discharged by means of either gravity or centrifugal force. Thus, two design methods are used for bucket discharge: low-speed gravitational discharge or high-speed centrifugal

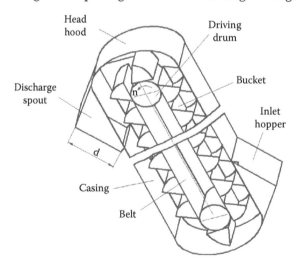

FIGURE 8.6
Components of a bucket elevator.

discharge. The first method allows low conveyor speeds (0.5–1.0 m/s), and it takes advantage of the material weight. The centrifugal discharge is the most common, and allows high conveying speeds (1.2–1.4 m/s); the separation distance between the buckets is two to three times the bucket height.

The weight w_m (kg) of the material conveyed by a bucket is

$$w_m = \psi V_b \rho_m \tag{8.13}$$

where:
  ϕ is the coefficient of bucket volume filling with material
  V_b is the bucket volume, m³
  ρ_m is the bulk material density, kg/m³

The bucket conveyor capacity q_c (kg/s) is then calculated as follows:

$$q_c = \frac{w_m v_c}{p} \tag{8.14}$$

where:
  v_c is the conveyor speed, m/s
  p is the bucket pitch, m

The standard pitch is $p = (2–3)\, h_b$, where h_b is the bucket height (m). For a flake bucket, $p = h_b$.

The *power P* (kW), which is required for driving the auger, can be calculated with Equation 8.6. The theoretical power should be increased by 10% to overcome the losses due to friction between conveyor parts, and power drive and transmission efficiency losses.

Buckets can be made of different materials, in various shapes and sizes. The common shape of a bucket is shown in Figure 8.7. Although the bucket elevators are less used in construction of combine harvesters, they are a good alternative to consider.

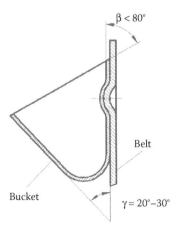

FIGURE 8.7
Common bucket configuration.

References

Dai, J., H. Cui, and J.R. Grace. 2012. Biomass feeding for thermochemical reactors. *Progress in Energy and Combustion Science* 38(5):716–736.

Owen, P.J. and P.W. Cleary. 2009a. Screw conveyor performance: Comparison of discrete element modelling with laboratory experiments. Presented at the Seventh International Conference on CFD in the Minerals and Process Industries CSIRO, Melbourne, Australia, December 9–11.

Owen, P.J. and P.W. Cleary. 2009b. Prediction of screw conveyor performance using the discrete element method (DEM). *Powder Technology* 193(3):274–288.

Rademacher, F.J.C. 1981. On seed damage in grain augers. *Journal of Agricultural Engineering Research* 26:87–96.

Shimizu, Y. and P.A. Cundall. 2001. Three-dimensional DEM simulation of bulk handling screw conveyors. *Journal of Engineering Mechanics* (9):864–872.

Verclyte, R. 2013. Mass and heat transfer modelling in screw reactors. Master's thesis, Ghent University.

Bibliography

Anderson B., R. Anderson, L. Håkansson, et al. 2011. *Computational Fluid Dynamics for Engineers*. Cambridge, UK: Cambridge University Press.

Boac, J.M., M.E. Casada, R.G. Maghirang, and J.P. Harner III. 2012. 3-D and quasi-2-D discrete element modeling of grain comingling in a bucket elevator boot system. *Transactions of the ASABE* 55(2):659–672.

Bolat, B. 2012. Increasing of screw conveyor capacity. *Journal of Trends in the Development of Machinery* 16(1):207–210.

Labiak, J.S. and R.E. Hines. 1999. Grain handling. In *CIGR Handbook of Agricultural Engineering*, ed. F.W. Bakker-Arkema. Vol. IV. St. Joseph, MI: American Society of Agricultural Engineers.

Ma, J. and M. Srinivasa. 2008. Particulate modeling in ANSYS CFD. Presented at the 2008 International ANSYS Conference, Canonsburg, PA.

Neculaiasa, V. and I. Danila. 1995. *Working Processes and Harvesting Machinery* [in Romanian]. Iasi: Romania: A92 Publisher.

Rosentrater, K.A. and G.D. Williams. 2004. Design considerations for the construction and operation of grain elevator facilities. Part II. Process engineering considerations. Presented at the ASAE/CSAE Annual International Meeting, Ottawa, ON.

9

Crop Residue Chopping and Spreading Processes and Equipment

9.1 Introduction

Modern, large combine harvesters discharge a higher flow of material other than grain (MOG) on a small-width swath relative to the header width. The weed and grain loss are thus concentrated behind the combine. The axial-flow harvesters process the vegetal material more intensively than the conventional combines; if the postharvest residues are left behind the combine and are shallowly incorporated into the soil, the accumulation of organic matter in the upper soil layer becomes a problem due to their insufficient decomposition and conversion to humus.

Field studies showed that distribution of nitrogen resulting from MOG left behind the combine is uneven, as shown in Figure 9.1.

Cereal crop residues (straw, husk, chaff, corn cobs, hulls, and some cornstalk pieces) are competitively regarded as very important organic fertilizers, organic materials for bedding and forage in the livestock industry, and for energy production from direct combustion and biogas production.

When not collected, managing the crop residue that results during cereal harvesting is very important for any consecutive seeding process, especially a reduced or no-tillage process. Heavy harrowing of the soil often groups the straw and corn husks in bunches that may not be properly incorporated into the soil; thus, all these residual MOG components, if not previously chopped and uniformly spread, do not disintegrate into the soil to release useful nutrients by mineralization. Success in uniformly distributing crop residue translates to better soil erosion protection and improved stand establishment.

It is believed that single-pass harrowing or cultivating of the soil is worse than no tillage at all because weed seeds are incorporating into the soil and loosen the stubble. Thus, failure to manage crop residue negatively influences the seeding process efficiency, leading to uneven distribution, hairpinning, plugging, lack of soil aeration and drying, and ultimately, low and nonuniform germination and emergence of the next crop seeds. Plants that emerge later than the neighboring plants may be outcompeted throughout their development, and crop yield may be reduced proportionately to plant emergence delay.

Modern combine harvesters are equipped with chopping–spreading equipment that chops and spreads the straw, husks, and chaff uniformly across the header cutting width.

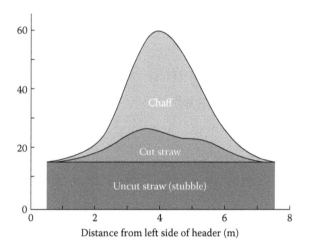

Distance from left side of header (m)

FIGURE 9.1
Amount of nitrogen (kg/ha) from decomposition of crop residue left behind the combine. (From Douglass Jr., C.L. et al., *Soil Science Society of America Journal* 56(4):1171–1177, 1992.)

Such a system has to satisfy a series of technological and process efficiency requirements, as follows:

- Produce consistent and uniform short chop length
- Uniformly spread the chopped material over the entire working width of the combine
- Generate a relatively low quantity of dust
- Consume low driving power
- Have a low level of maintenance

Material resistance to cutting depends on the crop MOG physical and mechanical properties, chopper knife design and kinematics, and feeding and discharging path of the material. The next section describes the construction of chopping–spreading systems that equip the combine harvesters.

9.2 Construction of Choppers and Spreaders

A chopping and spreading system may execute the process with a single working unit that chops and spreads the chopped material or may consist of two in-series units: chopper and spreader.

When the process is performed by a single unit, the process schematic is as shown in Figure 9.2. The straw released by the straw walkers or axial threshing unit is chopped into smaller pieces by the blades of a rotor or drum, which also generates airflow to help spread the chopped material behind the combine. The rotor could be equipped with four longitudinal rows of jointed blades (Figure 9.3), which are of two types: cutting blades and paddle blades. The straight, sharpened, and reversible blades finely cut the crop residue into uniform sizes. The lateral fan blades generate high air velocities (e.g., up to 145 km/h or

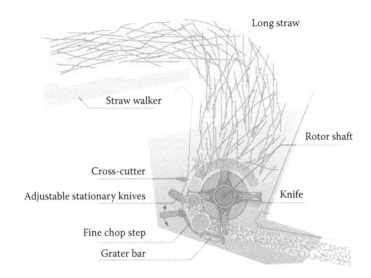

FIGURE 9.2
(See color insert.) Schematic of the chopping process. (Courtesy of CLAAS, Harsewinkel, Germany.)

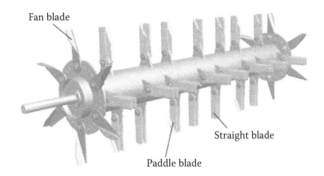

FIGURE 9.3
Active components of cutting and spreading rotor.

90 mph). The paddle blades increase the chopper airflow and spread width while improving the spreading uniformity across the combine working width. In this case, the chopped material is spread through special shrouds. Slightly different versions of this rotor are used in combine harvesters made by Case IH, Caterpillar, John Deere, New Holland, and Massey Fergusson. The usual number of blades on the rotor is 72–88. Case IH combine harvesters may also use a chopping rotor with multiknife (serrated) blades, which have opposite orientation and are grouped in pairs mounted on a helical path on the rotor barrel (Figure 9.4) to require a uniform torque over a complete rotation.

Regardless of the blade types, the blades are positioned within equally spaced transversal planes to the rotor axis to maintain a consistent cut length and for process and power efficiency reasons. To increase the cutting process efficiency, the cutting blades work in conjunction with stationary, yet adjustable counterknives. At the top, Figure 9.5 shows the counterknives of the chopping unit as well as a MOG spreading system composed of two spinning discs that rotate in opposite directions. Two spinning discs are needed for spreading the chopped material when the harvest swaths are greater than 7.5 m (25 ft);

FIGURE 9.4
Chopping rotor with multiknife (serrated) blades. (Courtesy of Case IH, Racine, WI.)

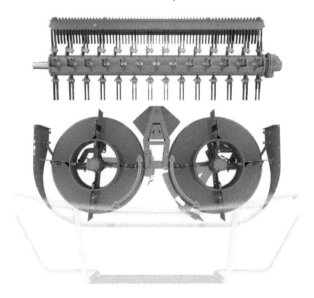

FIGURE 9.5
Chopping and spreading rotors. (Courtesy of CLAAS, Harsewinkel, Germany.)

otherwise, a single disc may be enough to perform this task. Straw choppers and spreaders are interchangeable on most combines, so the cereal growers can choose between different versions: one disc, two discs, or a combination of both. To increase material spreading width and uniformity, radial or curved vanes are attached on the spinning disc's surface.

The spreading width and distribution pattern are usually adjusted by

- Changing the fore–aft position where the material drops onto the spinning paddles
- Changing the angular position of deflectors or shields placed around the disc's border
- Changing the rotational speed of the discs

A list of crop residue chopping and spreading equipment can be found at the following link: http://pnwsteep.wsu.edu/tillagehandbook/chapter3/031997.htm.

9.3 Theory, Modeling, and Design of Crop Residue Chopping Process

The chopping process of crop residue, which is released by straw walkers or the axial threshing unit and cleaning shoe, is a cutting process of the material fibers by the edges of the cutting blades and counterknives of the chopper following physical phenomena such as bending, compression, and local crushing of the material. Such deformations occurring in the material depend on the form and orientation of the cutting edges and the kinematics of the process. Therefore, the cutting resistance of a material is strongly related to the shape and orientation of the cutting edges and process kinematics.

The material components have a random orientation when coming into contact with the active elements of the chopper. These components may be cut individually or in bundles instantly formed.

The cutting types of the material can be classified based on two criteria, as follows:

1. Position of the blade cutting edge relative to the straw (stalk) fiber direction
2. Direction of blade movement relative to the cutting-edge direction

Thus, based on the first criterion, there are two types of cutting: cutback and splitting (defibration), which is cutting along the straw fiber direction. In regard to the second criterion, the material can be sheared (when the direction of the blade speed is perpendicular to the cutting edge) or cut with sliding (the cutting edge slides over the material). This second type of cutting strongly decreases the power requirement for the chopping process.

Previous research on the plant cutting process extensively described the influence of different design and functional factors on the cutting process efficiency and equipment performance. Such factors include the blade speed, bevel angle, sharpness, thickness, edge serration, clearance between the edges, and physical and mechanical properties of cut material.

The cutting theory explained in Section 4.2.2.2, specifically Equation 4.42, is also applicable here. We emphasize here that the size of the angle between the blade edges during the cutting process is closely linked to the coefficient of friction between the blades and vegetal material.

In the following, we developed a theory that was not previously published. Let's assume the schematic of the chopper and process shown in Figure 9.6. On the chopper barrel rotating counterclockwise with a constant angular speed, the cutting blade is jointed at one end (point O_1); that is, it can freely rotate around it. The position of the axis OX of the associated coordinate system XOY corresponds to the radial orientation of the blade reaching the beginning of the cutting edge of the counterknife. Initially, the first pieces of material are sheared. This type of cutting requires a high and very nonuniform power due to irregular resistant torque created by material during the cutting process. Therefore, we are looking to determine a counterknife edge profile so that the power requirement and its nonuniformity will decrease. We aim to accomplish this desideratum by considering the blade clockwise rotation due to material resistance to cutting, and by designing the counterknife edge profile so that the rest of the material will be cut through a combination of shearing

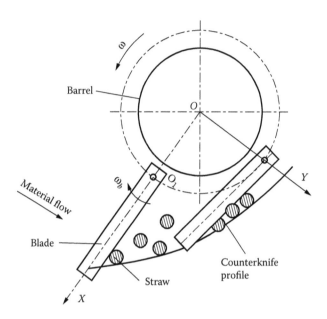

FIGURE 9.6
Schematic of the cutting process and counterknife edge profile.

and sliding. For mathematical modeling purposes, let us orient in a convenient way the coordinate system *XOY* and the counterknife edge profile as shown in Figure 9.7.

Let us consider an instantaneous position of the jointed point O_1 of the blade, when its positioning angle is α. The blade rotates with the instantaneous angular speed ω_b and meets the counterknife edge in point A. The distance polar coordinate of point A is R_c. At contact point A, the tangent to the curve that defines the counterknife edge shape makes the angle β with the horizontal (as considered by convention in this figure). Taking into account Equation 4.42, ideally the angle φ between the tangent to the curve at any point of the cutting operation and the blade edge should satisfy the following relationship:

$$\varphi \leq \varphi_1 + \varphi_2 \tag{9.1}$$

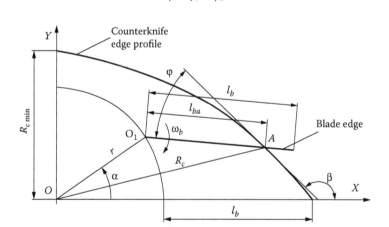

FIGURE 9.7
Schematic of shearing/cut by the sliding process and the counterknife edge profile.

where:

φ_1 is the friction angle of the straw (stalk) with the cutting blade, degrees

φ_2 is the friction angle of the straw (stalk) with the counterknife edge, degrees

Since the material of the blade and counterknife are similar, the values of these coefficients have equal or very close values, and they are considered as given design parameters.

We also assume that the following design parameters have already been established (chosen):

- r is the radial distance to the blade joint center on the chopper barel (base radius) (m).
- l_b is the length of the cutting blade edge (m).
- $R_{c\,min}$ is the minimum clearance for the blade (m).

We want to define an equation $y = f(x)$ of the counterknife edge profile that satisfies the following limit conditions:

$$\begin{array}{lll} f(x) = 0 & \text{if} & x = r + l_b \\ f(x) = R_{cmin} & \text{if} & x = 0 \\ f'(x) = \tan(-2\varphi) & \text{if} & x = r + l_b \end{array} \tag{9.2}$$

The third equation of the above system derives from geometrical interpretation of the first derivative of a function.

We particularly studied two reciprocally matching functions whose expressions were derived from basic functions, followed by reflection and sliding operations. They are as follows:

$$f_1(x) = R_{cmin} - ax^b \tag{9.3}$$

$$f_2(x) = R_{cmin}\left(1 - me^{nx}\right) \tag{9.4}$$

The coefficients/exponents a, b, m, and n must be determined from the above-mentioned limit conditions. The values of these coefficients of the above-defined two functions matching one another are given in Table 9.1 for a value of the considered design parameters $r + l_r = 0.25$ m.

The rotational speed of the blade varies from the base to the tip within the range, but is not limited to 20–50 m/s (65.6–164 ft/s).

This type of mathematical modeling has to be accompanied by a dynamic analysis of the blade rotation angle and angular speed or a specific design, material, and process parameter setting.

Usually, the cutting edge is formed by sharpening the blade on one side. The blade bevel angle (Figure 9.8) is a very important design and functional parameter. The bevel-angle value is usually within the range of 20–30°. A blade with a sharper angle will wear out and become duller faster and is more likely to be seriously damaged when subjected to foreign material (stones, lost metal parts). Besides, according to Chancellor (1988), when the bevel-angle value is decreased below 20°, there is very little reduction in the cutting energy

TABLE 9.1

Coefficient/Exponent Values of the Functions Expressed by Equations 9.3 and 9.4 when $r + l_r = 0.25$ m

Coefficient/ Exponent	Value	$R_{c\,min}$, m
a	17.36	0.158
b	3.392	
m	0.02323	0.165
n	15.112	

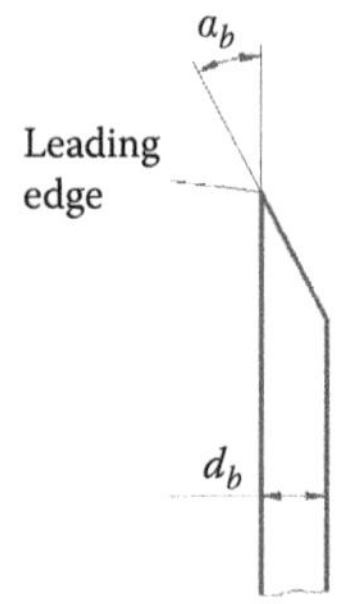

FIGURE 9.8
Design parameters of cutting blade.

requirement. An increase of the bevel angle greater than 35° leads to a higher requirement for cutting energy. For a better and longer use of blade material, the blades have two opposite cutting edges to allow reversing them when one cutting edge is irremediably worn out.

The thickness of the leading edge influences the magnitude of the cutting force and energy requirement. Thus, the cutting force is relatively constant when the edge thickness is 0.05–0.09 mm (0.002–0.0035 in.), but for a greater thickness, it increases significantly. This can be explained by the fact that a very sharp blade causes a more localized compression stress, that is, breaking failure in the material, while a dull blade exercises this action over a larger area, thereby requiring more energy for the same result.

Serrations on the blade-beveled area generate higher forces along the blade cutting edge and offer the advantage of a relatively larger angle between the shearing surfaces (blade and counterknife), while the material is not expelled. However, this translates to higher cutting forces and energy requirement because the material cutting occurs over a zigzag line instead of a continuous one.

9.4 Theory, Modeling, and Design of Crop Residue Spreading Process

The crop residue composed mainly from straw (stalks) and chaff can be spread in different ways as follows:

- Chaff is spread separately from the straw, which is chopped and then spread. The advantage of this process is separate setting of chaff- and straw-handling equipment, while the chaff does not overload the chopper.
- The chaff and straw are chopped and spread together over or into the stubble. The main advantage of the latest case is reducing the dust, that is, preserving the environment, and preventing the clogging of the cabin air conditioning filter and combine engine filter.

The spreading process is relatively simple: the chopped material is fed at a certain radial distance into the spreading disc that rotates uniformly in a horizontal or back-tilted plane. The effect of applied forces causes the material to move on the disc surface along a curved path until it meets the vane that acts as a guiding element until the particles leave the disc and fall based on the laws of physics.

For calculations of material particle dynamics, one needs to conventionally assess at least the size and average mass of a particle, Young's modulus, as well as the friction coefficients of the chopped material with the disc and vane materials. Young's modulus values as described in the literature are systematized in Table 9.2. The ASAE D251.2 standard gives values of friction coefficients for a few chopped forage plants; this is a good source to start with.

By applying *mutatis mutandis*, for data analysis, the methods of determining and expressing particle size described by ASAE S319.3 and ANSI/ASAE S424.1, we can assert that

- The analysis of mass percentage distribution of fragmented MOG in the chopper is based on the assumption that these percentages are logarithmic normally distributed.
- The size of MOG particles should be reported in terms of geometric mean length *lgm* and standard deviation *Sgml* by mass percentage, as given by Equations 5.109 and 5.110.

The mass m of a generic particle can be determined using the data published by Lam et al. (2008). Thus, the chopped straw particles with an average length of 9.078 mm (0.3574 in.) have an average mass of 0.008 g (0.0176 lb), while the particle density is 0.093 g/cm³.

Let us consider an inclined spreading disc at an angle α, and one of its straight vanes positioned at the initial radius r_0 and angle ψ_0, and being mounted with a forward inclination ψ with respect to the corresponding radial direction (Figure 9.9). The disc rotates uniformly with constant angular speed ω. A particle of mass m, initially discharged on the

TABLE 9.2

Young's Modulus of Individual Plant Stems

Material	Young's Modulus, GPa (individual stems)	Shear Strength, MPa (individual stems)	Author(s)
Alfalfa	0.79–3.99	0.4–18.0	Galedar et al., 2008
Barley	0.33–0.62		Tavakoli et al., 2008
Wheat	3.13–3.75		Esehaghbeygi et al., 2009
Wheat	0.90–1.80		Tavakoli et al., 2008
Wheat	4.76–6.58	4.91–7.26	O'Dogherty et al., 1995
Wheat	1.59–2.15	5.39–6.98	O'Dogherty et al., 1989

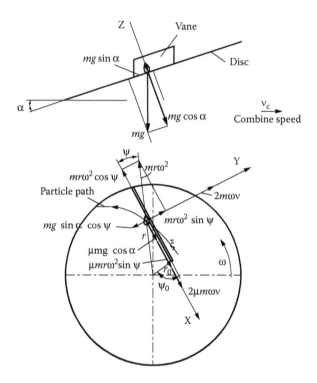

FIGURE 9.9
Acting forces on a particle interacting with the spreading disc.

disc at the interior end of the vane, is located at the radial position r after moving onto the distance ξ.

Consequently, we can write the following geometrical relationships:

$$r \sin \psi = r_0 \sin \psi_0 \tag{9.5}$$

$$r \cos \psi = \xi + r_0 \cos \psi_0 \tag{9.6}$$

Let us assign the coordinate system XYZ, which is oriented in a convenient way, as shown. The particle moves along the vane on the distance ξ with the speed v and acceleration a, respectively:

$$v = d\xi / dt = \xi' \tag{9.7}$$

$$a = d^2\xi / dt^2 = \xi'' \tag{9.8}$$

The following forces act on the particle:

- mg is the particle weight, which has two components sideways and perpendicular to the disc.

- $mr\omega^2$ is the centrifugal force oriented in the radial direction that passes through the weight center of the particle, with its two components depending on the angle ψ.
- $\mu mg\cos\alpha$ is the friction force of the particle with the disc surface; μ is the friction coefficient of the particle with the disc/vane material (galvanized steel). We neglect the friction force generated by the component $mg\sin\alpha$ because it is relatively very small.
- $2m\omega v$ is the Coriolis force.
- $2\mu m\omega v$ is the friction force of the particle with the vane material.
- $\mu\, mr\omega^2\sin\psi$ is the friction force of the particle with the vane that arises due to the Coriolis force component.

By applying the second law of dynamics along the axis X, we get

$$m\xi'' = mr\omega^2\cos\psi - \mu mg\cos\alpha - \mu mr\omega^2\sin\psi - 2\mu m\omega\xi' \tag{9.9}$$

If the vane is inclined backward, then the term $\mu mr\omega^2\sin\psi$ is negative; as a consequence, by considering both design options, Equation 9.9 becomes

$$m\xi'' = mr\omega^2\cos\psi - \mu mg\cos\alpha \pm \mu mr\omega^2\sin\psi - 2\mu m\omega\xi' \tag{9.10}$$

By dividing the last equation with the mass m and considering Equations 9.5 and 9.6, we can write

$$\xi'' + 2\mu\omega\xi' - \omega^2\xi = r_0\omega^2\cos(\psi_0\pm\phi)/\cos\phi - \mu g\cos\alpha \tag{9.11}$$

where the friction angle

$$\phi = \arctan\mu \tag{9.12}$$

Equation 9.11 is a heterogeneous differential equation of the second order; its characteristic equation has the following second-order form:

$$\lambda^2 + 2\mu\omega\lambda - \omega^2 = 0 \tag{9.13}$$

The unknown λ_1 and λ_2 expressions are

$$\lambda_1 = \omega\left(\sqrt{1+\mu^2} - \mu\right) \tag{9.14}$$

$$\lambda_2 = -\omega\left(\sqrt{1+\mu^2} - \mu\right) \tag{9.15}$$

In solving Equation 9.11, we followed the mathematical approach described by Bosoi et al. (1988) for a similar problem.

Let us introduce arbitrary constants C_1 and C_2 in Equation 9.11, which will take the following form:

$$\xi = C_1 e^{\lambda_1 t} + C_2 e^{\lambda_2 t} + r_0 \cos(\psi_0 \pm \phi) / \cos\phi - \mu g \cos\alpha / \omega^2 \tag{9.16}$$

These constants will be determined from the initial conditions of the particle movement; at the moment $t = 0$, the particle is discharged onto the disc at the position corresponding to the vane inner end (r_0 and ψ_0 coordinates). That means

$$\xi = \xi' = 0 \text{ at } t = 0 \tag{9.17}$$

Substituting for the constants C_1 and C_2 in Equation 9.16, the result is

$$\xi = \left[\frac{\mu g \cos\alpha}{\omega^2} - r_0 \frac{\cos(\psi_0 \pm \phi)}{\cos\phi} \right] \left[\frac{1}{\lambda_2 - \lambda_1} \left(\lambda_2 e^{\lambda_1 t} - \lambda_1 e^{\lambda_2 t} \right) - 1 \right] \tag{9.18}$$

From Equations 9.5 and 9.6 we get

$$\left(\xi + r_0 \cos\psi_0 \right)^2 = r^2 - r_0^2 \sin^2\psi_0 \tag{9.19}$$

From this equation we can express the current position r of the particle:

$$r = \sqrt{\left(\xi + r_0 \cos\psi_0 \right)^2 + r_0^2 \sin^2\psi_0} \tag{9.20}$$

The derivative of Equation 9.18 gives the particle velocity along its trajectory over the spreader's disc as follows:

$$\xi' = \left[\frac{\mu g \cos\alpha}{\omega^2} - r_0 \frac{\cos(\psi_0 \pm \phi)}{\cos\phi} \right] \left[\frac{\lambda_1 \lambda_2}{\lambda_2 - \lambda_1} \left(e^{\lambda_1 t} - e^{\lambda_2 t} \right) \right] \tag{9.21}$$

The above developed equations of the particle kinematics help a designer to simulate and optimize the movement of the particle in connection with the rotation and inclination of the disc so that most of the particles will leave the disc within a defined angular area and at a certain spreading distance.

References

ANSI/ASABE S424.1. 2006. Method of determining and expressing particle size of chopped forage materials by screening. In *ASABE Standards*, 619–621. St. Joseph, MI: American Society of Agricultural and Biological Engineers.

ANSI/ASAE S319.3. 2006. Method of determining and expressing fineness of feed materials by sieving. In *ASABE Standards*, 600–605. St. Joseph, MI: American Society of Agricultural and Biological Engineers.

ASAE D251.2. 2006. Friction coefficients of chopped forages. In *ASABE Standards*, 577–579. St. Joseph, MI: American Society of Agricultural and Biological Engineers.

Bosoi, E.S., O.V. Verniaev, I.I Smirnov, and E.G. Sultan-Shakh. 1988. *Theory, Construction, and Calculations of Agricultural Machines*. Rotterdam: A.A. Balkema.

Chancellor, W.J. 1988. Cutting of biological material. In *Handbook of Engineering in Agriculture*, 35–63. Vol. I. Boca Raton, FL: CRC Press.

Douglass Jr., C.L., P.E. Rasmussen, and R.R. Allmaras. 1992. Nutrient distribution following wheat-residue dispersal by combines. *Soil Science Society of America Journal* 56(4):1171–1177.

Esehaghbeygi, A., B. Hoseinzadeh, M. Khazaei, and A. Masoumi. 2009. Bending and shearing properties of wheat stem of Alvand variety. *World Applied Science Journal* 6(8):1028–1032.

Galedar, M.N., A. Tabatabeefar, A. Jafari, et al. 2008. Bending and shearing characteristics of alfalfa stems. *Agricultural Engineering International: The CIGR Ejournal* X, Manuscript FP 08 001.

Lam, P.S., S. Sokhansanj, X. Bi, et al. 2008. Bulk density of wet and dry wheat straw and switchgrass particles. *Applied Engineering in Agriculture* 24(3):351–358.

O'Dogherty, M.J., H.G. Gilbertson, and G.E. Gale. 1989. Measurements of physical and mechanical properties of wheat straw. In Fourth International Conference on Physical Properties of Agricultural Materials, Rostock, Germany, pp. 608–613.

O'Dogherty, M.J., J.A. Huber, J. Dyson, and J. Marshall. 1995. A study of the physical and mechanical properties of wheat straw. *Journal of Agricultural Engineering Research* 62:133–142.

Rekord. 2014. Spreading of chaff and straw on combines. Rekordverken, Öttum, Sweden AB. http://www.rekordverken.se (accessed February 17, 2014).

Tavakoli, H., S.S. Mohtasebi, and A. Jafari. 2008. Comparison of mechanical properties of wheat and barley straw. *Agricultural Engineering International: the CIGR Ejournal* 10:CE 12002.

Bibliography

Bitra, V.S.P., A.R. Womac, Y.T. Yang, et al. 2011. Characterization of wheat straw particle size distribution affected by knife mill operating factors. *Biomass and Bioenergy* 35:3674–3686.

Goering, C.E., R.P. Rohtbach, and A.K. Srivastava. 1993. *Principles of Agricultural Machines*. St. Joseph, MI: American Society of Agricultural Engineers.

Kumhala, F., Z. Kviz, J. Masek, and P. Prochazka. 2005. The measurement of plant residues distribution quality after harvest by conventional and axial combine harvesters. *Plant Soil Environment* 51(6):249–254.

Kushwaha, R.L., A.S. Vaishnav, and G.C. Zoerb. 1983. Shear strength of wheat straw. *Canadian Agricultural Engineering* 25(2):163–166.

Miu, P. 1999. Mathematical modeling of material other-than-grain separation in threshing units. ASAE Paper 99-3208. St. Joseph, MI: American Society of Agricultural and Biological Engineers.

Neculaiasa, V. and I. Danila. 1995. *Working Processes and Harvesting Machinery* [in Romanian]. Iasi, Romania: A92 Publisher.

Redekop. 2013. Put the NO in till. Saskatoon, Saskatchewan: Redekop. http://www.strawchopper.com (accessed February 17, 2014).

Sitkei, G. 1986. *Mechanics of Agricultural Materials*. Amsterdam: Elsevier.

Tabil, L., P. Adapa, and M. Kashaninejad. 2011. Biomass feedstock pre-processing: Part 1: Pre-treatment. In: *Biofuel's Engineering Process Technology*, ed. M. Aurelio and D.S. Bernardes. Rijeka, Croatia: InTech. http://www.intechopen.com/books/biofuel-s-engineering-processtechnology/biomass-feedstock-pre-processing-part-1-pre-treatment (accessed February 17, 2014).

Veikle, E. 2011. Modeling the power requirements of a rotary feeding and cutting system. PhD dissertation, University of Saskatchewan.

Veseth, R. 2014. The first no-till step–combine residue spreading. Pacific Northwest Conservation Tillage Handbook. http://pnwsteep.wsu.edu/tillagehandbook/chapter3/031997.htm (accessed February 27, 2014).

Zhang, Y., A.E. Ghaly, and B. Li. 2012. Physical properties of corn residues. *American Journal of Biochemistry and Biotechnology* 8(2):44–53.

10

Corn Ear Dehusking Process and Equipment

10.1 Introduction

Section 1.5.3 describes how the corn headers detach and collect the corn ears from the plants to be further threshed and cleaned by combine harvesters. Thus, a combine harvester delivers only corn kernels, and the husks are left as a swath or chopped and spread on the soil (see Chapter 9). In some countries, the corn is still harvested and preserved as cobs after removing the husks from the ears, while ensuring the integrity and quality of corn kernels.

Harvesting only the *corn cobs* has two big advantages: it provides the raw material for canning (e.g., sweet corn is the most important canned vegetable in the United States), and it allows a rapid drying of the corn kernels because the stored cobs allow a greater aeration flow, which can be naturally available in the storage bins exposed to the wind.

Harvesting the corn as cobs is done with a combine harvester of a different design, usually a pulled or self-propelled combine. The design of these combines is beyond the scope of this book. In the following, we will describe only the dehusking process of the corn ears and the afferent equipment (unit).

There are several husking unit designs, but they have in common pairs of rapidly rotating rollers, which catch the husks and pull them between the rollers while the cobs flow to the end of the rollers. Section 10.2 describes the roller construction and specifications.

10.2 Construction and Specifications of Corn Dehusking Units

The working unit that performs the corn ear dehusking process consists of a pair of cylindrical rollers (Figure 10.1), which rotate in opposite directions, have different working surface profiles, and are made of different materials. One roller is metallic (forged or cast steel), while the other one has cylindrical, rubber-made sections pressed over a metallic shaft. There are also combined rollers, the length of which the metallic and rubber surfaces alternate: a metallic portion, then a rubber one, and so on. At all times, the rollers are pressed one onto the other by helical compression springs, and their reciprocal contact surfaces are made of different materials, as stated above.

The outer surfaces of both rollers have different profiles (combinations of helical and longitudinal ribs, grooves, and spurs), whose main role is to develop nonuniform friction forces with the ear husks, to dislodge them, and to pull them down through the rollers.

FIGURE 10.1
Dehusking rollers made of steel (top) and rubber.

One roller is always placed at a slightly higher elevation to increase the difference between the friction forces developed by each roller with the ear husks, so that they can be easily dislodged and grabbed. Most of the dehusking units have a pushing mechanism, which exercises additional forces on the corn ears (pushing them forward and toward the rollers), so that the husks may be easily, uniformly, and eventually completely dislodged and removed from the ears, while the ears flow uniformly and fast over the rollers' surface. The pushing mechanism can be of a belt conveyor type with elastic paddles or rotary type with paddles or fingers (Figure 10.2). A dehusking unit is made of at least four rollers, which are arranged alternately: metal/rubber/metal/rubber. The rollers are inclined by 8°–15° with respect to horizontal to help cob movement along them.

The main design parameters of the rollers are outside diameter $D_r = 63.5$–80 mm (2.5–3.125 in.) and length $L_r = 0.8$–1.25 m (31.25 – 49.25 in.) when a pushing mechanism is present or $L_r = 1.45$–1.75 m (57–70 in.) otherwise. The dehusking units without a pushing mechanism are mainly used in stationary dehusking equipment. The peripheral speed of the rollers is 0.85–1.37 m/s (2.75–4.5 ft/s).

10.3 Dehusking Process Theory

The process of removing the husks on the cobs can be divided into three phases:

- Orienting the ears with their longitudinal axis parallel to the roller axis
- Dislodging the husks while the ears rotate and move along the rollers
- Dragging the husks through the rollers

Eventually, the cobs leave the rollers, and then they are collected and conveyed according to the material flow design of the combine.

The orientation of the ears is performed by the helical ribs of the metallic rollers in the section where material is fed. Due to different friction forces developed between the ears and each roller, the husks are one by one dislodged and pulled down through the rollers.

FIGURE 10.2
Rotary pushing mechanism with fingers.

Let us consider the dynamics of a cob between the rollers, as shown in Figure 10.3. The rollers are pushed against each other with the force P, at both ends, by pairs of helical compression springs. We noted the inclination angle of the rollers with respect to horizontal by α. Thus, the cob weight $G \cdot \cos \alpha$ has two components: G_1 and G_2. By applying the sinus theorem for these forces, we can write

$$\frac{G_1}{\sin\left[\frac{\pi}{2} - (\beta + \gamma)\right]} = \frac{G_2}{\sin\left[\frac{\pi}{2} - (\beta - \gamma)\right]} = \frac{G}{\sin 2\beta} \tag{10.1}$$

where γ is the positioning angle between the rollers' axes. By design, the value of this angle is chosen within 15°–20°.

Consequently, the cob weight components can be calculated as follows:

$$G_1 = \frac{G \cos \alpha \cos (\beta + \gamma)}{\sin 2\beta} \tag{10.2}$$

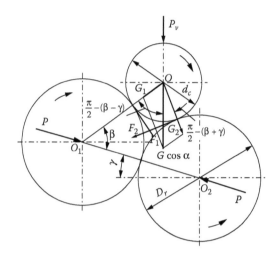

FIGURE 10.3
Dynamics of the cob between the rollers.

$$G_2 = \frac{G \cos \alpha \cos (\beta - \gamma)}{\sin 2\beta} \tag{10.3}$$

Due to the normal reactive forces on the cob that counteract the forces G_1 and G_2, the friction forces F_1 and F_2, respectively, occur at the contact surface between the cob and the rollers. Due to differences in their size, also determined by differences between friction coefficients μ_1 and μ_2, we can write $F_2 > F_1$. Explicitly, these forces can be expressed as follows:

$$F_1 = \mu_1 \frac{G \cos \alpha \cos (\beta + \gamma)}{\sin 2\beta} \tag{10.4}$$

$$F_2 = \mu_2 \frac{G \cos \alpha \cos (\beta - \gamma)}{\sin 2\beta} \tag{10.5}$$

In order to avoid cob catching and pulling by the rollers, the following two inequalities have to be satisfied:

$$F_1 \cos \beta < G_1 \sin \beta \tag{10.6}$$

$$F_2 \cos \beta < G_2 \sin \beta \tag{10.7}$$

If the angles of friction between the cob and the rollers are φ_1 and φ_2, then the value of the angle β, which characterizes the cob position with respect to the roller position, must satisfy the following requirements:

$$\beta > \varphi_1 \quad \text{and} \quad \beta > \varphi_2 \tag{10.8}$$

Based on the geometry of the rollers and cob, we can calculate the value of the positioning angle β as follows:

$$\beta = \arccos \frac{D_r}{D_r + d_c} \tag{10.9}$$

where
   d_c is the cob diameter
   D_r is the roller diameter

If we also consider a vertical pushing force P_v that is exercised on the cob by the pushing mechanism of the dehusking equipment, Equations 10.4 and 10.5 become

$$F_1 = \mu_1 \frac{(G \cos \alpha + P_v) \cos (\beta + \gamma)}{\sin 2\beta} \tag{10.10}$$

$$F_2 = \mu_2 \frac{(G \cos \alpha + P_v) \cos (\beta - \gamma)}{\sin 2\beta} \tag{10.11}$$

The cob rotates by its longitudinal axis due to the resultant moment ΔM generated by the friction forces, that is,

$$\Delta M = (F_2 - F_1) \frac{d_c}{2} \tag{10.12}$$

From the moment a husk is dislodged and grabbed, it will be pulled down by the rollers with a force T, which has to be higher than the breaking resistant force R_h of the husks from the cob, meaning that

$$T = (\mu_1 + \mu_2) P > R_h \tag{10.13}$$

The roller diameter is calculated so that pulling the smaller cobs through the rollers is not possible. Mathematically, from the condition $\tan \beta > \mu$, we draw the following design requirement:

$$D_r < \frac{d_c}{\sqrt{1 + \mu^2} - 1} \tag{10.14}$$

where μ value is chosen for each roller, depending on its material and surface profile. The force P is adjustable by pretensioning of the compression helical springs.

During pulling the husks, the clearance between the rollers varies within certain limits. Because the rollers are driven by spur gears, their teeth profile have to be adequately designed; the tooth is longer, while its base is thinner, when compared with the involute profile of a regular tooth.

Bibliography

Neculaiasa, V. and I. Danila. 1995. *Working Processes and Harvesting Machinery* [in Romanian]. Iasi, Romania: A92 Publisher.

Salunkhe, D.K. and S.S. Kadam. 1998. *Handbook of Vegetable Science and Technology: Production, Composition, Storage, and Processing*. New York: Marcel Dekker.

11

Power System of a Combine Harvester

11.1 Introduction

Modern self-propelled combine harvesters have evolved into complex machinery, which use a considerable amount of energy to move into the field and perform all the technological processes required for harvesting crops. The current marketplace for combine harvesters is substantially driven by advertised horsepower. The energy required by a combine harvester is generated and transmitted by the power system, which can be divided into the propulsion (traction) subsystem and process driving subsystem. Both subsystems share the energy generated by a diesel engine. A block diagram and components of the power system of a combine harvester are illustrated in Figure 11.1.

The engine converts the chemical energy of diesel fuel into mechanical power. Combustion of the fuel that occurs sequentially in the engine cylinders moves reciprocating pistons, which turn the engine crankshaft. The power from the crankshaft is transmitted to the power train and process drives via a flywheel.

The traction power is used for combine propulsion and steering. The drives for technological processes can be mechanical or hydrostatic as well.

The power train consists of a mechanical gearbox that receives the power via a clutch, or one or more hydrostatic pumps, or all together. If the power train contains a mechanical gearbox, then the final drives at the driving front wheels are also mechanical gearboxes with planetary spur gears. In this case, a differential gearbox correlates the individual rotation of each driving wheel when the combine turns in a direction or when dissimilar sliding of the wheels occurs at the contact with the ground.

Tables 1.1 and 1.2 indicate essential specifications of the engine, power train, and steering of modern conventional and axial combines, respectively.

11.2 Diesel Engine

Due to the complexity of phenomena occurring in an engine, and associated design and optimization problems, an extensive study of the diesel engine is beyond the scope of this book. There are books that are exclusively dedicated to the study of engines. In the following, we will just cover a few aspects of the diesel engine for a combine harvester. However, specific information on a particular engine should be required from the manufacturer.

The diesel engine of modern combine harvesters is a four-stroke engine with an electronically controlled fuel injection process; that is, the timing and duration of fuel injection

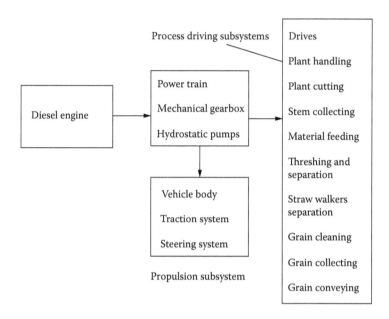

FIGURE 11.1
Components of the power system of a combine harvester.

are controlled by opening the associated solenoid valves. This allows optimizing generated power, while fuel consumption and exhaust emissions are diminished. Figure 1.43 shows a longitudinal cutaway of a diesel engine with in-line cylinders.

The *equivalent power of the fuel* P_{fe} that is generated by combustion into the engine chambers is given by the following equation (Goering, 1999):

$$P_{fe} = \frac{H_f q_f}{3600}, \text{ kW} \tag{11.1}$$

where:
   H_f is the specific heating value of fuel, kJ/kg
   q_f is the fuel consumption rate, kg/h

Depending on the boiling temperature range (160–260°C or 200–370°C), the heating value of diesel fuel is 45,500–45,700 kJ/kg, and corresponding densities are 0.832–0.834 kg/L.

Fuel combustion is possible in the presence of the air introduced into the engine cylinders at a certain air/diesel mass ratio.

For a stoichiometric composition, the combustion reaction of diesel fuel is represented as follows:

$$C_xH_y + \left(x + \frac{y}{4}\right)O_2 \rightarrow xCO_2 + \frac{y}{2}H_2O \tag{11.2}$$

where:
   x, y are the number of carbon and hydrogen atoms, respectively, in a diesel molecule ($C_{16}H_{34}$)
   z is the number of oxygen atoms in the reaction

The *air/fuel ratio* can be calculated with the following equation (Goering, 1999):

$$\frac{m_a}{m_f} = \frac{137.3\left(x + \dfrac{y}{4} - \dfrac{z}{2}\right)}{\phi\left(12x + y + 16z\right)} \tag{11.3}$$

where:
    m_a is the mass of air
    m_f is the mass of diesel fuel
    ϕ is the fuel equivalence ratio

The fuel equivalence ratio is reported in terms of a nondimension variable, which is the actual oxygen/fuel ratio normalized by the stoichiometric oxidant/fuel ratio. By definition, $\phi = 1.0$ corresponds to the stoichiometric condition. When the mixture has an excess of fuel, $\phi < 1$ and it is called lean. When the mixture has oxygen in excess, $\phi > 1$ and the mixture is called rich. In diesel engines, the m_a/m_f ratio is greater than 20.

Diesel engine rotation is limited by the amount of the fuel injected into the engine cylinders. In the cylinders there is always sufficient oxygen to burn the fuel. Consequently, by increasing the amount of injected fuel, the engine will attempt to increase the speed to meet the new fuel injection rate. If the engine load is very low, the engine can accelerate at a high speed that is limited by the engine governor, which controls the amount of the fuel.

An important design parameter of a diesel engine is the compression ratio, which is a measure of how much the pistons compress the gases in the cylinders. In diesel engines, the compression ratio takes values within the 14:1 to 24:1 range because mostly the air is compressed; the fuel is injected late in the compression stroke, and then starts evaporating and mixing with air during an ignition delay. *Compression ratio* r_c is calculated with the following formula:

$$r_c = \frac{V_d + V_c}{V_c} \tag{11.4}$$

where:
    V_d is the displacement volume, cm³ (in.³)
    V_c is the clearance volume, cm³ (in.³)

Displacement of an engine cylinder is given by the following equation:

$$V_d = \frac{\pi d_c^2}{4} s \tag{11.5}$$

where:
    d_c is the cylinder (bore) diameter, cm (in.)
    s is the piston stroke that equals two times the crankshaft crank (eccentricity length) when measured center to center, cm (in.)

Total displacement of an engine V_e with n cylinders equals the product nV_d.

The equivalent power of diesel fuel is divided into *useful power* (33%) that reaches the flywheel and *lost power* through radiation (7%), engine cooling (30%), and exhaust (30%). The mechanical power developed through fuel combustion in the cylinders is called the *indicated power* P_i and is given by the following equation:

$$P_i = \frac{p_{ime} V_e n_c}{60z}, \text{ W} \tag{11.6}$$

where:
  p_{ime} is the indicated mean effective pressure developed through fuel combustion, Pa
  V_e is the total engine displacement, m³
  n_c is the flywheel shaft speed, rpm
  $z = 2$ (for a four-stroke engine), $z = 1$ (for a two-stroke engine)

The indicated mean effective pressure is the difference between the peak combustion pressure (produced during the power stroke) and the peak compression pressure (produced during the compression stroke). Practically, variation of mean effective pressure is computer controlled by varying valve opening and closing.
    The available power at the engine flywheel (called *brake power* P_b) can be calculated with the following equation:

$$P_b = \omega_c M_b = \frac{2\pi n_c}{60} M_b = \frac{\pi n_c M_b}{30}, \text{ W} \tag{11.7}$$

where:
  ω_c is the angular speed of the engine flywheel shaft, s⁻¹
  M_b is the brake torque, Nm

The *mechanical efficiency* of the engine is the ratio

$$\eta_m = \frac{P_b}{P_i} = 1 - \frac{P_l}{P_i} \tag{11.8}$$

where P_l is the power loss through friction and heat dissipation.
    The *engine specific fuel consumption (SFC)* is a very important functional parameter that gives an indication of the average amount of fuel used by the engine to develop a certain level of power over a defined period of time:

$$SFC = \frac{\dot{m}_f}{P_b} \tag{11.9}$$

where $\dot{m}_f$ is the fuel consumption (kg/h).
    The *engine thermal efficiency* is given by the following formula:

$$\eta_e = 3600 \frac{P_b}{H_f q_f} \tag{11.10}$$

Engine thermal efficiency is the expression of fuel combustion efficiency calculated by comparing the heat energy potential of the fuel with the amount of usable mechanical work produced by the engine.

A typical combine harvester uses a single engine for powering the movement of the vehicle and all machine functions required to perform various technological processes of crop harvesting. Thus, the load on the engine varies in accord with the loads applied by various functional units, some of them operating continuously, while others are activated, operated, and deactivated during routine use of the combine. Usually, a typical combine operates at two or three forward speeds, and each speed requires a target engine rotation speed. The speed is set by the operator. Once the combine operator selects a speed, an electronic control unit (ECU) controls the engine speed, primarily by dynamically adjusting the amount of fuel injected into the engine cylinders. As a load is imposed, the engine rotation tends to decrease below a first predetermined speed; then the engine control unit adjusts the quantity of fuel being injected, and thus maintains the engine speed within a target range.

11.3 Construction of Power Train of Combine Harvesters

11.3.1 Construction of Mechanical Power Train

There are two kinds of power trains for combine harvester propulsion: mechanical and hydrostatic systems. Combinations of mechanical and hydrostatic components are also used. Mechanical components of a power train can be mechanically, hydraulically, or electromagnetically controlled. A simplified schematic of a mechanical propulsion system is shown in Figure 11.2. From the engine's flywheel, the power is transmitted through the clutch, mechanical gearbox, differential box, and spur gear final drives to the driving front wheels of the machine. The drive power is transmitted from the differential output shafts to the spur gear final drives using transaxles. A view of the power train designed and manufactured by ZF Friedrichshafen for a combine harvester is shown in Figure 11.3. In this figure, one can see the gearbox with the service brakes, transaxles, and final transmissions; the clutch is not shown.

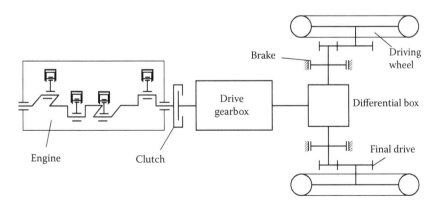

FIGURE 11.2
Mechanical propulsion system of a combine harvester.

FIGURE 11.3
Mechanical transmission assembly. (Courtesy of ZF Friedrichshafen AG, Friedrichshafen, Germany.)

The clutch provides a friction-dependent torque that is transmitted from the engine shaft from the moment the pressure of the driven disc on the driving disc (main body of the flywheel) exceeds a particular value. Once the clutch is left engaged, the torque is not friction dependent anymore; transmitted torque is synchronized with the driving torque developed by the engine. The cross section through a master clutch designed by the author is shown in Figure 11.4. The mechanical power from the engine flywheel is split into two paths: one for combine propulsion through the driven shaft and the other for combine propulsion through a trapezoidal belt pulley for driving the combine processing units. The clutch serves for combine propulsion by disengaging and engaging smoothly the power train for the gearbox. The pressure plate always pushes the friction disc with linings onto the flywheel. When the combine operator pushes the clutch pedal for disengaging the clutch for shifting the speed, the disengaging lever pushes back the pressure plate by overcoming the force of the compression springs. When engaging the clutch, the operator has

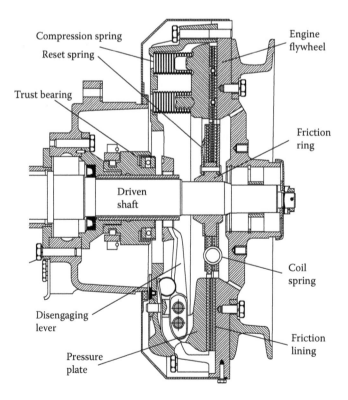

FIGURE 11.4
Cross section through a master clutch.

to progressively release the clutch pedal until there is enough contact friction among the linings on the disc, pressure plate, and flywheel surfaces.

Generally, the linings are not lubricated; they are made of *cerametallic* materials whose friction coefficient varies from 0.25 to 0.5 (Browning, 1978). Additional fluid coupling aids in a smoother transmission of the power through the clutch, increasing the comfort level and decreasing the power train vibrations. The disadvantage is the loss of energy due to occurring slip (lower friction coefficient) and required fluid cooling.

The range of forward speeds of a combine harvester is 0–25 km/h (0–15.5 mph). About three-fourths of the combine life is related to nominal speeds between 2.5 and 8 km/h (1.55 and 5 mph). The variation of forward speed can be continuous (when using hydrostatic transmissions) or by stepped-ratio transmissions using a gearbox with three or four forward speeds and one reverse basic speed. Figures 11.5 and 11.6 show the cross sections through a stepped-ratio gearbox, which was designed by the author. This gearbox has one input shaft, two intermediate shafts, and the output shafts from the differential. The parking/emergency brake is mounted on the second intermediate shaft.

Manual gear shifting can be done in three ways: with sliding gears, with a shift collar, or through a synchronized shift using a shift collar and a blocker with a friction surface. In all three cases, the gears or collars have a circular notch for the switch fork. The gearbox, shown in Figures 11.5 and 11.6, uses shift collars to engage every two adjacent gears.

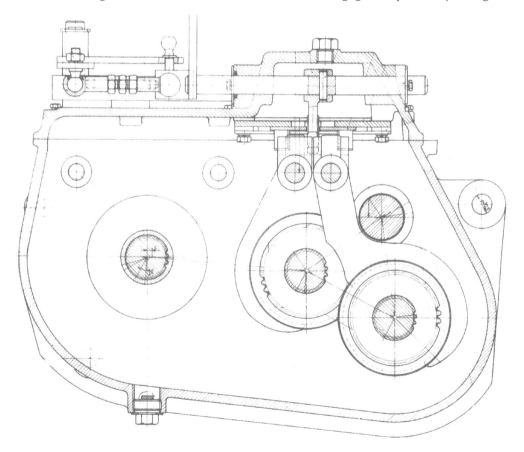

FIGURE 11.5
Cross section through a stepped-ratio gearbox with differential (I).

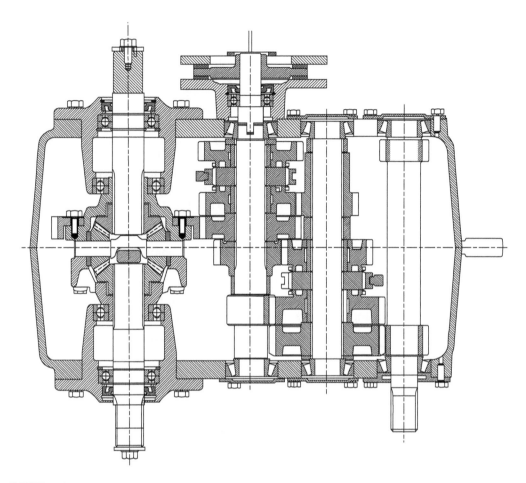

FIGURE 11.6
Cross section through a stepped-ratio gearbox with differential (II).

Because of the heat generated by the friction surfaces of the collars, gears, and bushings, the gearbox components are continuously lubricated by oil. The gears, collars, and bushings have holes in the radial direction, so that, due to centrifugal force, the lubricant oil can reach all surfaces in contact that have a relative motion to one other. The cross section through a shift collar designed by the author is displayed in Figure 11.7. The teeth have a semiconical shape to the exterior side to ease the collar engagement with the teeth of the adjacent gears. The external groove accommodates the fork that is handled by the combine operator through the speed-shifting mechanism.

Usually, the differential is compacted within the gearbox, and the service brakes are mounted outside of the gearbox, laterally, on the output shafts from the differential. A top view of such a gearbox is shown in Figure 11.8. The service brakes retard and fix the combine for on-road driving and for hillside operating; they are activated when the combine operator pushes the brake pedal to slow down or stop the combine. The most used type of service brake is the dry disc brake, whose caliper holds the brake paddles on lateral sides of the disc. Drum brakes are also used as service brakes.

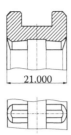

FIGURE 11.7
Cross section through a shift collar.

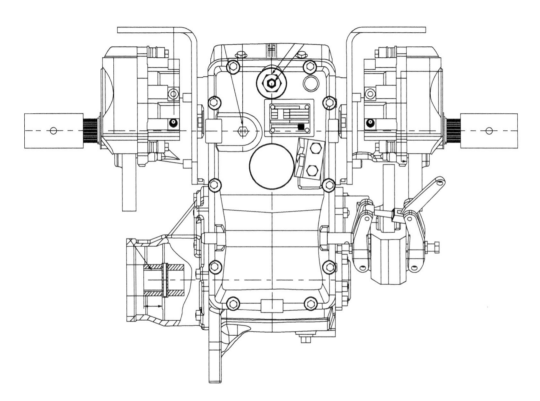

FIGURE 11.8
Top view of a gearbox with service brakes and couplings for transaxles. (Courtesy of ZF Friedrichshafen AG, Friedrichshafen, Germany.)

The output shafts end with couplings for the transaxles, which will transmit the driving power to the planetary or spur gear final drives (one on each side of the combine). The cross section through a spur gear final drive designed by the author is shown in Figure 11.9. The drive fits within the wheel rim that holds the tire. To minimize the assembly volume, the service brake is mounted on an additional shaft opposite to the input shaft of the final drive. This solution was adopted for a hillside combine to maximize the lateral inclination of the combine body with respect to the chassis that always remains parallel to the terrain.

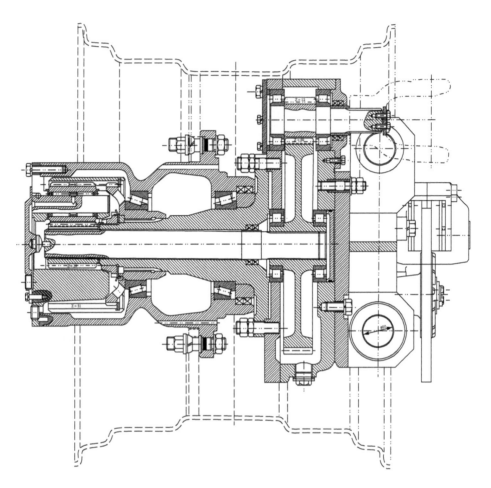

FIGURE 11.9
Cross section through a spur gear final drive.

Table 11.1 shows technical data of the combine power trains designed and manufactured by the German company Zahnradfabrik Friedrichshafen AG. The basic specifications for stepped-ratio transmissions of combine harvesters can be summarized as follows:

1. Forward speeds
   a. Bandwidth of nominal speeds according to market level: 0–30 km/h (1–18.6 mph)
   b. Ratio of neighboring speeds: 1.1–12 for main working range: 2.5–10 km/h (1.55–6.25 mph); higher ratios below 2.5 km/h, little higher above 10 km/h
2. Reverse speeds: Bandwidth of one nominal speed: 2–10 km/h (1.25–6.25 mph)
3. Durability
   a. Total transmission efficiency (from input shaft to wheels): Minimum 85% at full load for the main working range (2.5–10 km/h)
   b. Transmission life: minimum 6000 h

TABLE 11.1

Example of Technical Data of Combine Power Train

Technical Data	MD Series (Transmission)/S Series (Wheel Drive)			
	3 MD-40	S-800	3 MD-40	S-1050
Speed maximum (km/h)	25	25	25	25
Output torque maximum (Nm)	2,500	19,000	2,500	25,000
Ratio range	2.27–20.31	7.45	2.27–20.31	13.214
Gears (transmission gears)	3 front/3 rear	(1)	3 front/3 rear	(2)
Gear shifting	Dog clutch		Dog clutch	
Weight (kg)	190	340 (170*)	190	420 (210*)
Maximum load (kg)	18,000		24,000	

Source: Courtesy of ZF Friedrichshafen AG, Friedrichshafen, Germany.
* Per side.

As can be seen from Tables 1.1 and 1.2, most modern combine harvesters are equipped with hydrostatic transmissions; their construction is described in the next section.

To increase the adherence and traction on the hillside combines or combines working on paddy fields, they may be equipped with a self-leveling tracking system of wheels working in tandem (CLAAS, Figure 11.10).

The hydrostatic pump supplies the steering mechanism (Figure 1.45) with oil as well. In the hillside combine harvester, while the chassis remains parallel to the terrain, all wheels have to stand in a vertical position (Figure 1.49). This is done by a deformable, parallelogram-type bar mechanism. Figure 11.11 shows an example of this type of mechanism designed by the author. The steering bars are attached to a semicircular plate that rotates synchronously with the combine chassis.

11.3.2 Construction of Hydrostatic Power Train

A hydrostatic power train of a combine harvester consists of a hydrostatic pump driven by the combine engine, which can operate at a constant speed. The oil flow of the pump is controlled by the steering current. The pump is coupled with a hydrostatic motor that delivers the power to the driving wheels via a mechanical gearbox with differential. A hydrostatic

FIGURE 11.10
Self-leveling tracking system with tandem wheels. (Courtesy of CLAAS, Harsewinkel, Germany.)

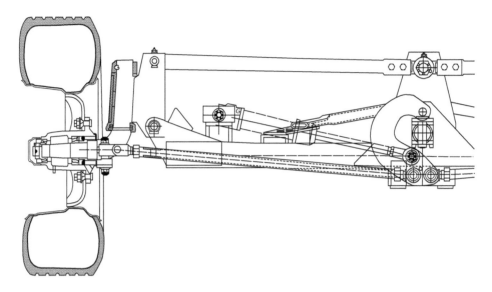

FIGURE 11.11
Steering wheel and bar along with parallelogram-type leveling mechanism in a hillside combine.

power train can also be made in an all-wheel-drive version (Figure 11.12). When the brakes are activated, the hydrostatic pump works as a power generator.

Inside the hydrostatic pump there is a rotating swash plate whose position determines whether the machine will move forward, backward, or is stationary. A number of pistons are connected to the swash plate. Depending on the swash plate angle, the pistons move back and forth with a stroke that is proportional to the swash plate angle. A pictorial circuit diagram of a hydrostatic transmission using an axial variable-displacement pump and a fixed-displacement motor is shown in Figure 11.13. The engine rotates the input shaft of the hydrostatic pump. The oil flow generated by the pump passes through a multifunction valve, which controls the fluid pressure; then, the motor converts the oil flow into the rotation of its output shaft. During this process, the oil is heated; thus, it may be cooled in a heat exchanger before flowing into the tank.

The multifunction valve also incorporates a charge pressure relief valve for safety operation and protection of the transmission components. The operator of the combine moves the control handle to adjust the displacement of the control valve of the pump. A detailed cross section through a pump with variable displacement is shown in Figure 11.14. The pump has an axial piston swash plate design that develops a nominal oil pressure of 250 bars. It is built in different sizes and has a displacement range between 18 and 100 cm³ (1.15 and 6.4 in.³). Consequently, the power varies within the range 16–96 kW (21.5–129 hp).

A hydrostatic transmission has two great advantages:

1. The choice of torque and speed of the machine can be reached with high accuracy, under load, for the full-speed range. In terms of design, for automatic speed control purposes, an output shaft at one front driving wheel must be available.

2. A hydraulic motor generates up to 10 times more power than an electrical motor with the same dimensions.

The overall hydrostatic transmission efficiency for full load is very near to the efficiency of stepped full-load transmissions.

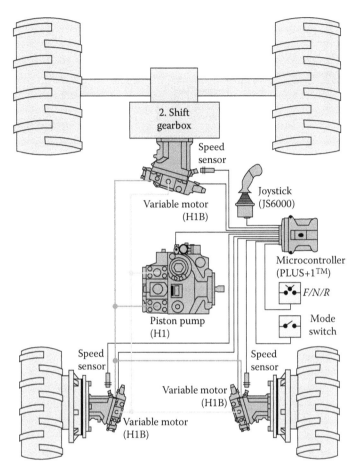

FIGURE 11.12
All-wheel-drive hydrostatic power train. (Courtesy of Sauer-Danfoss, Neumünster, Germany.)

11.3.3 Construction of Process Power Drives

The main technological processes performed by a combine harvester are the cutting and gathering of plants, grain threshing and separating, and grain cleaning and collecting into the combine tank. Consequently, process drives are required for driving the following process units: cutter bar, reel, header auger with retractable fingers (or draper), central conveyor, threshing/separating unit, grain pan (or multiple augers), straw walkers, cleaning unit, grain and tailing augers, and discharge auger. Modern combine harvesters incorporate a combination of mechanical, hydrostatic, and hydraulic drives. Figure 11.15 shows the distribution of these drives over the entire combine assembly. In summary, the characteristics of these drives are as follows:

- Power train hydrostatic motors with constant or variable displacement
- Hydraulic components with central, modular assembly
- Steering system and brakes for combine operator
- Fan drives in open or closed loops, with constant and variable displacement

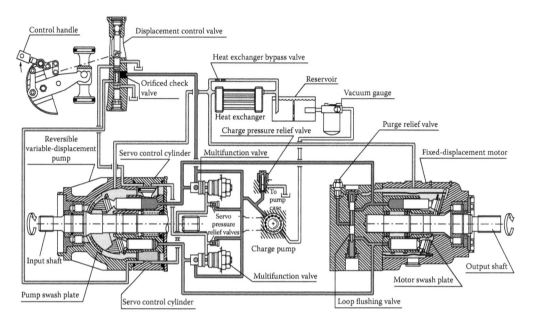

FIGURE 11.13
(See color insert.) Diagram of a hydrostatic transmission with a variable-displacement pump and a fixed-displacement motor. (Courtesy of Sauer Danfoss, Neumünster, Germany.)

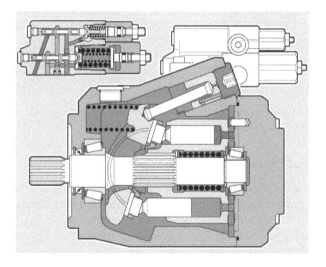

FIGURE 11.14
Hydrostatic pump with variable displacement. (Courtesy of Bosch-Rexroth Group, Lohr am Main, Germany.)

- Feeding auger and conveyor rotation that can be easily reversed
- Emergency stop function via the hydrostatic pump without additional valves
- Electronic management of the drive and implementation of hydraulics
- Control system with a controller area network (CAN) bus that can be networked as well

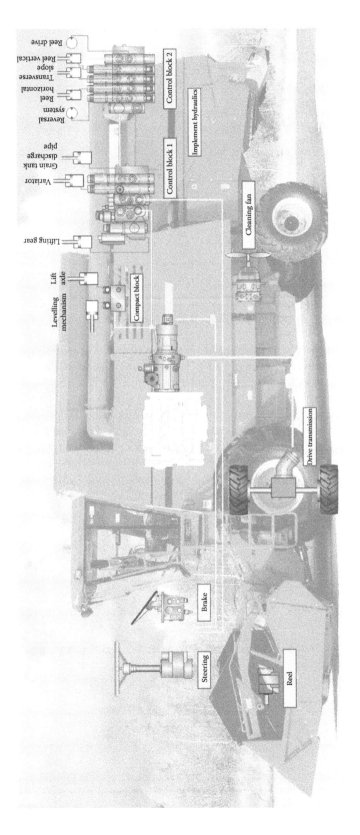

FIGURE 11.15
(See color insert.) Hydrostatic and implement hydraulics in a combine harvester. (Courtesy of Bosch-Rexroth Group, Lohr am Main, Germany.)

FIGURE 11.16
(See color insert.) Continuous variable transmission for an axial threshing unit. (Courtesy of Case IH, Racine, WI.)

Some of the mechanical drives have already been described in previous chapters. The mechanical continuous variable transmission for the axial threshing unit of a Case IH combine harvester is shown in Figure 11.16. It incorporates a planetary gearbox, a clutch, and a drum brake.

The transmission for driving a tangential threshing unit in a conventional combine can be composed of V-belt transmissions and a V-belt transmission with a variable sheave controlled by a hydraulic actuator. An example from a CLAAS combine is shown in Figure 11.17.

There are no ready receipts for combine power transmissions and process drives. Their design depends on several variables, such as machine type and complexity, power and torque for every rotational motion, physical design and location within the combine assembly, and type of process control and requirements—not to mention price considerations. The next section describes principles, methods, and mathematical equations that will help readers model the power system of combine harvesters.

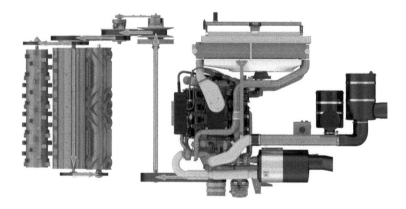

FIGURE 11.17
Power transmission for tangential threshing unit of a conventional combine. (Courtesy of CLAAS, Harsewinkel, Germany.)

11.4 Modeling of Power Train of Combine Harvesters

11.4.1 Modeling of Mechanical Power Train

The cumulative, effective tractive force F_{te} acting in the center of driving wheels of a combine equals the minimum between the tractive force F_t, which is generated by the engine and reaches the driving wheels, and the maximum friction force $F_{f\,max}$ allowed by friction between the tires and the terrain, that is,

$$F_{te} = \min\left(F_t, F_{f\,max}\right), N \tag{11.11}$$

The tractive force F_t that acts on the driving wheels can be calculated with the following equation:

$$F_t = 3600\eta_t \frac{P_b}{v_c}, N \tag{11.12}$$

where:
η_t is the transmission efficiency (minimum 0.85)
P_b is the engine brake power, kW
v_c is the combine speed, km/h

The maximum friction force that allows the vehicle to be pushed forward can be calculated as follows:

$$F_{f\,max} = 9.8066 M_{dw}\mu, N \tag{11.13}$$

where:
M_{dw} is the combine mass on driving wheels, kg
μ is the friction coefficient between the tire and the terrain

It is important to note that the combine mass on driving vehicles should be considered when the grain tank is full with grain whose relative bulk density is high (e.g., beans, sorghum) (see Table 3.5) while the combine moves in the field at full allowable speed. Certain terrain characteristics, such as terrain profile or disturbances, have to be considered as well.

Total resistance force R against combine movement results in a summation of the rolling resistance R_r and terrain grade resistance R_g. We will neglect the aerodynamic resistance because the combine harvester speed is not high in comparison with that of a car or truck. Thus, we can write

$$R = R_r + R_g, N \tag{11.14}$$

The rolling resistant force or rolling drag is the force resisting the motion of the combine tires on a surface. Rolling resistance, in a broad sense, is the force per vehicle weight unit that is required to move the vehicle on level ground, at a constant slow speed (so that the air resistance is negligible), and also where there are no traction or braking torques applied.

It is mainly caused by nonelastic effects in the tire and by the slippage that occurs between the wheel and terrain surface, and it is influenced by the vehicle speed. As a tire rotates under the corresponding fraction of combine weight, it is subject to repeated cycles of deformation and recovery, and it dissipates the hysteresis energy loss as heat. Analogous to with sliding friction, rolling resistance is often expressed as a rolling coefficient times the normal force that pushes the tire against the rolling surface.

Rolling resistance depends linearly on the combine mass M_c and machine speed v_c, as follows (Rakha et al., 2001):

$$R_r = 9.8066 C_r \left(c_1 v_c + c_2 \right) \frac{M_c}{1000}, \text{N} \tag{11.15}$$

where:

 C_r is the road surface coefficient that depends on road surface type and condition
 c_1, c_2 are the rolling resistance coefficients or rolling friction coefficients

When moving in field conditions, the values of the road surface coefficient vary within the range 2.5–16, corresponding to conditions varying from smooth but dirt surfaces to dunes with mud.

The rolling resistance coefficients depend on the tire type, as follows (derived from Fitch, 1994):

- For bias ply tires: $c_1 = 0.0438$ and $c_2 = 6.1$
- For radial tires: $c_1 = 0.0328$ and $c_2 = 4.575$

The grade resistance depends on the proportion of the combine mass that opposes the vehicle movement, and can be calculated with the following equation:

$$R_g = 9.8066 M_c \sin \alpha, \text{N} \tag{11.16}$$

where α is the terrain grade (slope).

Equation 11.16 shows that the machine runs faster downhill and slower uphill, depending on its mass and terrain slope. In field conditions, the combine engine faces a high load due to harvesting technological processes; therefore, combine mass and terrain slope will have a great role in machine dynamics.

The maximum acceleration a_c (m/s²) of the combine results from Newton's second law as follows:

$$a_c = \frac{F_{te} - R}{M_c} = \ddot{x}_c(t) \tag{11.17}$$

where x_c is the distance traveled by the combine.

Thus, we can write the first law that governs the combine motion:

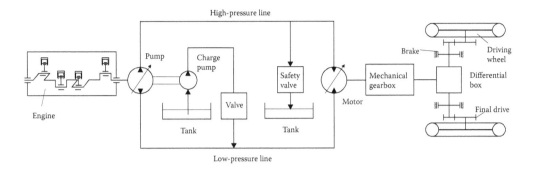

High-pressure line

FIGURE 11.18
Schematic overview of combine hydrostatic propulsion system.

$$F_{te} - R = M_c a_c \qquad (11.18)$$

For further development of the model, the reader is directed to Application 2.2 and Section 2.7.

11.4.2 Modeling of Hydrostatic Power Train

The hydrostatic propulsion system of a combine substantially alleviates the task of the operator while the machine is operating in the field. Furthermore, the use of a cruise control feature can help maintain a constant speed of the machine, whose motion is subjected to disturbances caused by the varying slope and total mass of the vehicle as the grain accumulates in the tank. A schematic overview of the hydrostatic propulsion system of a combine harvester is shown in Figure 11.18. The power delivered by the engine is transmitted via a drive shaft to the hydrostatic pump, whose oil flow is controlled by an electrical current. To ensure that oil is always available at a setup pressure, there is a charge pump connected to the same drive shaft. Inside the main pump, the swash plate incline determines whether the vehicle moves forward, backward, or remains stationary. The pressurized oil flow passes through a safety valve that adjusts the pressure if it is too high. The oil flow is used in a closed-loop circuit to drive the motor, which is connected to the mechanical gearbox. Finally, the power is transmitted to the driving wheels via the differential and final drive. The motor has also a swash plate. The swash plates of the engine and motor are used to control the desired combine speed by the operator. In the following, I propose a mathematical model of a hydrostatic power train that was derived from one for forest vehicles (Carlsson, 2006) and further improved by the author.

Figure 11.19 shows the mathematical model parameters and their association with physical components of the hydrostatic power train. Mechanical coupling of the engine with the pump is characterized by the following torque balance equation:

$$\left(J_d + J_p\right)\dot{\omega}_c = T_d - T_p - T_{pf} \qquad (11.19)$$

where:
J_d is the moment of inertia of the engine, Nms^2
J_p is the moment of inertia of the pump, Nms^2
ω_c is the engine crankshaft angular speed, s^{-1}

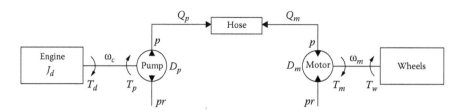

FIGURE 11.19
Model parameters and their association with hydrostatic power train components.

T_d is the torque developed by the engine, Nm
T_p is the resistant torque from the pump to the engine, Nm
T_{pf} is the torque developed by friction in the pump, Nm

The angular speed of the pump shaft is the same as that of the engine crankshaft due to their mechanical connection, usually through a shaft coupling.

The engine torque T_d depends on the mechanical efficiency η_m of the engine as follows:

$$T_d = \eta_m M_b \tag{11.20}$$

where M_b is the engine brake torque (Nm).

T_p is the resistant torque, which the engine senses from the pump; it is a function of pump displacement and the difference between charge and return pressure as follows:

$$T_p = \frac{(p - p_r)D_p}{\eta_{tp}} \tag{11.21}$$

where:
D_p is the pump displacement, m³/rad
p is the charge (high) pressure in the hose, Pa
p_r is the pressure in the return hose, Pa
η_{tp} is the torque efficiency of the pump, for example, $\eta_{tp} = 0.91$–0.930

The friction torque T_{pf} can be reasonably described by a sum of Coulomb friction and viscous friction as follows:

$$T_{pf} = T_{pfm} + T_v + T_c \tag{11.22}$$

where:
T_{pfm} is the torque due to internal mechanical friction in the pump, Nm
T_v is the resistant torque due to viscous shearing of the fluid in the pump, Nm
T_c is the constant friction torque (breakaway torque) that is independent of both fluid pressure and speed, Nm

The torque T_{pfm} can be calculated with the following equation:

$$T_{pfm} = \mu_p \left(p - p_r\right) D_p \tag{11.23}$$

where μ_p is the coefficient of mechanical friction between pump moving parts.

The torque T_v value can be determined experimentally. However, it may be calculated with the following equation:

$$T_v = c_v D_p \mu_f \omega_c \tag{11.24}$$

where:

c_v is the viscous shear coefficient
μ_f is the dynamic viscosity of the fluid, Pa·s

The fluid flow out from the pump is proportional to the pump displacement, shaft rotation, and volumetric efficiency of the pump η_{vp}, as shown in Equation 11.25:

$$Q_p = \omega_c D_p \eta_{vp} \tag{11.25}$$

The variation of the volumetric efficiencies of both pump and motor is large, depending on the displacement and the angular speed.

The equations describing the torque balance for the pump can be respectively adapted to describe the torque balance for the motor as follows:

$$\left(J_m + J_w\right) \dot{\omega}_m = T_m - T_w - T_{mf} \tag{11.26}$$

where:

J_m is the moment of inertia of the motor, Nms²
J_w is the moment of inertia of mechanical transmission (gearbox to wheels), Nms²
ω_m is the motor angular speed, s⁻¹
T_m is the motor torque delivered for the traction, Nm
T_w is the resistant torque of mechanical transmission (gearbox to wheels), Nm
T_{mf} is the torque developed by friction in the motor, Nm

The torque T_m delivered by the motor for machine propulsion is given by the following equation:

$$T_m = \left(p - p_r\right) D_m \eta_{tm} \tag{11.27}$$

where:

D_m is the motor displacement, m³/rad
η_{tm} is the torque efficiency of the motor, for example, $\eta_{tm} = 0.95 - 0.97$

Readers should note the difference between Equations 11.21 and 11.27, which takes into consideration the torque as a signal output versus input.

The friction torque T_{mf} can be reasonably described by a sum of Coulomb friction and viscous friction as follows:

$$T_{mf} = T_{mfm} + T_{vm} + T_{cm} \tag{11.28}$$

where:

T_{mfm} is the torque due to internal mechanical friction in the motor, Nm
T_{vm} is the resistant torque due to viscous shearing of the fluid in the motor, Nm
T_{cm} is the constant friction torque (breakaway torque of the motor), which is independent of both fluid pressure and speed of the motor

The torque T_{mfm} can be calculated with the following equation:

$$T_{mfm} = \mu_m \left(p - p_r \right) D_m \tag{11.29}$$

where μ_m is the coefficient of mechanical friction between pump moving parts.

The torque T_{vm} value may be calculated with the following equation:

$$T_{vm} = c_v D_m \mu_f \omega_m \tag{11.30}$$

where:

c_v is the viscous shear coefficient
μ_f is the dynamic viscosity of the fluid, Pa·s

The fluid flow in the motor is proportional to the motor displacement, shaft rotation, and volumetric efficiency of the motor η_{vm}, as shown in Equation 11.31:

$$Q_m = \frac{\omega_m D_m}{\eta_{vm}} \tag{11.31}$$

Note the difference between Equations 11.25 and 11.31; the reason to have η_{vm} in the denominator is that the flow is the input to the motor, while for the pump, it is the output.

The torque T_w from the wheels is not known. It has to be calculated based on design parameters as well as the type and slope of terrain, as shown in Section 11.4.1. Both torques T_d and T_w should be initially treated as disturbances of the system.

The equations of the propulsion system model should be rewritten based on the definition of the following system of column vectors:

$$x = \begin{pmatrix} \omega_c \\ \omega_m \\ \Delta p \end{pmatrix}, u = \begin{pmatrix} D_p \\ D_m \\ \dot{p}_r \end{pmatrix}, d = \begin{pmatrix} T_d \\ T_w \end{pmatrix} \tag{11.32}$$

and

$$\theta = \left(\mu_p \mu_m J_{dp} J_{mw} \theta_{\eta_{tp}} \theta_{\eta_{vp}} \theta_{\eta_{tm}} \theta_{\eta_{vm}} \right)^T \tag{11.33}$$

where:

$$J_{dp} = J_d + J_p \tag{11.34}$$

$$J_{mw} = J_m + J_w \tag{11.35}$$

The vector x represents the state vector, u is the input signal vector, and d is the disturbance vector, while the elements of vector θ are unknown.

For identification, a linear system can be written as follows:

$$\dot{x}(t) = A(\theta)x(t) + B(\theta)u(t) + d(t)$$
$$y(t) = C(\theta)x(t) + D(\theta)u(t) + h(t) \tag{11.36}$$

A further task is to estimate these parameters through different experiments.

At this point, readers can adapt/apply the above-mentioned model for each specific case. Further description of linear and nonlinear system identification methods is beyond the scope of this book.

References

Browning, E.P. 1978. *Design of Agricultural Tractor Transmission Elements*. ASAE Lecture Series 4. St. Joseph, MI: American Society of Agricultural Engineers.

Carlsson, E. 2006. Modeling hydrostatic transmission in forest vehicle. PhD dissertation, Linköpings University.

Fitch, J.W. 1994. *Motor Truck Engineering Handbook*, 4th ed. Warrendale, PA: Society of Automotive Engineers.

Goering, C. 1999. Engines. In *CIGR Handbook of Agricultural Engineering*, ed. B.A. Stout and B. Cheze, 41–53. Vol. III. St. Joseph, MI: American Society of Agricultural Engineers.

Rakha, H., I. Lucic, S.H. Demarchi, et al. 2001. Vehicle dynamics model for predicting maximum truck acceleration levels. *ASCE Journal of Transportation Engineering* 1:34.

Bibliography

Bosch-Rexroth Group. 2011. Drive and control systems for combine harvesters and forage harvesters. Lohr am Main, Germany: Bosch-Rexroth Group. http://www.boschrexroth-us.com/country_ units/america/united_states/sub_websites/brus_brh_m/en/markets_applications_jg/a_ downloads/re98071_1101.pdf.

Coen, T., I. Goethals, J. Anthonis, et al. 2005. Modelling the propulsion system of a combine harvester. Technical Report 05-47. Leuven: Belgium: ESAT-SISTA, Katholieke University Leuven.

Coen, T., J. Paduart, J. Anthonis, et al. 2006. Nonlinear system identification on a combine harvester. In *Proceedings of the 2006 American Control Conference*, Minneapolis, MN, pp. 3074–3079.

Gedeon, G. 1993. Diesel engine fundamentals. In *Department of Energy Fundamentals Handbook*. Vol. I. *Mechanical Science*. DOE-HDBL-1018/1-93. Washington, DC: Department of Energy.

Hatami, H. 2011. *Hydraulic Formulary*. Lohr am Main, Germany: Bosch-Rexroth.

Orthwein, W.C. 2004. *Clutches and Brakes*. New York: Marcel Dekker.

Renius, K.T. and H. Pfab. 1990. Traktoren 1989/90 [Agricultural tractors 1989/90]. *ATZ Automobiltechnische Zeitschrift* 92(6):334–346.

Sauer-Danfoss. 2011. Series 90 axial piston pumps. Technical information. Neumünster, Germany: Sauer-Danfoss.

Totten, G.E. 2012. *Handbook of Lubrication and Tribology*. Vol. I. *Application and Maintenance*. 2nd ed. Boca Raton, FL: CRC Press.

Waukesha, C.H. 2011. Waukesha pumps engineering manual. Delavan, WI: Wakesha Cherry-Burrell.

Xie, Y. 2013. Integrated plant and control design for vehicle–environment interaction. PhD dissertation, University of Illinois at Urbana–Champaign.

ZF Freidrichshafen. 2014. Drives for combines. Drives and chassis technology. Passau, Germany: Zahnradfabrik Freidrichshafen.

Zoz, F.M. and R.D. Grisso. 2003. Traction and tractor performance. Tractor Design 27. ASAE Publication 913C0403. St. Joseph, MI: American Society of Agricultural Engineers.

12

Dynamic Modeling of Material
Flow in a Combine Harvester

12.1 Introduction to Material Flow Dynamics in a Combine

After being cut from the field, vegetal material processed by a combine harvester evolves from its initial nonhomogenous, fibrous structure to unbound, multicomponent mixtures, which segregate and separate in different processing zones of the machine, following in series or parallel paths, in as many component flows. Cut material is formed of bendable/ fragile straws/stalks with ears, panicles, and so forth. *Multicomponent mixtures* of vegetal material consist of dynamically deformed or fragmented straw pieces, chaff, leaves, unthreshed ears, and grains of base crop, as well as weed plants and seeds, or even nonbiologic objects (e.g., soil granules). Characteristic of such *flows of multicomponent materials* is an uncertainty in the location and time of the particular constituents. Practically, it is not possible to exactly measure or predict the evolution of such component mixtures due to individual, unbound components of specific behavior, the complexity of component interactions, nonhomogeneity, and discontinuities of the mixture flow, as well as physical or simulation scaling. Besides, these components are not integrated within a fluid with relative predictable behavior—not to mention that high values of moisture released by vegetal material due to mechanical compression create additional disturbances by component adherence to active or passive elements of combine working units. As a consequence, a *stochastic approach* or an *ensemble averaging* concept may be used.

Vegetal material processed by a combine harvester should be regarded as a dynamic system; understanding its structure and behavior is essential for modeling and problem solving. Modeling of material behavior should only be considered in correlation with design and functional parameters of each working unit of the machine. Generally, mathematical modeling of dynamic flows requires taking a sequence of steps, as follows:

- Developing a dynamic hypothesis explaining the cause of a problem
- Creating basic causal graphs
- Augmenting each causal graph with necessary, available data
- Converting the augmented causal graph to a dynamic flow graph
- Translating the dynamic flow graph into mathematical equations
- Developing a simulation program and running it
- Validating the results
- Extracting applicable rules to be extrapolated to similar cases

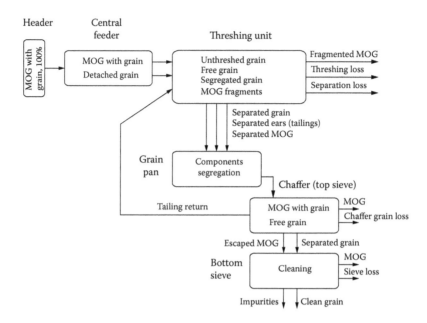

FIGURE 12.1
Schematic of realization of dynamic process flows (grain and MOG).

A casual loop provides insight into the structure of the system. Often, without experimental data, it might be difficult to infer the behavior of a system from its casual loop representation. As an example, an optimal linear model, which incorporates algebraic compensation for the internal return loop, makes estimation of the feedrate of grain flowing into a combine harvester possible. However, the estimation of material feedrate over a certain path is drastically influenced by the dynamics of the machine and by permanent evolution of the material mixture structure and component fractions of the material mixture, along with variation of their physical characteristics.

A comprehensive modeling of material dynamic flow can be developed by integration of individual working unit models. A generic Simulink realization of material dynamic flow (grain and material other than grain [MOG]) into an axial combine harvester is illustrated in Figure 12.1. By reviewing it, one can understand the multiple flows the material spreads into during its processing by successive working units of a combine harvester.

Connecting specific models of grain or MOG flows, as outlined in Chapter 5, produces a nonlinear, high-order characterization of overall combine process and design parameter influences. A set of material flow variations and nonlinearity of the dynamics of a machine's working unit affect the flows of both grain and MOG into the machine. Accumulations of material in different zones and delays over different paths have to be taken into consideration. The next section will introduce readers to the dynamic dimension of the systemic perspective, with a focus on the concepts of material stock, flow, and feedback.

12.2 Material Flow Dynamics Modeling

Causal loops are characterized by their ability to express relationships among the components of a system, for example, components of vegetal material processed by a combine harvester. But causal loops do not quantify the components of the system. Stocks and flows are concepts that account for such quantities.

A *stock* is an accumulation of material that has built up in a system over time. It is also known as a *level* that is a state variable. The stock value changes continuously over time by accumulation/reduction of rate integration, even when a rate is changing discontinuously.

A *flow*, or a rate, is expressed as an amount per time unit, reflecting the change of the stock (level) value. The value of the flow is not dependent on its previous evolution, but depends on the levels in the system, along with exogenous influences. The *inflow* F_{in} (or the input flow) is perceived as the rate at which a stock increases over time (the flow is going to), while the outflow F_{out} (or the output flow) is the rate at which a stock decreases (the flow is going out from). The net flow is the derivative of the total stock $S(t)$ with respect to time:

$$\frac{dS(t)}{t} = F_{in}(t) - F_{out}(t) \tag{12.1}$$

Consequently, the quantity inside a stock is given by the following integral:

$$S(t) = \int_0^t F_{tot}(s)\,ds \tag{12.2}$$

where F_{tot} is a function of the total flow of the system along the considered path.

If we take into consideration that initially, at $t = 0$, there is some stock, Equation 12.2 becomes

$$S(t) = \int_0^t \left[F_{in}(s) - F_{out}(s) \right] ds + S(0) \tag{12.3}$$

If there is an amount of material in transit and the quantity accumulates in *queue*, the outflow is the inflow shifted by a time delay d_t. But in combine harvesters, the order of outflow material does not necessarily depend on the inflow. In such a case, we have to consider an average delay time d_{at}. The outflow from a first-order material delay is linearly proportional to the stock by an average delay time. Mathematically, we can express this as follows:

$$\frac{d}{dt}S(t) = \frac{S(t)}{d_{at}} \tag{12.4}$$

In system as complex as a combine harvester, the material in different places cannot be considered stocks in cascade, for example, material accumulation on the grain pan that is generated by different inflows. To solve the flow for high-order delays, a system of differential equations is needed.

A *feedback loop* is a closed chain of causal connections from a stock, through a set of rules determined by physical laws or specified actions that are dependent on the level of stock, getting back through a flow to change the stock. A *negative feedback loop* is a balancing structure in a system; it represents a source of stability, that is, resistance to change.

12.3 Ensemble-Averaged Models of Multicomponent Vegetal Material Mixtures

Vegetal material processed in a combine harvester is composed of the following:

- Grains regarded as unbound, dry granular particles, weed seeds, or even nonbiologic objects, for example, soil granules
- Deformed or fragmented straw pieces, chaff, leaves, ears, and weed plants

This material mixture is not homogeneous because of the differences in shape, size, density, elasticity, brittleness, and morphology of the components. During different technological processes, grain segregation changes the spatial nonuniformity of the mixture. Efforts have been made to explain segregation from the nonequilibrium, dissipative nature of a vegetal material mixture; however, the mechanisms of segregation have not been fully understood; rather, they have been explained through a stochastic approach. The author has a contribution in this field, as described in Chapters 5, 6, and 7.

A vegetal material mixture is characterized by large variability due to the complexity of interfaces among its components, discontinuities in material flow properties, and physical scaling issues. When realization of a certain phenomenon, for example, grain segregation, is quantified, predictions in terms of *expected value* and *variance of grain accumulation* (stock) are interpreted as an average over all the events, that is, the *ensemble* of event realizations. A realization of material flow (grain flow) means the material motion between two levels. Accordingly, an ensemble of material flow realizations has a probability measure of realizations in constituent subsets. Ensemble averaging is the summation of the values of corresponding variables for each realization divided by the number of observations.

In this section, we do not give a complete overview of this subject; that would exceed a reasonable book size, as well as the author's expertise. Readers can find more information on this topic in the work written by Berry (2003) and indicated bibliography.

12.4 Modeling of Dynamic Grain Flow in a Combine Harvester

During operations, once the functional parameters of combine working units have been set, the combine moves on the field with a quasi-constant speed v_m; however, the speed is subject to adjustments based on combine operator training, skill, and working experience with a given combine. Due to inherent variations of material mass per field surface unit (m^2) and plant cutting height (even when controlled), the input material flow into the combine varies dynamically, and it is extremely difficult to continuously and accurately monitor it. A better way of estimating it is by sampling it at a certain period.

Let us assume that the instantaneous (at time t) grain yield on a given field surface unit is $m_u(t)$ (kg/m^2). If we measure it by sampling at period T_s, we get the data sequence $\{m_u(k)\}$.

The instantaneous value of grain input feedrate $q_{sin}(k)$ into a combine harvester can be estimated when the following two sampled parameters are known:

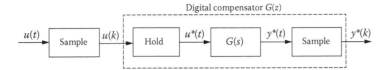

FIGURE 12.2
Schematic representation of digital compensation.

- Actual cutting width, $w(k)$
- Combine speed, $v_m(k)$

Thus, we can write

$$q_{sin}(k) = w(k)v_m(k)m_u(k) \qquad (12.5)$$

A combine harvester continuously performs its technological processes. To simulate the entire continuous technological process as a discrete-time system, one needs to determine the *discrete equivalent* of the overall combine process (preceded by a hold circuit and followed by a sampler) (Figure 12.2). The continuous and discrete equivalence is achieved by modeling the discrete compensation process.

In other words, the objective is to establish a transfer function $G(z^{-1})$ that corresponds to an ideal analog $G(s)$ such that the error is sufficiently small. The error depends primarily on the sampling period T_s and the holding method used for each input sample during this period.

If an overall process mathematical model $G(z^{-1})$ can be developed, then the grain output feedrate $q_{s\,out}$ is given by the following equation:

$$q_{sout}(k) = G(z^{-1})w(k)v_m(k)m_u(k) \qquad (12.6)$$

From Equations 12.5 and 12.6 we can write

$$q_{sout}(k) = G(z^{-1})q_{sin}(k) \qquad (12.7)$$

The model $G(z^{-1})$ is a discrete equivalent of a transfer function; if one translates the above equation into a frequency domain, then the following is obtained via numerical integration using the trapezoid rule (Tustin's method):

$$\frac{s(1+z^{-1})}{2} \leftarrow \frac{z^{-1}-1}{T_s} \qquad (12.8)$$

Under the trapezoid rule, the equivalent discrete-time system is stable if, and only if, the continuous-time system is stable.

The most difficult problem is to develop a global mathematical model of combine processes expressed as a Z-transform $G(s)$.

Let us consider an axial combine harvester whose axial threshing unit is axially fed with vegetal material. The combine cuts the plants at a predetermined cutting height. Therefore, for a given crop and cutting height setting, the material entering into combine is character-ized by a grain/MOG ratio ψ_{sp}:

$$\psi_{sp} = \frac{m_{sin}}{m_{pin}} = \frac{q_{sin}}{q_{pin}} \tag{12.9}$$

where m denotes the mass and q denotes the feedrate of grain (s) and MOG (p) entering (in) the combine.

Grains separated through the concave, grates, and cage fall onto the grain pan, which transports the grains with a constant speed v_{gp}. Let us assume that the length of the grain pan equals the length L of the axial unit.

Transportation of vegetal material within the axial threshing unit is done with the speed v_{ax} (Equation 5.23). The probability density function $s_d(x)$ of grain separation on the grain pan is given by Equation 5.47, and the cumulative separated grain $s_s(x)$ is given by Equation 5.48. Using Equations 5.102 and 5.103 in combination with Equation 12.9, we can write the dependence of threshing process parameters β and λ on the grain input feedrate as follows:

$$\beta = \frac{a \left(\dfrac{q_{sin}}{\psi_{sp}} \right)^b}{e^{\; c \frac{q_{sin}}{\psi_{sp}}}} \tag{12.10}$$

$$\lambda = \frac{ke^{\; m\frac{q_{sin}}{\psi_{sp}}}}{\left(\dfrac{q_{sin}}{\psi_{sp}} \right)^b} \tag{12.11}$$

The material path in the axial threshing unit is a cylindrical helix with the axial pitch given by Equation 5.24. Based on the helix parametric equations, the current position $x(t)$ can be calculated with the following equation:

$$x(t) = \frac{p_{am}}{2\pi} t = \frac{R_s}{abx^{b-1}} t \tag{12.12}$$

The speed of the material through the axial unit is higher than the grain speed on the grain pan. Then, the separated grain flow $q_{s\,out}$ (t) (kg/s) at the end of the grain pan can be modeled as follows:

$$q_{sout}(t) = q_{sin}s_s\left[x(t - \Delta T_d), q_{sin} \right] \tag{12.13}$$

where:
  ΔT_d is the mean transportation delay (s) for grain moving from the feeding zone of the threshing unit to the grain pan

$s_s(x)$ is the cumulative separated grain (Equation 5.48) to the grain pan over the length $0 < x < L$

The process parameters β and λ depend on many other parameters, as described by the author's model's equations (5.99 and 5.100). All these equations confirm the nonlinearity of threshing and separating processes, regardless of the type and design of the threshing unit.

A similar kind of nonlinearity characterizes the grain separation on the chaffer and bottom sieve of the cleaning unit, as expressed by Equation 7.21. For instance, the equation that models the grain flow q_{chout} separated through the chaffer, thus reaching the bottom (grain) sieve of the cleaning unit, can be written as follows:

$$q_{chout}(t) = q_{sout} u_s \left[x(t - \Delta T_{dc}), q_{sout} \right] \tag{12.14}$$

In the above-mentioned equation, we considered that the entire flow q_{sout} of separated grain from the threshing unit and conveyed by the grain pan will be fed to the cleaning unit chaffer. In this case, the delay time ΔT_{dc} is the mean transportation delay (s) for grain moving from the feeding zone of the chaffer to the bottom sieve of the cleaning unit. Equation 12.14 must be repeated for grain separation through the bottom sieve; obviously, the delay time will be different.

The dependence of grain separation through the chaffer on the grain feedrate is graphically shown by Beck (1999). Readers are advised that the author's work in the immediate future will be to develop mathematical models that quantify such dependence of the process parameters λ_c and c on the functional and design parameters of the cleaning unit.

The dynamic models of grain flow developed above can be used to develop, in a similar way, the models for MOG flow into the combine, based the author's models described in Chapters 5 through 7.

While this chapter lays out the foundation of dynamically modeling the grain and MOG flows in a combine harvester, there are a few more required steps to be taken, as described in Chapter 2: applying Z-transforms, filtering procedures, and inverse dynamics as compensation for the combine system dynamics. An example of such work can be found in the paper published by Maertens et al. (2001).

References

Beck, F. 1999. Simulation der Trennprozesse im Mähdrescher [Simulation of separation processes in combine harvesters], PhD dissertation, Hohenheim University, Forschritt-Berichte VDI, Reihe 14(92).

Berry, R.A. 2003. Ensemble equations for multiphase, multi-component, and multi-material flows. INEEL/EXT-03-01011. Idaho Falls: Idaho National Engineering and Environmental Laboratory, Bechtel BWXT Idaho.

Maertens, K., M. Reyniers, and J. De Baerdemaeker. 2001. Design of a dynamic grain flow model for a combine harvester. *Agricultural Engineering International: The CIGR Journal of Scientific Research and Development* III, Manuscript PM 01 005.

Bibliography

Choopojcharoen, T. and A. Magzari. 2012. Mathematics behind system dynamics. BSc thesis, Worcester Polytechnic Institute.

De Baerdemaeker, J., H. Ramon, J. Anthonis, et al. Advanced technologies and automation in agriculture. *Control Systems, Robotics, and Automation* XIX:284–320.

De Silva, C.W. 2009. *Modeling and Control of Engineering Systems.* Boca Raton, FL: CRC Press, Taylor & Francis.

Gent, S.P. 2009. Computational modeling of multiphase fibrous flows for simulation based engineering design. PhD thesis, Iowa State University.

Lenaerts, B., E. Tijskens, J. De Baerdemaeker, and W. Saeys. 2012. Simulation of grain-straw separation by a discrete element approach with bendable straw particles. Presented at International Conference on Agricultural Engineering, Valencia, article C0861.

Sun, J. 2007. Multiscale modeling of segregation in granular flows. PhD dissertation, Iowa State University.

Unger, T. 2004. Characterization of static and dynamic structures in granular materials. PhD dissertation, Budapest University of Technology and Economics.

Yohannes, B., K. Hill, and L. Khazanovich. 2009. Mechanistic modeling of unbound granular materials. Report MN/RC 2009-21. Department of Civil Engineering, University of Minnesota, Minneapolis, MN.

13

Sensors and Fault Diagnosis Systems for Combine Harvesters

13.1 Sensors

13.1.1 Introduction

High yields of agricultural crops are possible when the proper level of agricultural technology is achieved and climate and soil/terrain conditions are favorable for each crop. Soil conditions refer to the soil structure, texture, moisture, nutrient availability, and elevation. The agricultural technology level can be assessed by quantity, variety, working performance, ease of operation, and control of machinery and equipment. Spatial variability in properties and conditions of soil for agricultural use leads to local variations in crop yield. Shearer et al. (1997) reported that significant differences in measured yield exist between adjacent harvest swaths; such differences were attributed to potential machine/operator variability. Strubb et al. (1996) confirm that the accuracy of yield estimates depends greatly on the variability of harvesting conditions, the total field area in one grid, and the range of operating conditions over which a sensor is calibrated.

When a crop is harvested with combine harvesters, its process parameters have to be continuously adjusted to maximize machine and operator productivity and minimize loss, while the quality of the grain is very well preserved. The complexity of combine technological processes limits the ability of the operator to perform multiple tasks, and to make the best decisions in regard to process parameter settings. Moreover, combine operators are highly susceptible to fatigue because they have to monitor the crop flow intake by maintaining the row alignment, and threshing and separating processes, and oversee grain unloading without interrupting the operation. Thus, a high level of automation of combine harvester processes may lead to autonomous machines. However, this requires advances in sensor technology, integration of precision-adjusting mechanisms, real-time acquiring and processing of data, and implementation algorithms for process maximization, decision system, and machine control. This implies a mechatronic design of combine harvesters so that they can adapt to spatially variable crop yields, that is, performing precision agriculture.

The availability of material, grain, and material-other-than-grain (MOG) flow sensors in combine harvesters would allow real-time measuring of the grain yield during harvesting, an adaptive speed control of the machine, a dynamic adjustment of combine process parameters, and safe operation.

The sensors for combine harvesters must satisfy a series of critical performance criteria, as follows:

- Capability of measuring material flows up to 20 kg/s
- Less than 3% measurement errors of cumulative mass under dynamically varying step and ramp flow rates
- Less than 4% measurement errors of average instantaneous flow rates
- Measurement accuracy independence of variations in bulk properties of the material (grain or MOG)
- Low sensitivity to machine vibrations due to machine or working unit motion
- Large range of operating conditions
- Limited requirements for recalibration and maintenance
- Reasonable cost (low but not too low)

According to Thylen and Murphy (1996), the measurement errors were associated with low yield at the start of harvest, interruptions in crop intake, sudden changes in machine forward speed, variations in cutting width, and the delay time of material traveling from the cutter bar to the yield sensor.

Besides sensing a certain measurable parameter, a computational realization of a physics-based model is required that describes the relationship between the physical meaning of a parameter and the electrical signal generated by the sensor. This is usually done with Labview, MATLAB, or proprietary software.

As a general remark, although the combine process and mechanical design have changed a little in recent years, many advances in sensor and control technology have been incorporated into combine harvester functionality, thus improving the overall efficiency, work quality, and operating conditions.

13.1.2 Classification of Sensors for Combine Harvesters

There are several groups of sensors that can be used to a varying extent in a combine harvester:

- Sensors for material properties
- Process parameter sensors
- Mechanism/device adjustment monitoring sensors
- Machine-steering and speed control sensors
- Cab ergonomic sensors
- System functionality diagnostic sensors
- Sensors for multispectral remote sensing data

The *sensors that measure and monitor material properties* are *grain moisture sensors, grain protein sensors, grain gluten sensors, grain starch sensors, degree of plant maturity sensors,* and *proximal soil-sensing sensors*. The information on grain moisture variation helps in the adjustment of threshing, separating, and grain conveying process parameters, and dynamic correction of bulk density and physical density of harvested grains in regard to the grain flow.

The information about protein content of the grain is useful for on-farm sorting of grains by protein content and for field mapping protein. Creation of in-field protein distribution maps at harvest enables directing nitrogen-based fertilizers with a variable rate according to subsequent crop needs. Other applications take into consideration near-infrared sensors that dynamically analyze the contents of gluten and starch in grain.

The information delivered by the sensors for assessment of plant maturity, for instance, when harvesting maize for silos, is very useful for deciding on the magnitude of the pressure on the material as silo stock.

Although not related to the harvested crop, the proximal dynamic soil sensors mounted on a combine harvester deliver information during harvesting about the surfed topsoil water content, the total nitrogen and carbon content that can be used in selective grain harvesting technology, and the spatially variable grain quality versus environmental conditions (Wojciechowski and Czechlowski, 2013).

The *sensors for process parameters* measure *plant cutting height, cutting width, material flow* (*grain flow, MOG flow*), *grain loss, cleaning unit airflow,* and *air speed*. Their names unequivocally describe the physical quantity they sense.

The *sensors for monitoring mechanisms, device settings, and adjustments* are sensors of general use, such as *rotary sensors, proximity sensors, distance sensors, torque gauges,* and *hydraulic pressure* and *pneumatic pressure gauges*. A rotary transducer converts the rotation of a shaft into an electrical signal whose magnitude or frequency is proportional to speed. The proximity sensors are used in setting the clearance or position of different setting mechanisms, for example, setting the clearance between the cylinder and concave. A torque gauge monitors the required torque of a rotating unit (e.g., cylinder, rotor); this information is then used to adjust the fuel injection into the combine engine.

The *sensors for machine speed control and steering* are *rotary sensors*. Special sensors are *global positioning system (GPS)-based posture sensors* for machine guidance and orientation in the field. A posture sensing system measures machine displacement, velocity, and bearing relative to a selected reference frame.

The *cab sensors* monitor ergonomic and climate parameters, such as *temperature, humidity, noise,* and *mechanical vibration* parameters, to ensure ergonomic working conditions for the combine operator.

The *diagnosis of combine functionality* takes into consideration three classes of failures caused by malfunctioning actuators or sensors, unexpected/sudden process changes, and sudden changes of the combine process model parameters.

Multispectral remote sensing refers to mapping and monitoring the crop (grain yield), soil variability, and other conditions that affect plant health, quality of the crop, or weed spread.

In the following, detailed descriptions of most of the specific sensors used in combine harvesters are given.

13.1.3 Grain Yield Sensors

Grain yield sensors monitor the quantity of grain that is collected in the combine harvester by measuring the rate at which clean grain enters the grain tank. Usually, they report the data at small intervals of time (1–3 s). The grain rate can be measured either directly or indirectly. Direct measurement means the measurement of the actual mass or volume of grain. Indirect measurement is based on measuring some parameters whose values can be functionally correlated to the grain flow.

Besides the performance criteria mentioned in Section 13.1, researchers suggest additional features for yield sensors, as follows:

- Not impeding or blocking the grain flow; that is, the sensor should be of a nonintrusive type.
- Avoiding inaccurate indirect flow measurement, for example, strain gauge measuring or monitoring the torque of the grain conveying auger.

The volumetric flow rate or mass flow rate measured by the sensor must be corrected for the actual grain moisture. For instance, the mass flow rate of grain is a measure of conveyed grain mass considering that the bulk density remains constant.

The mass flow sensors can be classified by two types:

- *True mass flow meters* that sense the real mass flow of material through the device
- *Inferential mass flow meters* that multiply the instantaneous granular material concentration, that is, density, by the flow velocity

The measurement of true mass flow meters is not influenced by material density; it is the flow velocity that may influence the accuracy of measurements.

Table 13.1 shows the features of several commercially available yield sensors and their measurement principles.

13.1.3.1 Impact Force–Based Grain Yield Sensors

The installation (in New Holland combines), design, and dynamics of a grain yield sensor based on the impact force measuring principle are shown in Figures 13.1 through 13.3, respectively. This sensor performs indirect grain flow measurement. The device is composed of a jointed bar that has a sensing plate at one end and a counterweight at the other end that neutralizes the rubbing effect of the grain. A transducer senses the force exercised by the grain.

Let us assume that the grain exercises the force F_y on the sensing plate. The weight of the counterweight is G_{cw}. The force sensed by the transducer is F_s. From the equation of moment balance, we get

$$F_y = \frac{G_{cw}a + F_s c}{b} \tag{13.1}$$

TABLE 13.1

Characteristics of Commercially Available Grain Yield Sensors

Manufacturer	Measured Physical Quantity	Measurement Principle	Claimed Accuracy, %
New Holland	Mass	Impact force (impulse)	N/A
Ag-Leader	Mass	Impact force	±2
Micro-Trak	Grain mass	Impact force	±1–2
CLAAS	Volumetric flow	Displacement	±1
John Deere	Mass	Displacement	N/A
RDS Technology	Volumetric flow	Optoelectronics	±2
Massey Ferguson	Volumetric flow	Gamma radiation	±1

FIGURE 13.1
Installation of New Holland yield sensor. (Courtesy of New Holland.)

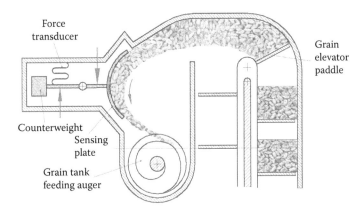

FIGURE 13.2
(**See color insert.**) Design impact force–based grain yield sensor.

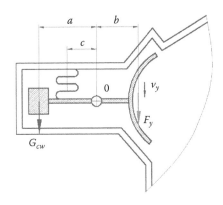

FIGURE 13.3
Dynamics of impact force–based yield sensor.

The grain impact on the plate generates the impulse H_y. Consequently, we can write

$$F_y = \dot{H}_y = \frac{d}{dt}\left(m_g v_y\right) = v_y \frac{dm_g}{dt} + m_g \frac{dv_y}{dt} \tag{13.2}$$

Since the speed of the grain is constant during two consecutive readings, the impact force is proportional to the grain mass derivative with respect to time and speed. Such grain yield sensors typically transmit a periodic report of the yield, for instance, at 1, 2, or 3 s intervals, as chosen by the combine operator or based on system presettings. A model of this type of sensor was proposed by Reinke (2010).

The reading of the sensor is independent of kernel mass and moisture content. There is no need for calibration of this sensor between different fields and crops. However, the accuracy of this type of sensor is ±1%–2%.

13.1.3.2 Volumetric Grain Flow Sensors

This type of grain flow sensor directly measures the volume of the grain passing through a paddle. To provide the mass flow, the information from a volumetric flow sensor must be multiplied by the grain bulk density corrected by real-time measurements of grain moisture content. An example of this type of sensor is the Claydon *Yield-O-Meter*, which was developed by J. Claydon, Norfolk, UK. The operation of this sensor is illustrated in Figure 13.4. The sensor is mounted between the outlet of the clean grain

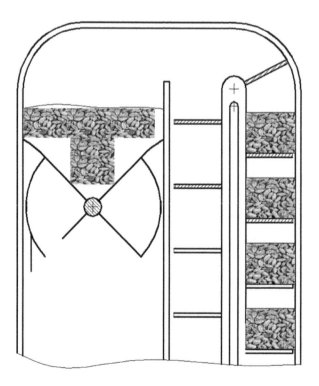

FIGURE 13.4
Schematic of Claydon Yield-O-Meter operation.

elevator and the inlet of the grain tank auger. The sensor consists of a paddle wheel and an optoelectronic sensor that senses the height of the grain in the paddle wheel. When the grain reaches the installation height of the level sensor, the paddle wheel is rotated so that the grain content is discharged. The grain volumetric flow is then obtained by multiplying the volume of the grain in the paddle by the number of paddle wheel revolutions during the time unit (e.g., rpm). As the filling of the space of the paddle wheel varies in both longitudinal and transversal directions, depending on the machine inclination, the information from the combine leveling sensor is integrated into the calculation of the grain flow. The manufacturer of the sensor and electronic monitor specifies that regular calibration of the entire sensing and measuring system is critical for reaching maximum accuracy.

Another volumetric grain flow sensor is the *RDS Ceres meter* (Figure 13.5). Patented by CLAAS, this meter can be adapted to any combine harvester. The functioning of this grain flow sensor is based on an indirect measurement of the height of the bulk grain lying on each paddle of the clean grain elevator. Thus, the near-infrared light emitter and detector are mounted as high as possible on opposite walls of the elevator. A computer converts the time the photosensor does not detect the light into a quantity equal to the height of the grain lying on the elevator paddle. The density of the bulk grain must be preentered into the program and real-time corrected based on the grain moisture.

In the following, a mathematical model to be used in connection with this type of sensor is proposed. We consider that the grain conveyed by the clean grain elevator flows into a chamber (with the base equal to the paddle constant dimensions). On one side, the grain mass changes in time by grain height change from one paddle to another. In other words, the height h of the grain in the chamber varies in time. On the other side, the grain mass varies because of grain density due to grain moisture variation.

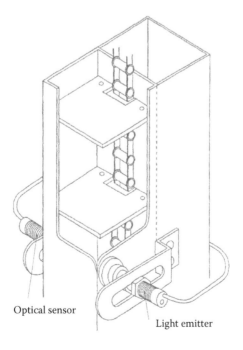

Optical sensor

Light emitter

FIGURE 13.5
Design of Ceres meter.

Thus, the flow rate at the grain elevator outlet m_o is a function of grain height h and density ρ_G, as follows:

$$m_o = m_{in} + A\frac{d(\rho_G h)}{dt} \qquad (13.3)$$

where:
m_{in} is the mass of grain entering the combine
A is the area of one paddle on which the grain lies

Thus, variation of grain flow in the combine is translated into variation of the height of grain lying along the elevator paddles; besides, the grain moisture also varies. Consequently, we can write

$$m_o = m_{in} + A\left(h\frac{d\rho_G}{dt} + \rho_G\frac{dh}{dt}\right) \qquad (13.4)$$

Reitz and Kutzbach (1996) carried out research for potential errors of this type of sensor by using two additional light barriers arranged alongside the driving direction of the combine. When a grain shift occurred due to combine harvesting on slopes, flow was measured more accurately. The relationship between measured light intensity and grain flow did not vary linearly; this relationship is also influenced by the crop type and species, moisture content, and combine inclination to the horizontal.

13.1.3.3 Displacement-Based Grain Mass Sensor

John Deere combine harvesters are equipped with a sensor that measures the mass of the grain exiting the clean grain elevator. Instead of a force transducer, it uses a spring potentiometer device that is connected to the curved impact plate. The resistance of the potentiometer varies linearly with the grain flowing on the impact plate. The grain impact plate is placed in the grain flow in a manner similar to that of Massey Fergusson and Case IH combines.

Yield sensors based on the mass detection principle compensate for variation in grain density, meaning that they are not susceptible to errors given by the extent of the field variation in crop-specific weight.

13.1.3.4 Gamma Radiation–Based Grain Flow Sensor

The gamma radiation–based Massey Ferguson yield sensor is mounted on top of the clean grain elevator as well (Figure 13.6). The gamma rays from the emitter are interrupted by the grain flow that is being discharged by the clean grain elevator. When the grain does not flow, the radiation detector receives the maximum number of gamma rays (approximately 30,000 counts/s). The detector senses the absorbed radiation, which varies exponentially with the density (layer thickness) of the grain flow, as described by the following equation:

$$I(\rho, x) = I_0 e^{-\mu\rho x} \qquad (13.5)$$

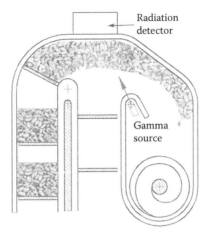

FIGURE 13.6
Gamma radiation–based grain flow sensor.

where:
- I is the intensity of detected radiation, Bq
- I_0 is the intensity of emitted radiation, Bq
- μ is the radiation absorption coefficient, m^{-1}
- ρ is the grain flow density, kg/m^3
- x is the distance of radiation travel, m

In the above equation, the distance and the absorption coefficient are constant. Consequently, the difference of gamma radiation intensity $I - I_0$ depends on the grain flow density only.

13.1.4 Grain Moisture Sensors

CLAAS, Case IH, John Deere, and other manufacturers install grain moisture sensors on the clean grain elevator of their combine harvesters. Grain moisture sensors continuously measure the capacitance variation between two electrodes when a controlled grain flow passes between them. One electrode is usually a tube, while the second one is installed in the middle of the tube. Figure 13.7 shows the longitudinal section through the grain elevator of a CLAAS combine that has a grain moisture sensor. Measurement samples are taken, for example, every 30 s, helping the operator make decisions on the setting of certain combine process functional parameters.

A grain moisture sensor also facilitates decisions such as whether to keep harvesting or whether the grain needs more time to dry or mature.

13.1.5 Plant Height Measurement Sensors

The average crop height is considered an important morphological characteristic due to the following:

- Direct relationship to the ears or the cob insertion height, thus allowing automated header height adjustment when harvesting the crop with a combine
- Direct relationship to the biomass yield, thus allowing correlation of soil fertilization needs for the next season

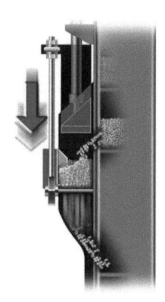

FIGURE 13.7
(See color insert.) Grain moisture sensor installation. (Courtesy of CLAAS, Harsewinkel, Germany.)

A real-time height sensor allows an indirect measure of biomass yield by correlation with individual stem diameters and the plant density per coverage area. A GPS-based system can georeference the biomass and grain yield for a yield map.

Using a laser scanner, the geometric model of plant height measurement is shown in Figure 13.8. The scanner is mounted on the header of the combine, which travels with the speed v_m. The mobile coordinate system XOY has an origin associated with the laser scanner beam light origin. The tangential resolution is R_t, and the corresponding angle is α. Using the notation in Figure 13.8, the equation that expresses the tangential resolution of the measurements is

$$R_t = \rho\left(\frac{\cos\gamma}{\cos(\gamma+\alpha)} - \sin\gamma\right),\, \text{m} \tag{13.6}$$

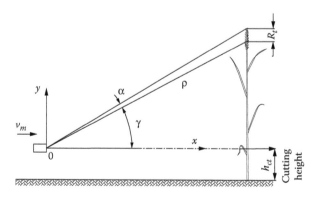

FIGURE 13.8
Laser scanner–based geometric model of plant height measurement.

where:
    ρ is the measurement distance, m
    γ is the elevation of the scanning direction to the horizontal

The usual value of the angle $\alpha = 0.5°$.

The measurement problems relate to the fact that the combine has a pitch roll due to the terrain slopes. Thus, the scanner position may change from downward to upward sloping. A downward-sloping-positioned laser scanner will cause a larger measured height, while an upward-sloping-positioned laser scanner will deliver a smaller measured height. For further details, practitioners may want to read the PhD thesis of Zhang (2011).

A system based on plants and stubble height measurement comparison maintains the combine header edge aligned at all times with the crop (Figure 13.9). The Laser Pilot (CLAAS) transmits pulsing beams of light that are reflected by the standing plants and the stubble. The reflection from the stubble takes more time than that from the standing crop. Based on the time difference, the system determines the edge of the crop and guides the combine along it. This offers the great advantage of maximizing the cutting bar working width, that is, a quasi-constant material feedrate at a constant ground speed of the machine.

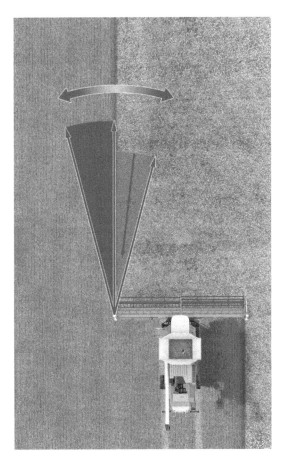

FIGURE 13.9
Laser pilot working mode. (Courtesy of CLAAS, Harsewinkel, Germany.)

13.1.6 Real-Time Sensors for Grain Content of Protein, Gluten, and Starch

Agricultural producers need to make informed management decisions in terms of soil fertilization needs (nitrogen) and crop rotation based on soil fertility data. At high spatial resolutions, soil sampling is a very expensive method. Grain yield monitors provide useful information; however, grain yield data does not comprehensively characterize the nutrition potential of the soil. An alternative method is proposed by New Holland and Textron Systems (Meier, 2004): real-time measuring of protein, oil, and moisture content of the grain using a near-infrared sensor that will be described in this section. When installed on a combine harvester, the information provided by such sensor can help producers claim protein premiums for their crops (Meier, 2004).

The functioning principle and main component parts are shown in Figure 13.10. Such a sensor consists of a photodiode array with at least 256 pixels that are sensitive to different wavelength rays ranging from the visible spectrum (400–700 nm) through near infrared (700–900 nm) to the short-wave infrared radiation spectrum (800–2500 nm). A tungsten/halogen lamp is the light source. The irradiated grain molecules absorb a fraction of

FIGURE 13.10
Near-infrared sensor for grain content of protein, gluten, and starch. (Adapted from Berzaghi, P. et al., On combine harvester grain tester using near infrared transmission, 2005, http://www.grainit.it/Grainit/RxGrains_file/nit_COMBINE_germany.pdf.)

radiation energy at wavelengths unique to each molecule type. The absorbance of the light energy is proportional to the concentration of the type of absorbing molecules. The rest of the energy is reflected back. Thus, the sensor infers that the absorbed energy corresponds to the chemical composition of the grain molecules. In practice, the influence of several factors has to be considered, as follows: grain sample thickness, light refraction, grain size and shape, temperature fluctuations, grain moisture, and the crystalline/amorphous fractions of the grain structure.

For sensor calibration, this method requires developing spectroscopic models that relate the amount and intensity of reflected light to each chemical component fraction (protein, gluten, starch, etc.) within the grain.

13.1.7 Sensor Integration System in a Combine Harvester

Figure 13.11 shows the interconnection of the sensors in a CLAAS combine harvester for calculating the grain yield. The information that is collected includes the following parameters: grain bulk density (preentered), cutter bar width, dynamic radius of the front wheel (for machine speed setting), machine inclination angle, volumetric grain flow, and grain moisture. Machine speed can be obtained by ground speed measurement from measuring the wheel rotation or radar mounted on the combine, or by using the GPS system.

All information is displayed on a terminal in the cab or on a memory card if the yield mapping equipment is installed and the GPS position is acquired. Grain moisture is monitored at a variable rate, every 10–50 s, depending on the grain feedrate. Every 5 s, the yield

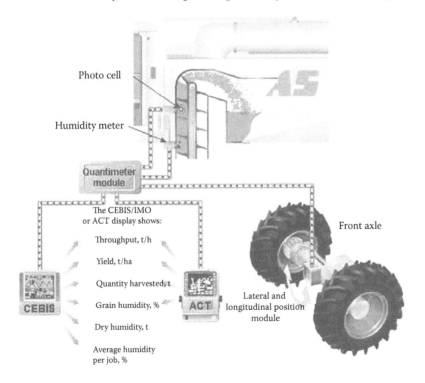

FIGURE 13.11
Integration of sensors for grain yield computing. (Courtesy of CLAAS, Harsewinkel, Germany.)

data and the current grain moisture are correlated with the GPS position to create grain weigh points onto a map.

Chapter 15 discusses in detail the GPS system, map projection and tracking, and geomagnetic direction sensor.

13.2 Fault Diagnosis Systems for Combine Harvesters

13.2.1 General

Although equipped with reliable control systems, modern combine harvesters are continuously operated by human operators who must drive the machine in the field and observe the process parameters, as well as distinguish different malfunctions or unusual events in an ever-increasing workload or transport conditions. Operating the machine in various environmental and uncertain functional conditions becomes a more and more difficult task, with possible costly outcomes in terms of safe operation of the machine or harvested crop quality.

High technical complexity and a multitude of process parameters of the combine harvester can be efficiently mastered by use of electronics and automation that incorporate concept and diagnostic solutions. Twenty years ago, the requirement to diagnose the faults of different combine functions was not on the machine specifications set. Today, combine function fault diagnosis has improved significantly, but it is still a challenge because of three comprehensive requirements: development of a powerful diagnostic system, adopting a user-friendly approach, and integration at a rational cost. Besides, consistent management and distribution of diagnostic data are mandatory for the combine manufacturer.

Different specific standards developed in the field are as follows:

- ISO 22900-1: Open Diagnostic Data Exchange (ODX) (ASAM MCD-2D)
- ISO 22900-2: Hardware interface (D-PDU API)
- ISO 22900-3: Interface between the runtime system and the test application (ASAM MCD-3D, D-Server API)
- ISO 14229-1: Unified Diagnostic Services on CAN

These standards represent the foundation for developing and implementing diagnosis systems at the necessary technological level and ensuring the interchangeability of the products, components, and data without restraining the competition among manufacturers.

Different diagnostic systems have been developed and implemented, such as DaVinci and CANdela Studio (Vektor Informatik), IQ-FMEA, and MATLAB/Simulink-based diagnostic design strategies. CADdela (CAN diagnostic environment for lean applications) was established in the development process at Daimler-Crysler, and later it was improved and implemented in the CLAAS combines (Figure 13.12). The diagnostic component CANdesc (CAN diagnostic embedded software component) can be automatically derived from the CANdela data (Frank and Schmidts, 2007).

A "fault" of a combine system is an unexpected or unpredicted deviation of an observed variable (parameter) outside of a defined or acceptable range of values or state

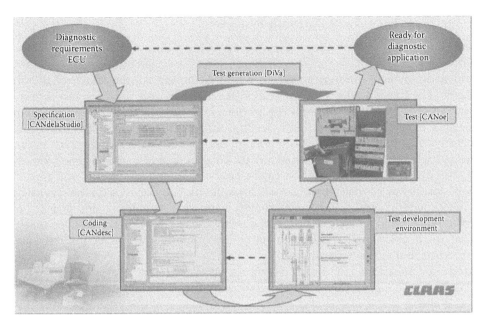

FIGURE 13.12
Diagnostic development at CLAAS using the CANdela tool chain. (Courtesy of CLAAS, Harsewinkel, Germany.)

of the considered system, such that the system operation either deteriorates or is demolished. Accordingly, we speak about either an incipient fault or a total failure of diagnosed system. A fault diagnosis system is capable of detecting the presence of faults in the monitored system, determining fault location, and indicating fault severity. Thus, a fault diagnosis system detects, isolates, and identifies the faults of a monitored system.

Detection of a fault or damage of combine harvester subsystems during working in the field is very important, especially when working on hillside fields with the tank full of grain or driving the vehicle on the road or in a field. In a combine harvester, one may deal with the failures of combine actuators, sensors, or process.

An *actuator fault* is the malfunctioning of an actuator, device, or subsystem that may affect the machine dynamics or process. A *sensor fault* is an abnormal indication in the observed signal output in the absence of any anomalous value of the monitored parameter. Sensors are used to either directly measure system states or generate state estimates for the control law of a system. Thus, the sensor faults may deteriorate state estimates and, consequently, determine an inaccurate control of the monitored system. Sensor faults occur due to inefficiencies in the manufacturing process, incorrect calibration, wear and tear with long-term usage, and mishandling.

Process faults are changes of the inner parameters of the combine technological process, such as an unmodeled change in crop harvesting conditions.

The faults may occur abruptly (involving discontinuities) or gradually (from incipient phase to a permanent state). The faults may be intermittent or continuous. In terms of consistency, the faults may be stable (consistent behavior) or heretical (randomly changing behavior).

When the combine is driven on a road, a fault may affect only the machine motion (dynamics). In such a case, any component of the power train, starting from the engine to the wheel final drive, may fail. During harvesting, failure of different components of the processing units or sensors for monitoring process parameters may become additional faults to those that can affect the machine dynamics.

13.2.2 Specifications of Fault Diagnosis Systems

Different diagnosis concepts and techniques have been developed and applied in the automotive industry. Fault diagnosis systems are software-based built-in-test systems that detect, identify (make a difference), isolate, and indicate (signal) the failure of the prime subsystems. The consideration and application of fault diagnosis solutions range from the development process of cars, through applications in the manufacturing process, to quality assurance in production, and finally to final diagnosis in a service garage. The fault diagnosis systems have not been given enough attention in agricultural machinery, but they are of high value at a supervisory control level for combine harvesters. Nowadays, system fault detection in combine harvesters, analysis of fault root cause, and repair are mainly performed by service technicians. Consequently, this depends on the service operator skills, experience, and diligence; moreover, erroneous conclusions and possibly decisions may intervene in a negative way and destroy buyer confidence in a given machine. Therefore, the development and implementation of fault diagnosis systems are not only very important, but also a necessity.

Desired specifications of fault diagnosis systems have been formulated by Venkatasubramanian et al. (2003). We further interpret and explain them as follows:

- *Balance between early detection of faults and sensitivity to measurement noise*: Quick detection versus generation of false alarms. While being sensitive to incipient faults, the diagnosis system should keep false alarms under healthy operational modes.

- *Identification and distinction of different types of failures*: The implementation and application of clear rules of failure classification. This is the capability of a diagnosis system to distinguish the origin of a fault among other potential fault sources or to locate a faulty component among different components of a diagnosed system. Isolability of a fault also depends on the fault effects on the diagnosed system output. Besides, modeling uncertainties and system disturbances affect the isolability degree of the faults.

- *Robustness*: The capacity of a diagnosis system to diminish the effects of model uncertainties, measurement noise, and system disturbances. Robustness essentially augments diagnosis system reliability and effectiveness.

- *Novelty identifiability*: The capacity of a diagnosis system to *not* wrongly classify novel malfunctions in the system as either an *a priori* known type of malfunction or a healthy operational mode. While detection of novel faults is relatively easy to achieve, their isolation is difficult to accomplish due to missing models of their unknown behavior.

- *Multiple fault identifiability*: The ability of the diagnosis system to identify and correctly classify multiple faults that may coexist in the system. This specification is difficult to achieve due to nonlinearities and coupling/interactions that usually exist between the states and faults of a diagnosed system.

- *Explanation facility*: The diagnosis system capability to explain where the fault originated and how it propagated through the system to the current status.

- *Classification error estimation*: The ability of the system to estimate in the beginning the expected errors in fault classification and explanation.

- *Adaptability*: The ability of a diagnosis system to intelligently adapt to system changes due to disturbances and maintain/improve its performance despite the hardware degradation in time.

- *Reasonable computational requirements*: The diagnosis algorithm and database of diagnostics classifiers should meet this specification. This has a high impact on required memory, storage capacity, and power requirement.
- *User friendliness*: This is required because in agriculture, one operator handles different types or makes of the machines. Thus, the time for adequate training is limited and obviously costly.

13.2.3 Fault Diagnosis Problem Formulation

A combine system is a dynamic, continuous-time system with multiple inputs and outputs, which can be described by a nonlinear mathematical model, as shown below:

$$\dot{x}(t) = Ax(t) + Bu(t) + \Phi_{d_i} f_{d_i}$$
$$\dot{y}(t) = Cx(t) + Du(t)$$

(13.7)

where:

$\dot{x}(t) \in \mathbf{R}^n$ is the state vector of n parameters; it is assumed that all system states are available for measurements
$\dot{u}(t) \in \mathbf{R}^q$ is the input vector of q parameters
$\dot{y}(t) \in \mathbf{R}^m$ is the output vector of m parameters
A, B, C, and D are the system matrices with measurable dimensions expressed in appropriate units
$\Phi_{di} \in \mathbf{R}^{m \times di}$ is the matrix of d_i fault components, where $1 \le d_i \le n$
$f_{di} \in \mathbf{R}^{di}$ is the vector of process fault magnitude, where $1 \le d_i \le n$; it may be interpreted as the channel over which the disturbances (fault consequences) are applied to the system

To represent a single fault, that is, $d_i = 1$, in the ith subsystem $\forall i \in [1, n]$, the fault component matrix is the ith column of the identity matrix $I_n \in \mathbf{R}^{n \times n}$.

Using a sampling period T_s (Lewis, 1986), Equation 13.7 may be described by the following nonlinear discrete-time state-space representation (Li et al., 2001) as follows:

$$x(k+1) = A_d x(k) + B_d u(k) + A_d \int_0^{T_s} e^{-A\tau} \Phi_{d_i} f_{d_i} (kT_s + \tau) d\tau$$

(13.8)

$$y(k) = Cx(k) + Du(k)$$

where the system matrices are defined as follows:

$$A_d = e^{AT_s}$$

(13.9)

$$B_d = \int_0^{T_s} e^{A\tau B d\tau}$$

(13.10)

and

$$\mathbf{f}_{di}\left(kT_s\right) = \mathbf{f}_{di}\left(k\right) \tag{13.11}$$

Under the full-state measurement assumption, matrices **C** and **D** in Equation 13.8 are the same as in Equation 13.7.

Assuming that $\mathbf{f}_{di}(t)$ is constant over the sampling interval T_s, this means that

$$\mathbf{f}_{di}\left(kT_s + \tau\right) = \mathbf{f}_{di}\left(k\right), \quad \forall \tau \in \left[0, T_s\right] \tag{13.12}$$

Consequently, the fault model in Equation 13.8 becomes

$$\mathbf{A}_d \int_0^{T_s} e^{-\mathbf{A}\tau} \mathbf{\Phi}_{d_i} \mathbf{f}_{d_i}\left(k\right) d\tau \tag{13.13}$$

Thus, in the discrete-time domain, we can write the fault model as follows:

$$\mathbf{A}_d \int_0^{T_s} e^{-\mathbf{A}\tau} \mathbf{\Phi}_{d_i} d\tau = \mathbf{A}_d \mathbf{A}\left(\mathbf{I}_n - \mathbf{A}_d^{-1}\right)\mathbf{\Phi}_d \tag{13.14}$$

and this depends on the system state represented by matrix **A**.

For designing a fault diagnosis system, it is very important to identify and understand the failure mechanisms in sensors and actuators and model their erroneous behavior.

13.2.3.1 Sensor Faults

Five generic types of sensor faults are *bias* (offset), *drift* (linear or not), *scaling* (gain, which is linear or not), *hard fault* (freezing, which implies loss or locking of the signal), and *accuracy degradation* (mainly due to calibration error). The effects of the above sensor faults on the measurements they deliver are shown in Figure 13.13.

For sensor fault mathematical representation, let $y_s(t)$ be the sensor output and $y_0(t)$ the real value (in the absence of a sensor fault or error) of a monitored parameter. Thus, the sensor faults can be generally described by the following equation:

$$y_s(t) = \alpha_s(t)\left(1 + \varepsilon_s\right)y_0(t) + \delta(t) + N \tag{13.15}$$

where:
$\alpha_s(t)$ is the scaling (gain) constant
ε_s is the attenuation coefficient
$\delta(t)$ is the bias (offset) value
N is the noise

For the above-mentioned sensor faults, the values of model parameters are given in Table 13.2. While the hard fault is easy to detect through the built-in test provided by the sensor manufacturer, the other faults are more difficult to detect, especially when the deviation $|y_s(t) - y_0(t)|$ is relatively small.

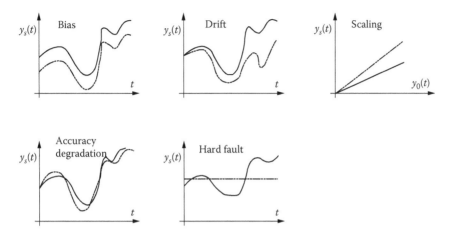

FIGURE 13.13
Sensor fault effects on their measurements.

TABLE 13.2

Model Parameter Values for Sensor Faults

Parameter	Bias	Drift	Scaling	Hard Fault	Accuracy Depreciation	No Fault
$\alpha_s(t)$	1	1	1	0	1	1
ε_s	0	0	$\neq 0$	0	$\neq 0$	0
$\delta_s(t)$	$\neq 0$ (constant)	$\neq 0$ (time varying)	0	Constant	$\neq 0$	0

13.2.3.2 Actuator Faults

An actuator uses the control signal to accordingly modify its actuation parameter (force, speed, torque, etc.) to drive the system toward the desired output. Thus, actuators are control effectors of the monitored system. Within the scope of fault detection application, each actuator should be considered a single device, which is monitored through the success rate and speed of achieving the desired output of the system.

Actuator faults usually depend on the actuator type. Four actuator fault modes can be identified: *lock-in-place* (freezing, jamming), *float*, *hard-over*, and *loss of effectiveness*. The effects of the above actuator faults on the actuation signal are shown in Figure 13.14.

For actuator fault mathematical representation, let $u_a(t)$ be the actuator output (signal), $u_c(t)$ the control signal, and u_f the real actuation level (signal). Thus, the sensor faults can generally be described by the following equation:

$$u_a(t) = \sigma_f k u_c(t) + \left(1 - \sigma_f\right) u_f \tag{13.16}$$

where:
σ_f is the fault-type coefficient
k is the actuator effectiveness, $0 < k < 1$

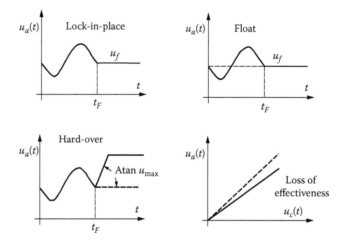

FIGURE 13.14
Actuator fault effects on actuator behavior.

TABLE 13.3

Model Parameter Values for Actuator Faults

Parameter	Lock-in-Place	Float (Oscillatory)	Hard-Over	Loss of Effectiveness	No Fault
σ_f	0	0	0	1	1
k	ϕ	ϕ	ϕ	$0 < k_f < 1$	1
u_f	$u_c(t_F)$ (constant)	Periodic	Constant (saturation)	0	0

For the above-mentioned actuator faults, the model parameter values are given in Table 13.3 after the time t_F of fault occurrence.

Besides actuators and sensors, every technical system comprises different other components, such as power sources, tanks, machine elements (bearings, hardware), lubrication system, and so forth. Often, it is difficult to mathematically model the faults of such components without extensive testing.

Nonetheless, other types of faults in a system may occur due to wear and tear of system components.

13.2.4 Model-Based Approaches for Fault Diagnosis

Fault diagnosis is a process that consists of system *fault detection, isolation,* and *identification.* This process involves designing and setting up a system, which uses software developed based on a certain method or algorithm.

Fault detection is the process by which the presence of a fault in the monitored system, along with the time of fault occurrence, is uncovered. By *fault isolation,* the type and location of the faults are determined. Finally, *fault identification* (or *estimation*) aims at determining the size and time-varying behavior of the faults.

Depending on the form of system information (or process knowledge), the analytical fault diagnosis approaches can be divided into *model-based* and *computational intelligence–based* approaches. The model-based *a priori* knowledge uses either *quantitative* historical data of a monitored system or *qualitative* information on the system, in the form of *if–then* rules. In quantitative models, the description of the monitored system or process has to be formulated in terms of mathematical relationships between the inputs and outputs of the system. The mathematical model is usually composed of differential or difference equations, and the outputs can be measured numerically. A group of computational intelligence–based approaches is based on *artificial immune system computational models*.

The architecture of model-based algorithms for fault detection in continuous variable systems combines a residual generator and a residual evaluation strategy that provides Boolean decisions on whether faults did occur. The schematic of such model-based architecture is shown in Figure 13.15. The most extensively studied residual generator was a nonlinear observer-based technique. Residual generation uses a model of the monitored system in which the control inputs and system outputs, as measured by the sensors, are used to predict the system behavior parameters, which are compared with the actual system parameters. A fault is indicated when the value of a residual is different than zero. The residual evaluation is required to distinguish between different signals, while decision logic helps in fault isolation. Most of the model-based fault isolation techniques belong to one of the following two groups: *structured residuals* and *directional residuals*.

The better the system model, the lower is the level of uncertainty in detecting and isolating the system faults. Various other sources of uncertainty are represented by the measurement noise and unmodeled exogenous disturbances that affect the system behavior.

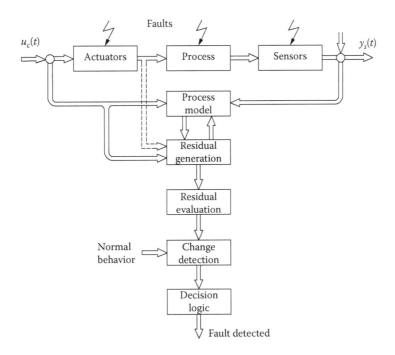

FIGURE 13.15
Architecture of model-based algorithms for fault detection.

13.2.5 Immune System–Based Computational Models/Algorithms for Fault Diagnosis

Design complexity and time-variant process nonlinearity of a combine harvester interacting with different biological materials make diagnosis techniques through quantitative modeling quite difficult, despite our best attempts to organize and generate knowledge throughout this book. Thus, important knowledge from experienced operators and engineers in the field can be further used in developing causal qualitative-based models for combine fault detection.

Important modeling approaches (not only for diagnosis system development) are based on the principles and actions of biological immune systems. The role of the biological immune system is to protect the biological being against pathogens and viruses. A biological immune system acts in two main ways: launch the primary response to the invading pathogen and launch the secondary immune response by remembering past encounters and responding faster the second time an infection occurs.

The immune system uses largely short- and long-term memory, recognizes patterns, identifies and classifies unexpected events, divides and controls the tasks in executing strategies, and continuously adapts to changing conditions the representing biological system encounters.

An *artificial immune system* (AIS) is a computational algorithm that is developed by combining *a priori* knowledge of an observed system with the adapting principles and defense mechanisms of biological immune systems to detect, identify, and isolate a broad spectrum of known and unforeseen faults of the monitored system. The aim of developing and implementing an artificial immune system is to apply the immune system functional rules and flow into the fault diagnosis of machinery or systems of a different nature. We need to specify that such immune-based computational algorithms are used in other techniques, for instance, pattern recognition, modeling, design, and control of various systems, including software for other applications, for example, antivirus software.

Using a finite number of discrete detectors derived from the body DNA, the adaptive immune system performs a succession of tasks, as follows: pattern recognition, detection, identification, optimization, and learning. The last task governs the adaptivity and efficiency of the immune system, which acts based on already available knowledge in the DNA molecules and continuously acquired knowledge from fighting the body intruders. Biological immune systems use a complex of cells (*lymphocytes*) to protect the living body against foreign invasions of microbes and viruses. There are two types of lymphocytes: *B-cells*, produced in bone marrow by clonal selection, and *T-cells*, processed in the thymus. The T-cells are first generated through a pseudorandom genetic rearrangement mechanism, which ensures high variability of the new cells in terms of biological identifiers; then a selection process occurs in the thymus by destroying the T-cells whose identifiers match the self. Only the different T-cells are allowed to leave the thymus and proliferate. While B-type lymphocytes secrete antibodies or serve as memory cells, the T-type cells interact with other cells concerning cellular immunity against foreign cells. The antibody diversity comes from somatic mutations, and the efficiency of antibodies increases due the selection process of high-affinity antibodies.

Development of artificial immune systems is based on their similarity with biological adaptive immune systems. In comparison with other computational intelligence–based algorithms, artificial immune systems offer several distinctive advantages, as follows:

- *Vast a priori knowledge* already acquired from the studies of DNA molecules in regard to body immunity.

- *Rapid convergence to an optimal solution*, which is achieved by rapid mutation and recombination of genetic representation of possible solutions to the problem.

- *Uniqueness* and *variety*: Each artificial immune system responds in a specific way—hence the variety of them. However, the solution to the problem must be the same regardless of the algorithm type used for solution convergence.

- *Learning about new encounters* from the interaction between the population of antibodies and antigens that leads to self-organizing network structures.

Other features of artificial immune systems are specified in computing terms: memory of past encounters, highly distributed/parallel processing, multilayered, robust, and self-tolerant.

Timmins (2000) defined an artificial immune system to be "a computational system based upon metaphors of the natural immune system." Figure 13.16 shows a diagram of the immune system–level concept, which was adapted from KrishnaKumar (2003).

Substances created by the immune system that are capable of triggering a specific immune response are commonly referred to as antigens, which include viruses, bacteria, fungi, and so forth. The immune recognition is based on the complementarity between the binding region of the antibody receptor and a portion of the antigen. Each antibody can recognize a single antigen.

The idea of shape space (Castro and Timmins, 2002) is that the features of a receptor molecule that are necessary to characterize its binding region with an antigen are called generalized shape. This can be represented by an attribute string of a certain length L in a generic L-dimensional space called shape space.

The immune memory is acquired and maintained by (1) *clonal expansion and selection* and (2) the *immune network*. The clonal expansion is the immune system's capability of storing high-affinity antibody-producing B-cells for subsequent encounters. The theory of immune networks states that B-cells costimulate each other via portions of their receptor molecules in such a way as to mimic antigens. A network of mutually, highly stimulated B-cells survive, while the weak ones are suppressed from the system. This is a control function for regulating the overstimulation of B-cells and maintaining a stable memory of the immune system.

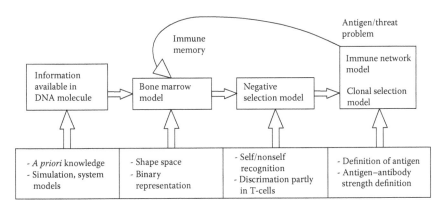

FIGURE 13.16
A diagram of immune system–level concept.

Analogous with the responsive patterns and actions of biological immune systems to different threats, the following immune system–based computational models/algorithms have been developed:

- Bone marrow model
- Negative selection algorithm
- Clonal selection algorithm
- Immune network algorithm
- Immunized computational algorithms
- Somatic hypermutation model

In the following subsections, we present details of these immune system–based computational models.

13.2.5.1 Bone Marrow Models

The B-cells in the bone marrow collectively form an immune network; they remain in the immune system as long they are required. Some of the first work to explore antibody diversity generation was performed by Stephanie Forrest and her colleagues (Hightower et al., 1995).

In a bone marrow algorithm, gene libraries are used to create antibodies through a random concatenation of segments. Thus, a random segment is picked from each gene library and joined together with those chosen from the others. For instance, one segment of 16-bit size is chosen from each gene library. Thus, a newly generated molecule from four gene libraries will be expressed through a 64-bit sequence genome (Figure 13.17). This is then compared to a set of antigens, and an antigen-matching score is assigned to each antigen based on how well the antibody matched each one. This process is repeated many times for many different antibodies (population of individuals). Finally, the fitness of each antibody (individual) is calculated as the average of all antigen-matching scores.

But the gene libraries may not contain genes that can define antibodies for every possible antigen. Instead, gene fragments from particular regions in the gene library of a parent cell may be randomly recombined in every new antibody genome, allowing a diverse range of antibodies.

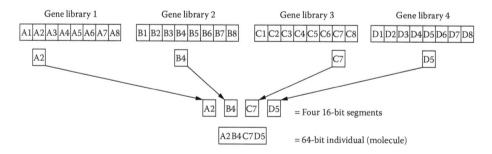

FIGURE 13.17
Random concatenations of elements to express an antibody molecule.

13.2.5.2 Negative Selection Algorithm

Since B-cells produce antibodies from the gene libraries (randomized template), sometimes they may generate the wrong thing. This means that the immune system has the potential to attack self-cells by mistake. The immune system has a special mechanism to prevent this: *negative selection* takes place in the thymus gland. The B-cells undergo a maturation process by being exposed to self-proteins in a binding process. If the cell is activated through the binding process, it is killed; otherwise, it is allowed to enter the lymphatic system.

The negative selection principle inspired the algorithm developed by Forrest et al. (1994, 1997) to detect data manipulation by computer viruses. The basic idea is to generate detectors and use them to classify new (unseen) data as self (not manipulated) or nonself (manipulated data). The algorithm is split into two parts: *censoring* and *monitoring* (Figure 13.18). In the first part, a set **D** of detectors is generated from a set **S** of elements of length l in shape space. These detectors are then used to continuously monitor the data stream for any changes by continually matching (based on an affinity function) the detectors in **D** against the data stream. The affinity function is based on checking for the similarity of r consecutive characters at any point in the detector and self-strings; this is called the r-contiguous matching rule.

Negative selection algorithms have been criticized for difficulties encountered in defining self-set and certain scaling problems (Timmins et al., 2008).

13.2.5.3 Clonal Selection Algorithm

When a pathogen invades and multiplies within a human body, the immune system produces the right kind of B-cells that generate the antibodies to kill that pathogen. Thus, the clonal selection principle, which applies in such a case, describes the basic property of the immune system to respond to a pathogenic stimulus by multiplying only the cells capable of recognizing that pathogen. Clonal selection operates on both B- and T-cells, when the antibodies produced by them bind with an antigen, becoming activated and differentiated into plasma or memory cells.

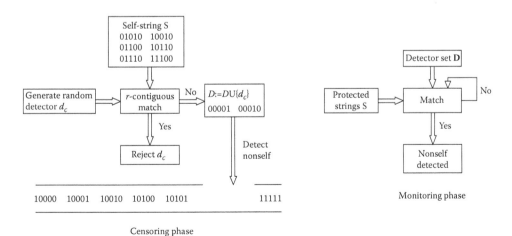

FIGURE 13.18
Censoring and monitoring parts of a negative selection algorithm.

Therefore, the immune system stores high-affinity antibody-producing cells following the first infection, thus enhancing the action speed and effectiveness for subsequent encounters.

This theory inspired the development of a clonal selection algorithm that differs from a genetic algorithm in terms of reproduction and evaluation. The sequence of a generic clonal selection algorithm is as follows:

```
Input: S—set of patterns to be recognized having n worst elements to be
selected for removal
Output: M—set of memory detectors, which are proficient in classifying
unseen patterns
Begin
     Create an initial randomized set A of antibodies
     for all patterns in S, DO
        Determine the affinity with each antibody from A.
        Generate clones of a subset of the antibodies in A with the
        highest affinity.
        (The quantity of an antibody clone is proportional to its
        affinity.)
        Mutate attributes of these clones inversely proportional to
        their affinity. Add these clones to the set A.
        Place a copy of the highest-affinity antibodies into the memory
        set, M.
        Replace the n lowest affinity antibodies in A with new randomly
        generated antibodies.
        end
     end.
```

Other published works are based on the alternative theory of how immune systems function. Instead of algorithms resembling genetic algorithms, such works led to the development of immune network algorithms, which are described in the next section.

13.2.5.4 *Immune Network Algorithm*

According to the immune network theory, the immune system consists of a regulated network of molecules and cells, which recognize each other, self-organize, and develop memory even in the absence of antigens. The recognition of an antigen by an antibody (cell receptor) leads to network activation and antibody proliferation, while the recognition of an idiotope (marker of a cell receptor) leads to network suppression. This is modeled by eliminating all but one of the self-recognizing cells.

The original theory did not explicitly account for the results of network activation or suppression; consequently, various immune network algorithms have been developed and published. Due to the specific purpose of this review, we will just present the description of an immune network algorithm that uses the clonal selection algorithm, as developed by Castro and Zuben (2001). The algorithm runs as follows:

```
Input: P—set of antigenic patterns to be recognized
Output: M—set of memory detectors, which are proficient in classifying
unseen antigenic patterns
```

Begin
1. Create an initial randomized set A of antibodies.
2. **For all** patterns in P, **DO** (apply the clonal selective algorithm that will return a set **M** of memory detectors and their coordinates for the current antigen).
   a. Determine the affinity (degree of matching) among all the antibodies in **M***.
   b. Eliminate all but one of the antibodies (individuals) in **M*** whose affinity is greater than a given threshold.
   c. Concatenate the remaining individuals from the previous operation with the remaining individuals found for each antigenic pattern presented.
   d. Determine the affinity of all population M individuals and suppress all but one of the self-recognizing elements. Repeat this cycle until a predefined stopping criterion is met, such as minimum pattern recognition or classification error.
   end
end.

A big advantage of artificial immune systems in comparison with traditional fault-tolerant approaches is the evolutionary property of the immune system. While the first technique generates static detectors that have to be updated periodically, the artificial immune systems enable the development of adaptable fault-tolerant systems, in which fault detectors may evolve over time.

It is expected that this review of immune system–based computational algorithm applications in fault detection and diagnosis will represent a tangible knowledge base for readers who are interested in the subject and will stimulate them to carry out further research on this topic.

References

Berzaghi, P., J.C. Ferlito, and L. Serva. 2005. On combine harvester grain tester using near infrared transmission. http://www.grainit.it/Grainit/RxGrains_file/nit_COMBINE_germany.pdf (accessed July 3, 2014).

Castro, L.N. and J. Timmins. 2002. Artificial immune systems: A novel paradigm to pattern recognition. In *Artificial Neural Networks in Pattern Recognition*, ed. J.M. Corchado, L. Alonso, and C. Fyfe, 67–84. SOCO-2002. Paisley, Scotland: University of Paisley.

Castro, L.N. and F.J. Zuben. 2001. aiNET: An artificial immune network for data analysis. In *Data Mining: A Heuristic Approach*, ed. H.A. Abbas, R.A. Sarker, and C.S. Newton, 231–259. Idea Group Publishing.

Forrest, S., S. Hofmeyer, and A. Somayaji. 1997. Computer immunology. *Communications of the ACM* 40(10):88–96.

Forrest, S., A. Perelosn, and A.L. Cherukuri. 1994. Self-nonself discrimination in a computer. In Proceedings of the IEEE Symposium on Research Security and Privacy, Oakland, CA: pp. 202–211.

Frank, H. and U. Schmidts. 2007. Vehicle diagnostics: The whole story. Harsewinkel, Germany: CLAAS.

Hightower, R., S. Forrest, and A.S. Perelson. 1995. The evolution of emergent organization in immune system gene libraries. In *Proceedings of the Sixth International Conference on Genetic Algorithms*, ed. L.J. Eshelman, 344–350. San Francisco: Morgan Kaufmann.

KrishnaKumar, K. 2003. Artificial immune system approaches for aerospace applications. Presented at the 41st Aerospace Sciences Meeting and Exhibit, Reno, NV.

Lewis, F. 1986. *Optimal Estimation*. New York: John Wiley & Sons.

Li, W., H. Raghavan, and S. Shah. 2001. Fault diagnosis-relevant subspace identification of continuous time systems. Presented at the 4th IFAC Workshop, Seoul, Korea.

Meier, C.G. 2004. Protein mapping spring wheat using a mobile near-infrared sensor and terrain modeling. MS thesis, Montana State University.

Reinke, R. 2010. Self-calibrating mass sensor. MS thesis, University of Illinois at Urbana–Champaign.

Reitz, P. and K.H. Kutzbach. 1996. Investigations on a particular yield mapping system for combine harvesters. *Computers and Electronics in Agriculture* 14(2–3):137–150.

Shearer, S.A., S.G. Higgins, S.G. McNeil, and G.A. Watkins. 1997. Data filtering and correction techniques for generating yield maps from multiple-combine harvesting systems. ASAE Paper 97-1034. St. Joseph, MI: American Society of Agricultural Engineers.

Strubb, G., B. Missoten, and J. De Baerdemaeker. 1996. Performance evaluation of a three-dimensional optical volume flow meter. *Applied Engineering in Agriculture* 12(4):403–409.

Thylen, L. and P.L. Murphy. 1996. The control of errors in momentary yield data from combine harvesters. *Journal of Agricultural Engineering Research* 64:271–278.

Timmins, J. 2000. Artificial immune systems: A novel data analysis technique inspired by the immune network theory. PhD dissertation, University of Wales, Aberystwyth.

Timmins, J., A. Hone, T. Stibor, and E. Clark. 2008. Theoretical advances in artificial immune systems. *Journal of Theoretical Computer Science* 403(1):11–32.

Venkatasubramanian, V., R. Rengaswamy, K. Yin, and S.N. Kavuri. 2003. A review of process fault detection and diagnosis: Part I: Quantitative model-based methods. *Computers and Chemical Engineering* 27:293–311.

Wojciechowski, T. and M. Czechlowski. 2013. Proximal soil sensing unit for cereal combine harvester. *Proceedings of the Institute of Vehicles* 4(95):179–188.

Zhang, L. 2011. Sensor development for estimation of biomass yield applied to *Miscanthus* Giganteus. PhD dissertation, University of Illinois at Urbana–Champaign.

Bibliography

Aickelin, U. and D. Dasgupta. 2004. Artificial immune systems. In *Search Methodologies: Introductory Tutorials in Optimization and Decision Support Techniques*, ed. E.K. Burke and G.K. Kendall. http://arxiv.org/ftp/arxiv/papers/0803/0803.3912.pdf (accessed July 13, 2014).

Arslan, S. 2008. A grain flow model to simulate grain yield sensor response. *Sensors* 8:952–962.

Balaban, E., A. Saxena, P. Bansal, et al. 2009. Modeling, detection, and disambiguation of sensor faults for aerospace applications. *IEEE Sensors Journal* 9(12):1907–1917.

Berzaghi, P., J.C. Ferlito, and L. Serva. 2006. On combine harvester grain tester using near infrared transmission. Research Report presentation, University of Padua/Grainit S.R.L., Padua, Italy.

Burks, T.F., S.A. Shearer, J.P. Fulton, and C.J. Sobolik. 2001. Influence of dynamically varying rates on clean grain elevator yield monitor accuracy. ASAE Paper 01-1182. St. Joseph, MI: American Society of Agricultural Engineers.

Castro, L.N. and F.J. Zuben. 1999. Artificial immune systems: Part 1: Basic theory and applications. Technical Report TR-DCA 01/99, Campinas, Brazil.

Castro, L.N., F.J. Zuben, and G.A. de Deus. 2003. The construction of a Boolean competitive neural network using ideas from immunology. *Neurocomputing* 50:51–85.

Craessaerts, G., J. De Baerdemaeker, and W. Saeys. 2010. Fault diagnostic systems for agricultural machinery. *Biosystems Engineering* 106(1):26–36.

Dasgupta, D., K. KrishnaKumar, D. Wong, and M. Berry. 2004. Negative selection algorithm for aircraft fault detection. *Artificial Immune Systems, Lecture Notes in Computer Science* 3239:1–13.

De Baerdemaeker, J. De, H. Ramon, J. Anthonis, et al. 2004. Advanced technologies and automation in agriculture. In *Control Systems, Robotics, and Automation*. Vol. XIX. www.eolss.net/sample-chapters/c18/E6-43-35-04.pdf (accessed June 14, 2014).

Garcia, L.R., L. Lunadei, P. Barreiro, and J.I. Robla. 2009. A review of wireless sensor technologies and applications in agriculture and food industry: State of the art and current trends. *Sensors* 9:4728–4750.

Hofmeyr, S.A. and S. Forrest. 1999. Architecture for an artificial immune system. *Evolutionary Computation* 7(1):45–68.

Hongbing, W., L. Peihuang, and T. Dunbing. 2013. Adaptive dynamic clone selection neural network algorithm for motor fault diagnosis. *International Journal on Smart Sensing and Intelligent Systems* 6(2):482–504.

Isermann, R. 2005. Model-based fault-detection and diagnosis: Status and applications. *Annual Reviews in Control* 29:71–85.

Jennett, A.T. 2012. Decision support system for sensor-based autonomous filling of grain containers. Paper 12979. Graduate theses and dissertations, Iowa State University, Ames.

Jiang, L. 2011. Sensor fault detection and isolation using system dynamics identification techniques. PhD dissertation, University of Michigan.

Long, D.S., R.E. Engel, and M.C. Siemens. 2008. Measuring grain protein concentration with in-line near infrared reflectance spectroscopy. *Agronomy Journal* 100(2):247–252.

Martinet, P., Bonton, P., Gallice, J., et al., 1998. Automatic guided vehicles in agricultural and green space fields. In *Proceedings of the 4th French-Israeli Symposium on Robotics, FIR'98*. Besançon, France, pp. 87–92.

Marzat, J., H. Piet-Lahanier, F. Damongeot, and E. Walter. 2012. Model-based fault diagnosis for aerospace systems: A survey. *Journal of Aerospace Engineering* 226(10):1329–1360.

Metternicht, G. 2006. Use of remote sensing and GNSS in precision agriculture. Presented at the UN-Zambia-ESA Regional Workshop in Sub-Saharan Africa, Curtin University of Technology.

Moore, M. 1998. An investigation into the accuracy of yield maps and their subsequent use in crop management. PhD dissertation, Cranfield University.

Norng, S. 2004. *Statistical Decisions in Optimizing Grain Yield*. Brisbane: Queensland University of Technology.

Sobhani-Tehrani, E. and K. Khorasani. 2009. *Fault Diagnosis Nonlinear Systems Using Hybrid Approach*. Heidelberg: Springer.

Stafford, J.V., B. Ambler, and M.P. Smith. 2001. Sensing and mapping grain yield variation. In *Automated Agriculture for the 21st Century*, 356–366. St. Joseph, MI: American Society of Agricultural Engineers.

Taylor, J., B. Whelan, L. Thylen, et al. 2005. Monitoring wheat protein content on-harvester: Australian experiences. In *Precision Agriculture*, ed. J.V. Stafford. Wageningen, The Netherlands: Academic Publishers.

Timmins, J. 2007. Artificial immune systems—Today and tomorrow. *National Computation* 6:1–18.

Timmins J., T. Knight, L. de Castro, and E. Hart. 2004. An overview of artificial immune systems, In *Computation in Cells and Tissues: Perspectives and Tools for Thought*, eds. R. Paton, H. Bolouri, M. Holcombe, J.H. Parish, and R. Tateson, pp. 51–86. Natural Computation Series, Springer.

Wang, Y. and D.H. Zhou. 2006. Sensor gain fault diagnosis for a class of nonlinear systems. *European Journal of Control* 5:523–535.

Yule, I. 2008. Investigation of remote sensing technology to optimize applied nitrogen management in arable crops. Lincoln, Canterbury, UK: Foundation for Arable Research.

14

Cab, Controls, and Human–Machine Interface

14.1 Introduction

The combine harvester is a complex system, and human interaction with it becomes complex to the same extent. While operating the combine, the operator may face numerous health hazards, including noise, whole-body vibrations, temperature extremes, dust, bright lights, diesel exhaust, and awkward postures due to placement and operation of the controls and monitor, as well as psychosocial health risks, such as time pressure, extended working time, and irregular shift work. During harvesting periods, it is quite common to operate a combine for 16 h/day. Furthermore, several operators may alternately operate the combine.

The cab of the combine harvester provides the operator with the operating environment, seating, and interface with the controls for driving and for the technological process of the machine. Designing an effective cab layout with a large visual envelope, controlled environment conditions, a comfortable seat with suspension, and easily reached operating controls reduces operator stress and health hazards, and allows the operator to mainly concentrate on the harvest process and optimize combine working capacity. In the cab, different accessories and equipment for acquiring crop yield and moisture data can also be mounted, as well as communication equipment for improving harvest management.

The design team of the combine harvester cab must have thorough knowledge of the combine functions, control methods and devices, operator–machine interface, human factors engineering, safety standards, and integration methods within design, as well as mechanical and electrical design skills and experience. Design practitioners benefit from engineering standards, data, and recommendations published by the American Society of Agricultural and Biological Engineers (ASABE), Society of Automotive Engineers (SAE), and International Organization for Standardization (ISO); many of them are cited in this chapter. When standard provisions for a combine cab (e.g., from a certain country) differ from those stated in general standards, the provisions of the specific standards take precedence over the provisions from general standards.

The views of two combine harvester cabs are shown in Figures 1.9 and 14.1. Their design ensures wide windscreens, that is, visibility, rational positioning/sliding of the controls for easy access and intuitive utilization by the operator, an air conditioning system, and an optional global positioning system (GPS) satellite system. In the following section, we will discuss in detail the cab design specifications in an attempt to guide the designer (practitioner) on the right path by following a proven methodology while using the right resources.

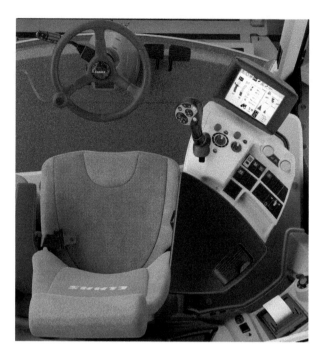

FIGURE 14.1
(See color insert.) Top view of a cab. (Courtesy of CLAAS, Harsewinkel, Germany.)

14.2 Specifications of Combine Harvester Cab

No matter the combine operation mode, it is desirable for the operator to unobstructively observe the road and traffic conditions, field crop, and whole header while interacting with the crop from the cab. Therefore, it is necessary to design a cab that will offer the operator maximum comfort and flexibility. This will allow the operator to remain mentally alert on the job for the long time intervals required for crop harvesting within the optimum period while weather conditions may not always be favorable.

In the following description, forward and rearward directions are related to the normal direction of combine travel; upward and downward orientations are related to the operating floor of the cab.

Typically, the cab is located behind and just above the header, and in front of the grain tank.

As seen in Figure 14.2, a combine cab is an enclosure usually made of a metal frame (steel, aluminum) that is rigidly mounted on the machine chassis, although a vibration damping system may instead be considered. Flat or curved tempered-glass windshield and windows are fitted to the frame using special gaskets for protection against water, wind, temperature, dust, and noise. The operator enters or exits the cab through a lateral door using a ladder and handrails.

Located in the center of the cab is the steering wheel; laterally, there are brake pedals, a clutch pedal, and an acceleration pedal, although the last may be place somewhere else according to the design team's decision. The operator's seat is located behind the steering wheel. The seat design is constrained by human engineering data, control location and

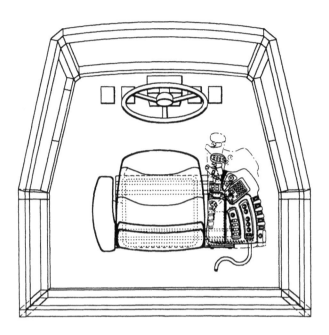

FIGURE 14.2
Top view of a cab with sliding control console. (From EPO, Operator seat sliding control console, European Patent Application EP 0 841 211 A2, Int. Cl. B60N 2/38, Zedelgem, Belgium, 1997.)

accessibility, ergonomics data, and budget or pricing restrictions imposed by market study predictions. The operator must be provided with a process control console that is easy to operate and maintain, and adjustable for different location and direction needs for a wide range of individual sizes. A process control monitor (e.g., Figure 1.55) may also be fitted in the cab. A highly functional monitor can provide the visualization of various functional and performance parameters of the machine.

During cab design, the following organizing steps may be followed:

- Establishment of the purpose of the cab—to what extent the cab should satisfy certain criteria beyond the basic ones: operator and control sheltering, minimum comfort for the operator, safety operation and protection against hazards, and so forth
- Identification of hazards, health, and safety concerns for the operator
- Identification and compliance with applicable standards, regulations, rules, and key features
- Design for minimization of the severity of harm or damage, should an incident/accident occur

Once design specifications have been established, hardware is selected based on engineering calculations, and once the design concept and project management schedule are in place, the design team may fully proceed with assembly and design details. However, back and forth iterative steps may be required when interfacing all the components of the cab.

14.2.1 Hazards for Combine Operator

The combine operator may be exposed to significant hazards, significant hazardous situations, and events, as listed in Annex A of ISO 4254-1 (2008). This is a comprehensive list that cannot be integrated into this text due to space limitations. It is very important for the cab design team to analyze all potential hazards carefully and develop its own specific list. Beyond this step, practitioners should consider every possible inadvertent exposure of the operator to a hazard resulting from the person's action during normal operation or service of the combine.

The cab should be designed according to the principles of risk reduction specified in ISO 12100-1 (2003), Clause 5.

14.2.2 Environmental Specifications for Operator Comfort

The comfort zone for a human is defined as the range of environmental conditions in which the human body can achieve thermal comfort; this is affected by climate, work rate, and state of acclimatization. The thermal comfort of a human is defined as a mental condition, which is based on the perception of noticeable changes in temperature (due to the above-mentioned factors), resulting in a personal expression of satisfaction with the environment to which the human is exposed.

When considering design of a combine harvester cab, the comfort specifications should accommodate the 5th percentile female to the 95th percentile male when seated, for environmental parameters as specified in Table 14.1. Thermal comfort for a person at rest (case of combine harvester operator) is best at air temperature $T = 22 \pm 2°C$, relative humidity $RH = 50 \pm 2\%$, air speed $v_a < 0.2$ m/s, and mild radiation exchange.

The *air temperature*, measured with a dry-bulb thermometer, is the average temperature of the air surrounding the occupant (e.g., air temperature measured in a combine harvester cab). The temporal average of air temperature is based on 3 min intervals, with temperature measured in at least 18 equally spaced points in the surrounding area.

Relative humidity (RH) is a measure of the actual moisture that is present in the air relative to the total moisture the air could hold at the actual air temperature. Relative humidity is expressed as a percentage. High levels of relative humidity affect the evaporative cooling effects of sweating, leaving the human body prone to overheating. Figure 14.3 shows the comfort zone dependence on the relative humidity and dry-bulb temperature of the air.

A comfort function, developed by Fanger (1972) and adapted in ISO 7730, is a model of correlation between the subjective human perception, expressed through the vote of comfort (*PMV*) on a scale ranging from –3 (very cold) to 3 (very hot), and the equation of thermal balance for the human body, as follows:

TABLE 14.1

Limits of Environment Conditions for Humans

Environment Parameter	Comfort Limits		Tolerable Limits	
	Lower Limit	Upper Limit	Lower Limit	Upper Limit
Temperature, °C (°F)	19 (66)	24 (75)	–1 (30)	39 (102)
Relative humidity, %RH	30	70	10	90
Ventilation, m³/min (ft³/min)	0.3 (10.6)	0.55 (19.4)	0.15 (5.3)	1.25 (44)
Air speed, m/s (ft/min)	0.2 (39.4)	0.4 (78.7)	0.1 (19.7)	1.00 (197)

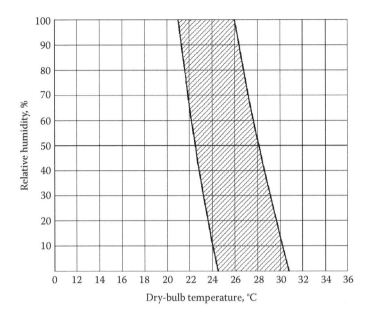

FIGURE 14.3
Air temperature–relative humidity chart and comfort zone.

$$PMV = \left(0.303e^{-2100M} + 0.028\right)\left[(M-W) - H - E_c - C_{res} - E_{res}\right] \tag{14.1}$$

where:
  PMV is the *predictive mean vote* of comfort
  M is the human metabolic rate, W/m²
  W is the effective mechanical power, W/m²
  H is the sensitive heat losses
  E_c is the heat exchange by evaporation on the skin
  C_{res} is the heat exchange by convection in breathing
  E_{res} is the evaporative heat exchange in breathing

A finer function known as *predicted percentage of dissatisfied people* (PPD) is defined by the following equation:

$$PPD = 100 - 95e^{-0.03353PMV^4 - 0.2179PMV^2} \tag{14.2}$$

The curve of Equation 14.2 has the appearance of an inverted Gauss probability density function.

The standard ISO 7730 indicates the limits of all data required to perform calculations of the above-described parameters.

ASAE S525-1.1 describes the requirements, test methods, and safety practices for environmental *air quality* in the agricultural cabs. The air intake filter should provide with a 99% or greater efficiency the air filtration against 3.0 μm (aerodynamic diameter) particulates. The acceptance criterion is based on the minimum efficiency measured in the 2.5–5.0 μm (aerodynamic particulate diameter) size range. The above-mentioned standard provides further information on particle concentration measuring and performance assessment, sealing efficiency against air leakage through all joints on the cab pressurization system, and other valuable information for cab design.

The *auditory comfort* depends on the sound pitch range as well as the sound pressure, measured in decibels (dB). The range of sound frequency a human can hear is 20–20,000 Hz. The sound pressure level we can sense is between 0 and 130 dB. The above-given ranges diminish as the body ages or due to a preponderance of certain sound levels during day-to-day work activities. The human ear is sensitive to sounds whose characteristics are within the human voice. The perception of loudness is related to the sound pressure level, duration of sound, and changes in sound frequency. Figure 14.4 shows the dependence of the human ear on equal-loudness levels expressed in phones on the sound frequency and pressure level. The human ear is most sensitive to sounds whose frequency is between 2 and 5 kHz. The range of human voice is between 80 and 2.5 kHz. In terms of human exposure to extended noise levels, the standards vary. Typically, 90 dB is the noise threshold value of human exposure for an 8 h work shift. Per Occupational Safety and Health Standard 1910.95 of the U.S. Department of Labor, the allowable time T_a (h) for exposure to noise with a weighted sound pressure level L_w (dB) can be calculated with the following equation:

$$T_a = 8/2^{(L_w-90)/5} \qquad (14.3)$$

In a combine harvester, the noise is generated by multiple sources; the most important are the engine and power train, cutting bar, threshing unit, straw walkers, cleaning unit, hydraulic motors, and all conveyors. A six-cylinder engine rated at 2100 rpm generates a sound with a frequency of 210 Hz (Goering et al., 2003). The noise reaches the operator's

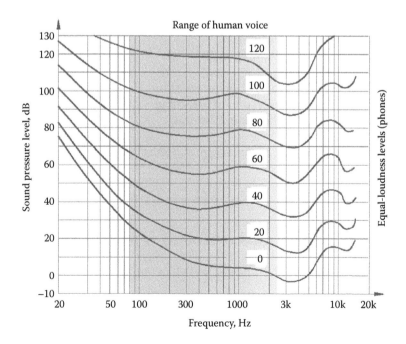

FIGURE 14.4
Equal-loudness curve dependence on sound frequency and pressure level.

ears through the air and cab structure. The cab floor and walls can be treated with sound-barrier materials. The noise pressure level at the operator's level, in the cab of modern combine harvesters, is around 75 dB (e.g., CLAAS, New Holland), measured per the ISO 5131 standard.

Practically, all state-of-the-art combine harvesters expose the operator through the seat and cab structure to *vibrations* that are not healthy. The vibrations are mainly generated by the engine, cutting bar, threshing unit, straw walkers, and cleaning unit; other types of vibrations are generated by the road and field profiles while the machine moves on.

Vibrations are oscillatory motions of mechanical systems, as mentioned above. Mechanical vibrations are characterized by frequency (Hz), amplitude (m [in.]), peak-to-peak displacement (m [in.]), velocity (m/s [in./s]), power spectral density (G²/Hz), and phase (degrees). The term G is actually G_{RMS}, where RMS means root mean square, and its normalized instantaneous value is

$$G = \frac{a}{g} \tag{14.4}$$

where:
  a is the vibration instantaneous acceleration, m/s² (in/s²)
  g is the gravitational acceleration, m/s² (in/s²)

In a combine harvester, the vibration spectrum can be described as a series of sinusoidal tones superimposed on a background of random vibrations (sine-on-random).

Taking advantage of our extensive experience in physical testing and vibration simulation of different products, we present here a minimum of knowledge practitioners should know about vibration theory and, for practical reasons, about the conversion between sinusoidal vibration and random vibration.

The characterizations of sinusoidal vibration and random vibration are based on two distinctly different sets of mathematical formulae. *Sinusoidal vibration* is based on a sinusoidal wave as a function of time t (Figure 14.5), as given below:

$$x = A \sin \omega t \tag{14.5}$$

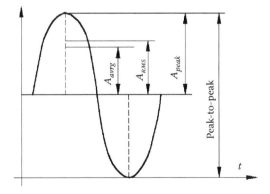

FIGURE 14.5
Constants for true sinusoidal wave.

where:

    x is the displacement
    A is the peak amplitude
    ω is the angular frequency of vibration, rad/s

We need to remember that 2π rad/s = 1 Hz.
The constants for true sinusoidal vibration are as follows:

$$A_{avrg} = 0.637 A \tag{14.6}$$

$$A_{RMS} = \frac{1}{\sqrt{2}} A \tag{14.7}$$

The severity of the vibration is characterized by the upper and lower limits of frequency, sweep duration, and peak acceleration of the sinusoidal wave vibration.

Random vibration is an oscillatory motion whose instantaneous amplitudes occur as a function of time following a normal (Gauss) distribution curve. In fact, random vibration is based on a continuous spectrum of different frequencies.

The amplitude of any discrete frequency could be very large or small at any given instant time. Random vibration can be represented in the frequency domain by a *power spectral density (PSD)* function (as adopted by industry), although the term *acceleration spectral density* is much more appropriate. Equation 14.7, attributed to John W. Miles (1954), defines the random vibration as follows:

$$G_{RMS} = \sqrt{\frac{\pi}{2} PSD f_n Q} \tag{14.8}$$

where:

    G_{RMS} is the root mean square acceleration of vibration, G
    PSD is the power spectral density, G^2/Hz
    f_n is the natural frequency, that is, the first resonance frequency of the vibrated system, Hz
    Q is the transmissibility (or amplification) factor

The transmissibility Q is defined as a function of the critical damping ratio ς, as follows:

$$Q = \frac{1}{2\varsigma} \tag{14.9}$$

G_{RMS} is determined by the square root of the area under the PSD curve; an example of such a curve is given in Figure 14.6.

Miles' equation (14.7) can be used in both design and testing of a product that is subject to vibrations. For instance, if the natural resonance frequency of a product has been determined (using ANSYS or COMSOL), then Miles' equation can be used to estimate the loads due to random vibration. In testing, the accelerations of a multiple degree of freedom system due to random vibrations can be approximated using Miles' equation.

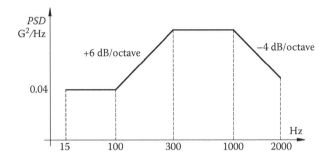

FIGURE 14.6
Random vibration spectrum per MIL-STD-810G. (From U.S. Department of Defense. Department of Defense test method standard: Environmental engineering considerations and laboratory tests. MIL-STD 810G, Section 514. Washington, DC: U.S. Department of Defense, 2008.)

Often, it is very useful to convert random vibrations to a sinusoidal vibration, and vice versa, because some standards only list sinusoidal vibration requirements. The following method relates random and sinusoidal vibration based on the effective damage theory (Wang, 2014).

To convert sine sweep or dwell to random vibration, the power spectral density is

$$PSD = A_{peak}^2 / (\pi f_n) \tag{14.10}$$

The duration of random vibration t_{rd} can be calculated by dividing the total vibration cycles N of sinusoid dwell by the dwell frequency:

$$t_{rd} = \frac{N}{f_n} \tag{14.11}$$

The duration of random vibration t_{rd} can also be expressed as a function of transmissibility Q and the rate of sweep R as follows:

$$t_{rd} = \frac{\ln \dfrac{1 + \dfrac{1}{2Q}}{1 - \dfrac{1}{2Q}}}{R \ln 2} \tag{14.12}$$

To convert the random vibration to sinusoidal vibration, we consider that the static accelerations in both types of vibrations are equal. Then, the peak acceleration is

$$A_{peak} = \sqrt{\pi PSD f_n} \tag{14.13}$$

The following formula will calculate the number of sinusoidal wave cycles:

$$N = t_{rd} f_n \tag{14.14}$$

Then, the sine sweep rate R (octave/s), can be calculated using the following equation:

$$R = \frac{\ln \dfrac{1 + \dfrac{1}{2Q}}{1 - \dfrac{1}{2Q}}}{t_{rd} \ln 2} \tag{14.15}$$

The method described above is applicable for sinusoidal vibration with a constant peak acceleration sine sweep.

During work and transportation, a combine operator encounters *whole-body vibrations* through the seat, *hand–arm vibrations* through the steering wheel and levers, and *footrest vibrations* through the cab floor.

In regard to the exposure of workers to the risks arising from vibration, European Council Directive (EEC) 2002/44/EC defines the whole-body vibration as "the mechanical vibration that, when transmitted to the whole human body, entails risks to the health and safety of workers, in particular lower-back morbidity and trauma of spine." The human body usually has a natural frequency between 4 and 5 Hz and a low tolerance for amplitudes between 0.5 and 1.5 G. Table 14.2 shows the natural frequencies of the human body parts.

EEC 2002/44/EC stipulates daily vibration exposure limit values for whole-body vibration and hand–arm vibration. The *vibration exposure level* is defined as a function of the vibration magnitude perceived by a person over a period of time t, as follows:

$$a_w = \left[\frac{1}{T} \int_0^T a_w^2(t) \, dt \right]^{1/2} \tag{14.16}$$

where:
  $a_w(t)$ is the frequency-weighted acceleration over the time period (a_{rms}), m/s²
   T is the measurement period, s

TABLE 14.2

Natural Frequencies of Human Body Parts

Body Part	Natural Frequency, Hz
Trunk	3.6
Eyeball	20–25
Chest	4–6
Thorax	3
Spine	3–5
Heart	4–5
Shoulders	2–6
Head	30
Stomach	4–7
Colon	20–25

TABLE 14.3

Frequency Weighting and Multiplying Factors Specified by ISO 2631-1

Vibration Direction	Frequency Weighting	Multiplier k
Longitudinal axis (X)	W_d	1.4
Transversal axis (Y)	W_d	1.4
Vertical axis (Z)	W_k	1

If the human is exposed to vibrations of different magnitudes during i periods of time, the frequency-weighted equivalent acceleration A_{eq} is given by the following equation:

$$A_{eq} = k \left[\frac{\sum a_{wi}^2 T_i}{\sum T_i} \right]^{1/2} \tag{14.17}$$

where:

T is the length of time period i, s

k is the orthogonal axis multiplier (Table 14.3)

Based on Scarlett et al.'s (2005) study, the vibration levels for agricultural machinery operators vary widely and approach or exceed the exposure limit value. The *daily* (8 h) *vibration exposure level* A(8) (m/s²) may be derived from the following equation:

$$A(8) = A_{eq} \sqrt{\frac{t}{8}} \tag{14.18}$$

where:

t is the daily exposure time, h

A_{eq} is the frequency-weighted equivalent acceleration (Equation 14.17)

If the frequency-weighted equivalent acceleration is not known, then the daily vibration exposure level can be calculated using Equation 14.16, as follows:

$$A(8) = k \left[\frac{1}{T_0} \int_0^T a_w^2(t) dt \right]^{1/2} \tag{14.19}$$

where T_0 is the reference time (8 h = 28,800 s).

The daily *vibration dose value* (VDV) (expressed in m/s$^{1.75}$) a person is exposed to can be calculated with the following formula:

$$VDV = k \left[\int_0^T a_w^4(t) dt \right]^{1/4} \tag{14.20}$$

where:

$a_w(t)$ is the frequency-weighted acceleration over the time period (a_{rms}) at a vibrated, supporting surface, m/s²

T is the duration of exposure to vibration within a 24 h period, s

k is the orthogonal axis multiplier (Table 14.3)

The European Council Directive (EEC, 2002) also provides a derogation that allows averaging the daily exposure to vibration when occasional exposure levels are greater than the *exposure limit value* (ELV). In such occasions, the weekly average vibration exposure is normalized to a reference of 40 h. For more information on this, readers may read Research Report 321, which was conducted at Silsoe Research Institute by Scarlett et al. (2005).

14.3 Operator–Machine Interface and Controls

14.3.1 Introduction

The *operator–machine interface* is the factual interaction of the human body and brain with an engineering system (hardware and software subsystems), to control the system for a desired output, with certain efficiency, during a rational work time over the machine life expectancy, while the human person's safety and health are preserved based on ergonomics principles. The human body interacts with the machine by its senses and muscular power, while the human brain acts by exercising knowledge, experience, and decision-making command.

Our definition of the operator–machine interface implies not only a correlation between human engineering data and specific controls of the machine, but also consideration of safety, comfort, and convenience for the operator, while the machine is used within its operating parameter range, for maximum output, over its predicted life of use.

The main ergonomics principles refer to the following:

- Designing for human capabilities and skills
- Easy/efficient to use (the interface with the machine)
- No overstraining of the human body and brain
- Maintaining a safe and healthy environment (in the combine cab)

Therefore, from the perspective of the human operator, it is important to consider the seat design and the operator's interaction with pedals, levels, buttons, speakers, and the software interface (e.g., a screen, speakers) and other instruments, while satisfying ergonomics principles for adjustable working position and envelope, visibility, hearing, and touch senses.

Human engineering data defines acceptable ranges of different human characteristics and parameters that represent design considerations for the human–machine interface and controls. In the cab workspace design, a designer must also consider the human–environment interactions, as well as physical and cognitive limitations. The ergonomics design flow of the human–machine interface is based on a series of interaction priorities, as follows: the primary visual tasks and their controls, emergency controls, control/display relations, functional/sequential grouping, and frequency of use.

The design reference points and zones refer to the seat reference points, arm rotation points, eye reference points or zones, visual envelope, mobility, and comfort adjustment ranges that define the operator working envelope. In terms of physical interactions, the dynamic measures for the human body refer to the range and strength, grip, grasp, and exerted forces for pushing and pulling in different directions.

Human physical data, as outlined by SAE J833 (1989), "Vehicle Surface Recommended Practice," can be used as the main anthropometric data of adult humans. Table 14.4 displays selected physical dimensions of adult humans. Figure 14.7 shows the distribution of weight of the U.S. male and female populations. Such data is useful for designing the operator seat, as shown in the following section. The rule of using the 5th and 95th percentile values should be applied in a differentiated way; for example, the lower percentile is used to ensure a short human arm can reach a control button behind the steering wheel.

TABLE 14.4

Selected Physical Dimensions for Human Adults

Body Measurements	5th Percentile		50th Percentile		95th Percentile	
	mm	in.	mm	in.	mm	in.
Standing height with shoes	1550	61.02	1715	67.52	1880	74.02
Sitting height	798	31.42	879	34.61	960	37.80
Knee height sitting	495	19.49	555	21.85	614	24.17
Horizontal seat height	398	15.67	440	17.32	482	18.98
Elbow sitting height	189	7.44	214	8.43	239	9.41
Thigh height sitting	533	20.98	598	23.54	662	26.06
Knee-to-buttock length	524	20.63	586	23.07	648	25.51
Buttock-to-calf length	409	16.10	457	17.99	505	19.88
Elbow-to-elbow width	386	15.20	443	17.44	500	19.69
Hip width sitting	325	12.80	360	14.17	395	15.55
Hip pivot to horizontal seat height	64	2.52	80	3.15	96	3.78
Hip pivot to floor (leg length)	816	32.13	916	36.06	1016	40.00
Arm span	1564	61.57	1742	68.58	1920	75.59
Hip pivot width	168	6.61	177	6.97	188	7.40
Shoe width	90	3.54	103	4.06	116	4.57
Foot width	84	3.31	94	3.70	104	4.09
Weight, kg/lb	48	105.8	73	160.93	98	216.05

Source: SAE J833, Human physical dimensions, vehicle surface recommended practice, in *SAE Handbook*, Vol. 3, Society of Automotive Engineers, Warrendale, PA, May 1989.

Notes: 5th percentile data is for female population. 95th percentile data is for male population. African descendants can have 2% longer arms and 4% longer legs than the dimensions provided. Oriental descendants can have 7% shorter arms and 10% shorter legs than the dimensions provided.

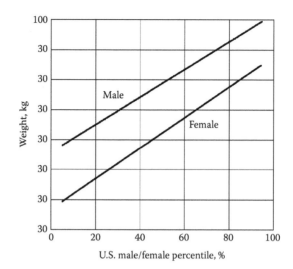

FIGURE 14.7
Statistics of the weight of the U.S. male and female populations.

14.3.2 Operator's Seat and Controls

One of the most important ergonomics aspects is related to the operator's seat design and its location with respect to the cab floor, pedals, steering wheel, and all controls. The seat location depends on the *operator working envelope* and seat movements: floating up and down due to its suspension and moving forward and rearward to accommodate all size ranges of the designated operator. In addition to human anthropometrics, the operator working envelope also depends on the *human body postures* and their ranges (Figure 14.8).

The seat movements relate to a central point called the *seat index point* (SIP). Figure 14.9 shows the SIP of the seat as well as ISO 4253 working zones. The seat should allow at least 150 mm (6 in.) fore and aft adjustment, 75 mm (3 in.) height adjustment, and 150 mm (6 in.) vertical stroke. The seat lumbar support angle to the vertical should allows 5°–20° rotation toward the rear direction. A modern seat (e.g., KAB 85-K6 made by KAB Seating; the measurements are shown in Figure 14.10, and the seat is shown in Figure 14.11) has additional features, such as adjustments of the seat cushion length, seat cushion tilt, and backrest extension, an adjustable suspension stroke limiter, and a ride indicator for proper suspension. The armrests are adjustable and foldaway.

An interesting mathematical model of the seat suspension system for vibration control was developed by Koszewnik and Gosiewski (2013). The passive system of the seat suspension consists of a cross-mechanism, a hydraulic damper, and a pneumatic spring. Such a mechanism reduces the vibration transmitted from the vehicle platform to the operator.

The design of the seat and placement of the controls around it should be based on a *user-centered approach* (Pheasant, 1996). This generally accepted approach derives from the acknowledgment that the operator has particular needs within the context of the considered human–machine interface, but the adaptation of its controls is not always useful or practical for a combine operator. The user-centered design approach considers equally well both physical (human strength) and cognitive integration of the human system with the engineered system at hand.

Human strength for operating a combine harvester is needed for steering the machine, actuating the clutch, braking (actuating the foot pedal), setting the safety brake on or off, and actuating different levers, as the case may be, while the permanent need to "feel" the

FIGURE 14.8
Human posture categories and ranges. (From Jorgensen, M.J. et al., Repeatability of a checklist for evaluating cab design characteristics of heavy mobile equipment, National Institute for Occupational Safety and Health, Spokane Research Lab, Spokane, WA, 2005.)

steering, clutch, and brake should be fully considered. Frequently used controls should be placed such that the operator will use them without undue exertion or fatigue.

In regard to the foot pedal actuating, the following specific data should be considered during combine seat control design:

- The full range of the knee angle is 160°–165°. The most comfortable range during knee flexion adduction is 115°–125°, with a minimum knee angle of 90°.

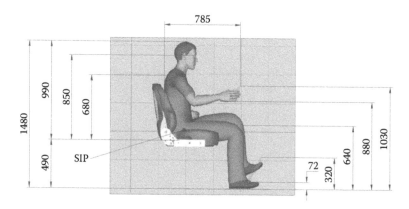

FIGURE 14.9
Operator seat SIP and ISO 4253 zones (dimensions in mm). (Mannequin from Dooley, W.K., Ergonomics and the development of agricultural vehicles, presented at the 2012 Agricultural Equipment Technology Conference, Louisville, KY, 2012.)

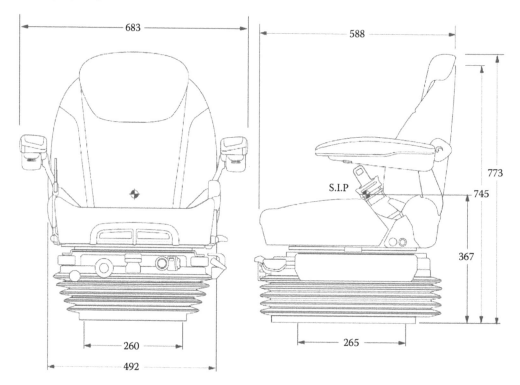

FIGURE 14.10
Design and dimensions (mm) of a KAB seat for combine harvesters (Courtesy of KAB Seating, Northhampton, England.)

- The ankle angle during foot operations (ankle flexion) should be within the 90°–115° range.
- The optimum lateral offset of clutch and brake pedals from the body midline should be 220–260 mm (8.65–10.25 in.).
- The minimum pedal travel distance is 30 mm (1.18 in.), and the maximum force response should be considered 300 N (67.42 lb).

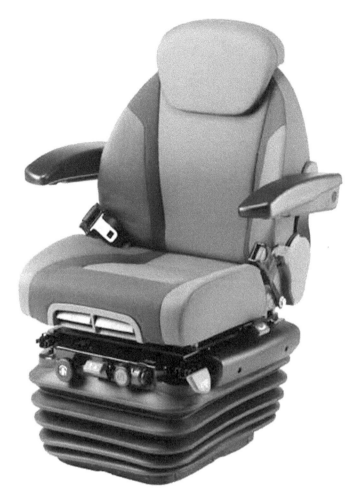

FIGURE 14.11
KAB 85-K6 seat made by KAB Seating, Northampton, England (dimensions given in Figure 14.10).

As a general rule, the *foot reach* is defined by a hemisphere of 800 mm (31.5 in.) radius centered on the centerline at the front edge of the cushion and extending downward. In regard to the *hand controls*, *hand reach* is defined by a sphere of 1000 mm (39.4 in.) radius centered on the seat front centerline, 60 mm in front of and 580 mm above the SIP of the seat. However, the cab may limit the space as defined above and shown in Figure 14.12. For a hand-operated lever, the minimum operating distance is 100 mm (4 in.), while the maximum actuation force is 80 N (20 lb).

The standards referenced below specify a series of other criteria for operator platform, handrails, step requirements, and so forth.

14.3.3 Cognitive Ergonomics and Human–Machine Interaction Modeling

During driving and operating a combine harvester, the operator continuously perceives, processes, and decides if and how to react to a wide variety of information presented in different forms: visual, auditory, and tactile. Consequently, when concurrent decisions have to be made and multiple tasks need to be performed, the operator's mental workload

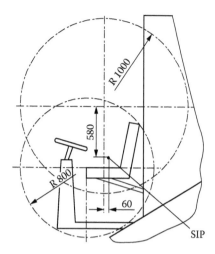

FIGURE 14.12
Arm- and foot-reach definition. (Data from ISO 4254-7, Agricultural machinery—Safety: Part 7: Combine harvesters, forage harvesters and cotton harvesters, International Organization for Standardization, Geneva, 2008.)

may approach his or her capability limits. At the mental level, a trained combine operator interacts with the machine in different ways, which can be generally classified as *perception* through sensory organs and *operation* with hands and feet. Between perception and operation, the operator has to make decisions on if, how, and when to react. The perception denotes the operator's aesthetic pleasure to interact with the machine, perception of machine complexity (quantity, variety, and interconnections), perception of machine quality, and assessment of its reliability. The operation of a machine implies using a set of operation skills (evaluated through the fitness for task) by processing the perceived information, identifying controls, and acting by overcoming any difficulty in using them (control usability), while continuously maintaining a certain level of situational awareness by monitoring the system output.

According to Bridger (1995), the operator tasks in machine operation involve using short-term memory processes; this infers a limited storage capacity of the memory and a short interval of time for retaining the perceived information in regard to five to nine distinct tasks. Wickens and Hollands (2000) introduced a model of information processing by humans that illustrates the perception and cognition mechanisms, their relationships with short- and long-term memories, and the response creation (Figure 14.13).

Dooley (2012) gives the example of monitoring the information displayed on the monitor screen of a combine harvester; he classifies such information into three categories: frequently viewed (grain loss, engine load), occasionally viewed (rotor speed, fan speed, crop moisture and yield), and seldom viewed (header elevation, fuel level, coolant temperature). It is agreed that each operator has his or her own habits in attention distribution, information processing time, and reaction time. Individuals establish filters based on their background, experience, customs, and so forth. Some of the screened factors may be considered irrelevant (not necessary truth) and thus ignored. The filtering mechanism changes in certain conditions related to decision-making stress, tiredness, psychological state, and overload. Consequently, modeling human behavior is nowadays a necessary condition to eliminate, or at least reduce, machine operating errors.

When solving the assignment (e.g., soybean harvesting), the operator's behavior characteristics as the combine controller depend on four groups of variables: task variables

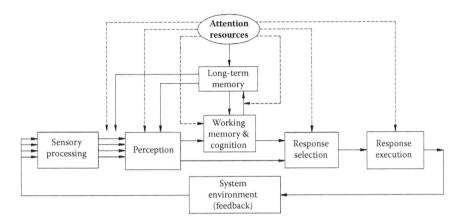

FIGURE 14.13
Model of information processing by humans. (Modified from Wickens, C.D. and J.G. Hollands, *Engineering Psychology and Human Performance*, 3rd ed., Prentice Hall, Upper Saddle River, NJ, 2000.)

(Figure 14.14), environment variables, operator-centered variables, and procedural variables. The task variables include all system inputs and control system elements external to the operator; these variables directly and explicitly affect the operator assignment. The *operator assignment* resides in *driving* the combine in the field with the crop to be harvested and *controlling the combine working processes* so that the machine will work at full capacity, while the grain loss is minimized and the grain quality is preserved. Thus, the operator workload is affected by two groups of combine harvester–handling qualities: *vehicle handling (guidance)* on a path in the field and working *process handling (process stabilization)* qualities. Process stabilization is the most demanding workload for the operator requiring frequent control inputs to maintain a constant combine load and good quality of grain.

The operator's dynamics is affected by the following task variables: forcing function, display, manipulator, and controlled element. The *system forcing function* is a random or randomly appearing time function that has stationary or quasi-stationary properties. Depending on how the display presents the operator with a visual stimulus, the display may operate as either a compensatory or a pursuit tracking system. A compensatory tracking task is one in which the output signal from the controlled machine is fed back and compared with the input signal, resulting in an error the operator is required to at least

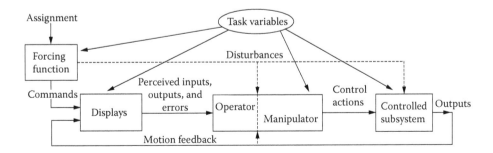

FIGURE 14.14
Task variables affecting the operator–machine system. (Adapted from McRuer, D.T. and H.R. Jex, *IEEE Transactions on Human Factors in Electronics* 8(3):231–248, 1967.)

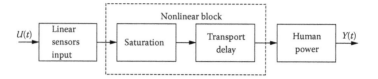

FIGURE 14.15
Possible model of human operator behavior in controlling a combine harvester.

minimize. An example of such a system is the implementation of a grain loss sensor, and the operator attempts to reduce the error.

The environment variables, external to the operator, include vibration, temperature, humidity in the cabin, and so forth. The operator-centered variables characterize operator motivation, stress, workload, training, fatigue, and so forth. The procedural variables are more difficult to quantify; they relate to instructions, practice, and order of presentation, as well as experiment design.

Human behavior in the process of system control can be described by relevant elements of an automatic control system. Different researchers (McRuer and Krendel, 1974; Cameron et al., 2003) proposed alternative mathematical models of human behavior in a dynamic control system. The behavior of a human operator derives from Figure 14.14 and can be described by a block diagram, as shown in Figure 14.15. The *linear block* delivers the model input signals obtained by quantifying the information captured by human sensory organs. The nonlinear component consists of the *saturation model* and *information transport delay* blocks. The saturation model takes into account the limited reaching range of human arms and legs (human manipulator characteristics). The information transport delay is the time for human response to outer stimuli. The last block is the *model of muscle active and passive characteristics* for actuating the machine controls.

The *Precision Operator Model* represented by Equation 14.21 is an extension of the *Crossover Pilot Model* (McRuer and Krendel, 1974). This model is a single-input, single-output (SISO) model expressed in the Laplace domain as follows:

$$Y_p = K_p \frac{T_L s + 1}{T_I s + 1} \frac{e^{-\tau s}}{(T_n s + 1)\left(\dfrac{s^2}{\omega_n^2} + \dfrac{2\varsigma_n}{\omega_n} s + 1\right)} \tag{14.21}$$

where Y_p is the *overall output* of the operator's behavior and action, and K_p is the *operator gain*, that is, operator habits for a given type of machine. The operator gain represents the operator ability to respond to an error in the magnitude of a controlled variable. T_L is the *lead time constant*, which depends on the operator's experience. It quantifies the operator's ability to predict a control input to avoid the occurrence of a possible situation or event. This ability results from the highest level of situational awareness that comes from knowledge and experience of the controlled system state and dynamics. T_I is the *lag time constant*, which describes the ease with which an operator executes the required control input. The lag constant depends on the degree of learned stereotypes and routines applied by the operator in exercising machine control. τ is the *time constant* that characterizes the *delay* of the brain response to the operator's muscles and eye perception. This parameter represents the period between the decision to change

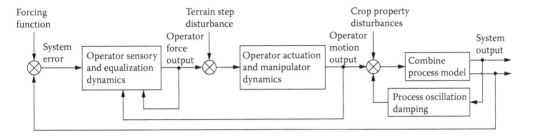

FIGURE 14.16
Integration model of human operator–machine control system.

a control input and the act itself. Such a delay depends on the actual state of the neuromuscular system, as well as on one's physical and mental condition. T_N is the *neuromuscular lag time constant*, which represents the operator's action delay to trigger the muscle from the moment the signal is sent from the brain. ω_n is the *undamped natural frequency* of the neuromuscular system. ζ_n is the *damping ratio* of the neuromuscular system.

The terms

$$K_p \frac{T_L s + 1}{T_l s + 1}$$

in Equation 14.21 are collectively described as "human equalization characteristics," while the last terms are described as "inherent human limitations" (Cameron, 2002).

The integration model of the human operator–combine harvester control system can be achieved as illustrated in Figure 14.16. The human operator model can be represented by Equation 14.21. The combine process model is a complex model that can be built at least based on the information outlined by this text. It could also be considered the model of combine vehicle kinematics while steering (see Application 2.2). The loop of process oscillation damping represents the first negative feedback; its aim is to reduce the reactive process oscillations, and it can be designed as a linear member with amplification less than 1.

For more information on this topic, readers are primarily directed to the References and Bibliography, specifically Szabolcsi (2007), Cameron (2002), and McRuer and Krendel (1974).

References

ASAE S525-1.1. 1997. Agricultural cabs—Environmental air quality: Part 1: Definitions, test methods, and safety practices. St. Joseph, MI: American Society of Agricultural Engineers.

Bridger, R.S. 1995. *Introduction to Ergonomics*. New York: McGraw-Hill.

Cameron, N. 2002. Identifying pilot model parameters for an initial handling qualities assessment. PhD disseration, University of Glasgow.

Cameron, N., D.G. Thomson, and D.J. Murray-Smith. 2003. Pilot modelling and inverse simulation for initial handling qualities assessment. *The Aeronautical Journal* 511–560.

Dooley, W.K. 2012. Ergonomics and the development of agricultural vehicles. Presented at the 2012 Agricultural Equipment Technology Conference, Louisville, KY.

EEC 2002/44/EC. 2002. Council directive on the minimum health and safety requirements regarding the exposure of workers to the risks arising from physical agents (vibration). *Journal of the European Communities* OJ L 177.

EPO. 1997. Operator seat sliding control console. European Patent application EP 0 841 211 A2, Int. Cl. B60N 2/38, Zedelgem, Belgium. Inventor: D.B. Stauffer, New Holland, PA.

Fanger, P.O. 1972. *Thermal Comfort*. Copenhagen: Danish Technical Press.

Goering, C., M.L. Stone, D.W. Smith, and P.K. Turnquist. 2003. Human factors and safety. In *Off-Road Vehicle Engineering Principles*, 421–462. St. Joseph, MI: American Society of Agricultural Engineers.

ISO 4253. 1993. Agricultural tractors: Operator seating accommodation. Geneva: International Organization for Standardization.

ISO 4254-1, B-1050. 2008. Agricultural machinery—Safety: Part 1: General requirements. Geneva: International Organization for Standardization.

ISO 4254-7. 2008. Agricultural machinery—Safety: Part 7: Combine harvesters, forage harvesters and cotton harvesters. Geneva: International Organization for Standardization.

ISO 4254-7. 2009. Agricultural machinery—Safety: Part 7: Combine harvesters, forage harvesters and cotton harvesters. Geneva: International Organization for Standardization.

ISO 5131. 2012. Acoustics—Tractors and machinery for agriculture and forestry—Measurement of noise at the operator's position—Survey method. Geneva: International Organization for Standardization.

ISO 7730. 2005. Moderate thermal environments—Determination of the PMV and PPD indices and the specifications of the conditions for thermal comfort. Geneva: International Organization for Standardization

Jorgensen, M.J., N.K. Kittusamy, and P.B. Aedla. 2005. Repeatability of a checklist for evaluating cab design characteristics of heavy mobile equipment. Spokane, WA: National Institute for Occupational Safety and Health, Spokane Research Lab.

KAB. 2014. KAB Seating—Agricultural seating range. Queensland, Australia: Acacia Ridge.

Koszewnik, A. and Z. Gosiewski. 2013. Modelling of the seat suspension system for the vibration control system. *Mechanics and Control* 32(3):97–101.

McRuer, D.T. and H.R. Jex. 1967. A review of quasi-linear pilot models. *IEEE Transactions on Human Factors in Electronics* 8(3):231–248.

McRuer, D.T. and E.S. Krendel. 1974. Mathematical models for human pilot behavior. AGARD-AG-188. Paris: NATO Advisory Group for Aerospace Research and Development.

OSHA 1910.95 App A. 2014. Noise exposure computation. Washington, DC: U.S. Department of Labor, Occupational Safety and Health Standard. https://www.osha.gov/pls/oshaweb/owa-disp.show_document?p_table=STANDARDS&p_id=9736 (accessed September 27, 2014).

Pheasant, S. 1996. Bodyspace: Anthropometry, ergonomics, and the design of work. 2nd ed. Philadelphia: Taylor & Francis.

SAE J833. 1989. Human physical dimensions. Vehicle surface recommended practice. In *SAE Handbook*. Vol. 3. Warrendale, PA: Society of Automotive Engineers.

Scarlett, A.J., J.S. Price, D.A. Semple, and R.M. Stayner. 2005. Whole-body vibration on agricultural vehicles: Evaluation of emission and estimated exposure levels. Caerphilly, UK: HSE. hseinformationservices@natbrit.com (accessed September 15, 2014).

Szabolcsi, R. 2007. Modeling of the human pilot time delay using Pade series. *AARMS* 6(3):405–428.

U.S. Department of Defense. 2008. Department of Defense test method standard: Environmental engineering considerations and laboratory tests. MIL-STD 810G, Section 514. Washington, DC: U.S. Department of Defense.

Wang, F.F. 2014. Relating sinusoid to random vibration for electronic equipment. *COTS: Journal of Military Electronics and Computing*. http://www.cotsjournalonline.com/articles/view/100314 (accessed September 30, 2014).

Wickens, C.D. and J.G. Hollands. 2000. *Engineering Psychology and Human Performance*. 3rd ed. Upper Saddle River, NJ: Prentice Hall.

Bibliography

ANSI/ASAE S279.17. 2013. Lighting and marking of agricultural equipment on highways. St. Joseph, MI: American Society of Agricultural Engineers.

ASAE S318.10. 1992. Safety for agricultural equipment. St. Joseph, MI: American Society of Agricultural Engineers.

Boduch, M. and W. Fincher. 2009. Standards of human comfort, relative and absolute. Presented at the Seminar in Sustainable Architecture, University of Texas at Austin.

Boril, J., R. Jalovecky, and R. Ali. 2012. Human–machine interaction and simulation models used in aviation. Project UO-K206: Complex Electronic System for UAS UDeMAG (University of Defence, MATLAB Group). Brno, Czech Republic.

CEN TC 122. 2008. Ergonomics. Guidance on the application of the essential health and safety requirements on ergonomics set out in Section 1.1.6 of Annex I to the Machinery Directive 2006/42/EC.

Dreyfuss, H. and A.R. Tilley. 2002. *The Measure of Man and Woman*. Rev. ed. New York: John Wiley & Sons.

EN ISO 14738, B-105. 2008. Safety of machinery: Anthropometric requirements for the design of workstations at machinery. Geneva: International Organization for Standardization.

Gaynor, R., T. Willkomm, and W.E. Field. 1986. Hand controls for agricultural equipment. Plowshares 2. Breaking New Ground Technical Report. Purdue University.

Gniady, J. and J. Bauman. 1991. Active seat isolation for construction and mining vehicles, 1–6. SAE Technical Paper Series. Warrendale, PA: Society of Automotive Engineers.

Inoue, E., O. Kinoshita, J. Kashima, and J. Sakai. 1993. Distribution and spectrum analysis of noise on a combine harvester. *Journal of Faculty of Agriculture* 38(1–2):137–145.

ISO 14738. 2002. Safety of machinery—Anthropometric requirements for the design of workstations at machinery. Geneva: International Organization for Standardization.

Kim, H.J., M.H. Kim, E. Inoue, T. Okayasu, and D.C. Kim. 2013. Ergonomic study for comfort operating the pedals of the agricultural tractors. *Journal of Faculty of Agriculture* 58(2):329–338.

Klooster, S.J. 2004. Vibration suspension and safety seat motion design of a hyper-active seat. MS thesis, Georgia Institute of Technology.

Makhsous, M., R. Hendrix, Z. Crowther, E. Nam, and F. Lin. 2005. Reducing whole-body vibration and musculoskeletal injury with a new car seat design. *Ergonomics* 48(9):1183–1199.

Miller, L. and C. Gariepy. 2008. Heavy mobile equipment: Ergonomics and the prevention of musculoskeletal injuries. Edmonton, Canada: EWI Works International.

Milles, J.W. 1954. On structure fatigue under random loading. *Journal of Aeronautical Sciences* 753–762.

Sümer, S.K., S.M. Say, F. Ege, and A. Sabanci. 2006. Noise exposed of the operators of combine harvesters with and without cab. *Applied Ergonomics* 37(6):749–756.

15

Guidance and Control of Autonomous Combine Harvesters

Following a task assigned (i.e., programmed) by a human supervisor, an *autonomous combine harvester* should be capable of the following while maintaining permanent communication with the human supervisor:

- Automatic navigation on the road and in the field
- Self-acting and reacting to its environment (other vehicles, obstacles, road conditions, etc.)
- Self-regulating the harvesting process of a crop

Before becoming autonomous, a combine harvester must have the entire technological process completely automated and self-controlled for maximizing the working capacity and preserving the quality of grain, which may be harvested from dispersed areas featuring certain crop variability. This primary requirement for an autonomous combine harvester can be achieved by using the mathematical models described in this book, control theory, and other specific knowledge in conjunction with in-house acquired design experience and quantified in-field experience.

Then, an autonomous guiding system that controls the vehicle direction, speed, and acceleration has to be implemented. Consequently, a camera-based system and a global positioning system (GPS)-based system need to be coupled in an autonomous combine with vehicle motion control. Thus, the camera system can be used without an *a priori* map of the crop field and can also be used for obstacle detection. The GPS system helps in navigating the vehicle on the road and prevents positioning errors from accumulating indefinitely. In addition, each system may mutually help compensate for the failure modes of the other system. Thus, the GPS eliminates the problems of the camera-based system due to poor lightning and sparse crop conditions, while the camera-based system may compensate GPS complications due to multipath choices or occluded satellites. There are also cases of using geomagnetic position sensors for implementing the noncontact location of the vehicle.

The following sections describe the principles, methods, specific algorithms, and equipment useful in autonomous vehicle development and, in particular, for integration within combine harvesters as complex mechatronic systems.

15.1 Precision Harvesting, Geospatial Dispersion, and Variability of Crops

The location of a crop field, weather, soil features and heterogeneity, and associated recent history of crop cultivation on that field have a substantial influence on the growing and maturing of actual and subsequent crops. The large geospatial dispersion of cereal crops and the farm size associated with different ownership make it difficult to manage crop cultivation in a particular area without collecting, interpreting, and using current and historical data about the crops grown in that area and surrounding region.

Precision farming is an information and technology-based management system whose aim is to improve crop production efficiency by adjusting farming operations to specific conditions within each area of a field, for a certain crop, or crop rotation.

The term *precision harvesting* derives from the precision farming concept for facilitating the farming management based on data collection during harvesting of different crops, whose variability (biomass development, yield, grain quality, genetic and organic matter content, and distribution) is inherently characterized by spatial and temporal components. Moreover, further relationships may be derived from association of the above-mentioned crop variables with the influence of terrain location, its topography, and soil chemical composition (nitrogen, magnesium, potassium, etc.). This requires the synchronization of an on-the-go measurement and data acquisition system mounted on the combine harvester with a navigation system for the field. Collected data may also be used for representing the crop growing process and crop harvesting operations through mathematical models that explain and predict the possible evolution of the considered process. Models of fertilization and water flow throughout a field, associated with historical yield data acquired during crop harvesting, may be cued to in-field growth models. For instance, in organic farming, the model of nitrogen (N_2) entering crop rotation via symbiotic leguminous–rhizobium N_2 fixation may be correlated with site-specific conditions and spreading of organic manure.

Chapter 13 describes a series of sensors that can sense the variation of different crop variables and deliver this information to an electronic communication and storage system.

Correlating comprehensive and accurate data in real time during harvesting with the features of the terrain, as well as with local farming history, became possible and affordable by the development and implementation of reliable *global navigation satellite systems* (GNSS), such as the U.S. GPS NAVSTAR (Global Positioning System—Navigation System by Time and Range), Russian GLONASS (Global Navigation Satellite System), and GALILEO from the European Union. China is also in the process of developing and expanding the Compass navigation system by 2020. Both GPS and GLONASS work on the same principle. The next section gives readers an in-depth introduction to GPS.

15.2 Introduction to GPS, Coordinate Systems, and Latitude and Longitude

GPS provides continuously specially coded satellite signals that can be processed with a GPS receiver (located on a vehicle of choice on the earth) for computing and displaying, at any time, the position and velocity of a moving or stationary vehicle. The current system, which became fully operational in June 1993, is administered by the U.S. Air Force for the U.S. Department of Defense. The GPS system is based on three segments: *space segment*, *control segment*, and *user segment*. The space segment comprises 24 geostationary satellites (4 satellites on each orbital plane, with 55° and 60° inclination between adjacent orbital planes) that orbit our planet at the altitude of 20,183 km, once every 12 h (Figure 15.1). Each satellite has its own identification number and sends messages for its identification, positioning, timing, and ranging data, as well as its orbit parameters. The control

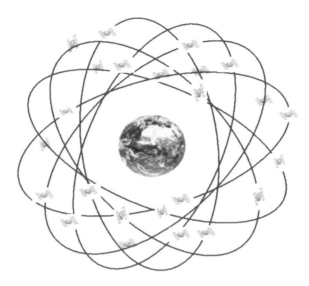

FIGURE 15.1
(See color insert.) GPS constellation of 24 satellites (6 orbital planes, each with 4 satellites).

station—a network of earth-based facilities—controls the space segment. The user segment consists of a receiver that tracks the satellite signals, processes the information based on an embedded algorithm, and displays the map and user position on the map and its speed. Information processing consists of time synchronization (by computing the time offset), satellite data decryption, computation of the pseudorange of each satellite, and position coordinate computation.

The signals emitted by a minimum of four GPS satellites are required for computing three-dimensional (3D) positions and the time offset. Each satellite emits a unique signal. The receiver calculates the distance by recording the time difference between the moment when the satellite starts its signal and the moment when the signal reaches the receiver. To ensure very accurate time measurement, the satellite atomic clock has an accuracy of 3×10^{-9} s. The GPS NAVSTAR provides autonomous position information with accuracies within $\pm(10-15)$ m. For higher accuracy, a differential GPS (DGPS) from a base station calculates the errors of the runtime and pseudorange measurement by comparing the calculated position with the real position. Thus, the position accuracy of the mobile receivers can be improved to $\pm(2-5)$ cm. DGPS systems require a differential signal from either a free service such as Wide Area Augmentation System (WAAS) or Coast Guard Beacon, or a commercial service such as OmniStar or John Deere's StarFire system. A comparison of such systems, their accuracy, and advantages is shown in Table 15.1. WAAS provides free GPS differential correction data for visible satellites. The phase differential real-time kinematic (RTK) systems are the most precise GPS systems, with centimeter accuracies, but they are also expensive. The accuracy depends on the receiver design, time spent on measurement, data postprocessing, relative positions of satellites, technically known as position dilution of precision (PDOP), and clock time offset. The U.S. Department of Defense controls the selective availability (S/A) feature of GPS, that is, the intentional degradation (referred to as dithering) of the standard positioning service (SPS) signals by a time-varying bias to limit accuracy for non-U.S. military and government users; this feature can be reactivated at any time without prior notification.

The coordinate system used by GPS is shown in Figure 15.2. Both latitude and longitude use the center of the earth as the vertex, but their zero reference is different.

TABLE 15.1

Comparison of DGPS Systems

Specification	WAAS	Submeter	Decimeter	Centimeter, RTK
Price range, US$	100–500	500–2,500	2,500–7,500	15,000–50,000
Source of differential signal	WAAS	Coast Guard, Omnistar VBS, StarFire I, local differential services	OmniStar HP, StarFire II (requires dual-channel receiver)	Real-time moving systems need a base station within 9.5–16 km (6–10 miles)
Accuracy, m (in.)	1–3 (39–118)	1–3 (39–118)	0.075–0.3 (3–12)	0.025–0.05 (1–2)
Advantage	Low-cost, small handheld unit	Better accuracy	Best accuracy without RTK	Highest accuracy and repeatability
Use	Mapping, yield monitor	Mapping, yield monitor, limited guidance	Guidance (close to row crop)	Precise guidance, elevation and survey-grade mapping

Source: Stephens, S.C. and V.P. Rasmussen, High-end DGPS and RTK systems, Utah geospatial extension program, USU/NASA space grant, 2010.

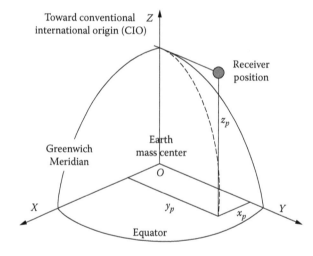

FIGURE 15.2
Coordinate system used by GPS.

The principle of object positioning is based on the following observation equations:

$$p_i = \sqrt{\left(x_p - x_i\right)^2 + \left(y_p - y_i\right)^2 + \left(z_p - z_i\right)^2} + \Delta p + e_i \tag{15.1}$$

$$\Delta p_i = c\,\Delta T_i \tag{15.2}$$

where:
- i is the satellite number, $i = 1, ..., 4$
- p_i is the pseudorange of satellite i (computed distance from four satellites)
- x_p, y_p, z_p are the GPS receiver coordinates
- Δp is the offset due to satellite error

ΔT is the receiver clock correction time
e_i is the additional error
c is the light speed

The GPS time is very precise; thus, we can substitute an atomic clock with a GPS receiver. There are some factors that may influence the determined position accuracy, as follows:

- Receiver design (e.g., number of channels, different tracked systems).
- Due to the way the satellites orbit, there might not be four satellites "visible" at all times. Consequently, when planning the use of GPS, the satellite availability should be checked. Moreover, if the number of required channels is greater than the number of available channels, the measurements are less accurate.

It is good to know some more aspects in regard to GPS codes, their travel, GPS signal modulation, code phase ranging, and so forth; however, they are beyond the scope of this text.

15.3 Map Projection

The latitude indicates the position of a place on the earth ranging from 0° to 90° north or south of the equator. Around the poles, the earth's circumference is approximately 40,007.9 km (24,859.82 miles). That means every degree latitude is the equivalent of 111.133 km (69.05 miles).

The earth map is divided into 60 Universal Transverse Mercator (UTM) zones of latitude, each 6° wide at the equator. They extend from 84° north to 80° south. The numbering of these zones begins at 180° longitude and is performed consecutively eastward.

The longitude indicates the position of a place on earth ranging from 0° at the Greenwich meridian (prime meridian or 0° longitude) to +180° eastward and −180° westward. At the Equator, the earth's circumference is approximately 40,075.06 km (24,901.55 miles). That means, at the equator, every degree latitude is the equivalent of 111.32 km (69.17 miles).

A UTM conformal map gives the location of a place on the earth using a two-dimensional (2D) Cartesian coordinate system, independently of vertical position. It uses a secant transverse Mercator projection in each of the 60 UTM zones. As an example, the UTM zones of U.S. territory are shown in Figure 15.3. The representation of a region on map allows determining the distance between every two points using Pythagoras's theorem because a UTM map is a grid-based system.

Besides vehicle-operated or autonomous navigation, the UTM maps and GPS technique can be used to track the spatial variability of soil properties, crop properties, and yield obtained during harvesting.

15.4 Geomagnetic Direction Sensing

The *geomagnetic field*, that is, the earth's magnetic field, is mainly generated in the earth's conducting fluid outer core through the self-exciting dynamo effect of motion of the iron-rich

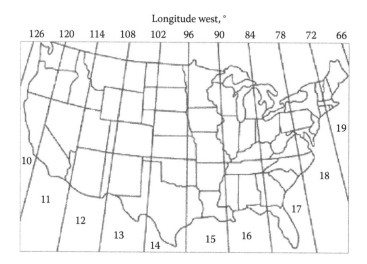

FIGURE 15.3
UTM zones of U.S. territory.

molten material (Figure 15.4). Other magnetic fields produced by magnetized rocks and electric fields, for example, in the ionosphere and magnetosphere, superimpose on the main field. The intensity of the earth's magnetic field is between 31 μT at 0° latitude and 65 μT. For comparison, the intensity of the strongest continuous magnetic field produced in a laboratory (Florida State University, National High Magnetic Field Laboratory) was 45 T.

The magnetic field at any point in space can be represented as a vector with direction and magnitude. The magnitude B, which is the total intensity of the magnetic field, can be

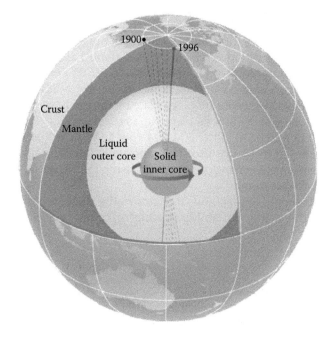

FIGURE 15.4
Origin of the earth's magnetic field. (Courtesy of NASA, Washington, D.C.)

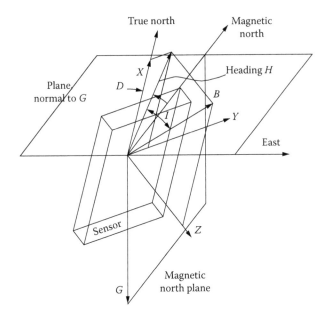

FIGURE 15.5
Geomagnetic field parameters.

represented on three orthogonal components on X, Y, and Z axes, as shown in Figure 15.5. The angle between the horizontal component H (heading) and true north (i.e., geographic north) is called the (magnetic) declination (D). The angle between magnitude B and the horizontal plane is called the (magnetic) inclination (I) or magnetic dip. D and I are measured in degree units, positive cast for D and positive down for I, with respect to Figure 15.5.

The magnetic field of the earth is uniform over a wide area (e.g., several kilometers). A vehicle (made by ferrous metals) creates a local disturbance in this field, whether moving or standing still. An anisotropic magnetoresistive (AMR) sensor can detect a change in the earth's magnetic field due to a disturbance generated by a vehicle. AMR sensors can sense direct-current (DC) static fields as well as the strength and direction of the field. An AMR sensor (Figure 15.6) is made of a nickel-iron (permalloy) thin film deposited on a silicon wafer, and it is patterned as a resistive strip in a Wheatstone bridge configuration. Because they are a thin-film technology product, AMR sensors can be easily integrated in commercial integrated circuit packages.

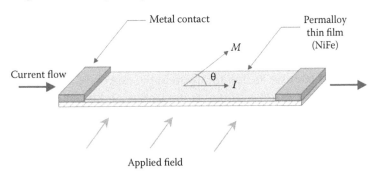

FIGURE 15.6
Anisotropic magnetoresistive element. (Courtesy of Honeywell, Plymouth, MN.)

The AMR sensors represent a very convenient way to measure both linear and angular displacement, as well as the position of an object subject to the earth's magnetic field. Thus, they can be successfully used for vehicle detection, movement direction and speed, and compass navigation (Caruso and Withanawasam, 2000).

15.5 Computer Vision

U.S. lawmakers have recently passed legislation that allows fully autonomous vehicles to share public roads. The demonstrations at the DARPA Grand Challenges and by industry leaders have established that the core technical barrier to achieving autonomous vehicles is road scene understanding. There is a perception that a wide gap exists between what is needed by the industry to successfully deploy camera-based autonomous vehicles and what is currently possible using computer vision techniques. This section attempts to explain the challenges associated with the computer vision field and outline a summary of how computer vision is related to an autonomous combine harvester.

Computer vision is the combination of science and technology for developing and applying computer algorithms for image data processing and analysis, pattern recognition, photogrammetry, and computer graphics to produce numerical or symbolic information, for example, in the form of decisions. The image data can be provided as video sequences, views from multiple cameras, or multidimensional data. Computer vision is the construction of an explicit, meaningful description of the properties of a 3D world from 2D images.

Computer vision deals with automating and integrating a wide range of methods, representations, and computing for vision perception. Subdomains of computer vision include event detection, video tracking, learning, indexing, motion estimation, and image restoration.

According to Marr (1982), there are three levels of description of a visual information processing system: physical level (hardware), representations and algorithms, and computational level. Sensors are used to obtain the images from the 3D environment. A static scene is in general 3D, while a dynamic scene is four-dimensional, represented by spatial–temporal information. The input to the system is a camera image from which, through a multilevel process, a representation of the environment scene that caused the image is obtained. The conceptual framework, within which image processing and understanding is explained, is represented in Figure 15.7. There are four consecutive, task-specific processes whose outcome is materialized into specific intermediate representations of the acquired image. Each process is possible if certain types and appropriate levels of knowledge are applied. The first process is the radiometric feature identification and segmentation of a digital raster image. Through the low-level vision process, mapping of 3D scene elements is performed to extract the real-world features as immortalized in the processed image.

As the name suggests, object recognition is the process of identification and extraction of the image objects by applying particular knowledge of object models, regardless of the angle of camera view. The latest process—high-level vision—results in detection and interpretation of object/time-transcending connections, such as object configurations, special situations, and articulate motion sequences. Image understanding can be achieved by developing and applying various matching algorithms. That means the results from the low level of processing are compared with those from the high representation level. Such interconnections are suggested by the bidirectional arrows shown in Figure 15.7.

Type of knowledge	Representation level	Processing level
Common sense Situation models Process models	Processes Situations Object configuration ⇕ Objects	High-level vision
Object models	Trajectories ⇕ Scene elements	Object recognition
Projective geometry Photometry Physics	3D surfaces, volumes or contours ⇕ Image elements:	Low-level vision
Real-world features	edges, regions, texture, motion folw ⇕ Digital raster image (rough image)	Feature extraction Segmentation

FIGURE 15.7
Hierarchy of knowledge-based processes of image understanding. (Adapted from Neumann, B., Bildverstehen-ein Überblick, in *Einführung in die künstliche Intelligenz*, ed. G. Görz, 559–588, Addison-Wesley, Bonn, 1993.)

Although very coherent and suggestive, the above-described image understanding processes may not always be accurately followed.

15.5.1 Image Formation and Acquisition

The image forms on an analog film or digital silicon, when the light from a scene passes through the lens of a camera. According to the physical law of image formation through the lens, the relationship between the distance z_0 to an object and the distance behind the lens at which a focused image z_i is formed can be expressed as follows:

$$\frac{1}{z_0} + \frac{1}{z_i} = \frac{1}{f} \tag{15.3}$$

where f is the focal length of the lens.

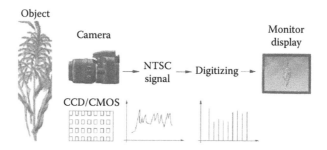

FIGURE 15.8
(See color insert.) Image processing by an analog-to-digital (A/D) converter sampling the National Television System Committee (NTSC) signal.

Due to the fact that the index of refraction in glass varies with wavelength, a simple lens may be subject to *chromatic aberration* because the light at different wavelengths is focused at slightly different focal lengths f', and hence a different depth x_i'. This effect is reduced by compound lenses.

The lenses also have a *vignetting effect* (natural or mechanical), which is the effect of image brightness falling off toward the edge of the image. Natural vignetting occurs due to foreshortening in the object surface, projected pixel, and lens aperture. Mechanical vignetting occurs due to internal occlusion of the light rays near the lens periphery.

An image acquired by a camera may be displayed on a monitor using a signal-digitizing card according to the process schematic shown in Figure 15.8.

An imaging sensor picks the light using an active sensing area, and then the signal is passed to a set of amplifiers. There are two main types of such sensors: *charge-coupled device* (CCD) and *complementary metal oxide on silicon* (CMOS). For every set of the scene points with coordinates (X_s, Y_s, Z_s), we get image points (x_i, y_i). If the scene and image points are represented by homogenous vectors, and the coordinates of the image point are

$$x_i = f \frac{X_s}{Z_s}$$

$$y_i = f \frac{Y_s}{Z_s}$$

(15.4)

then the central projection (Figure 15.9) is a linear transformation expressed as follows:

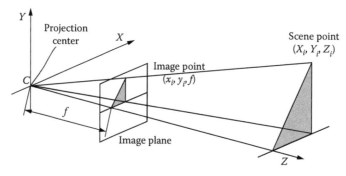

FIGURE 15.9
Formation of the projection center on the imaging sensor.

$$
\begin{bmatrix} u \\ v \\ w \end{bmatrix} = \begin{bmatrix} f & 0 & 0 & 0 \\ 0 & f & 0 & 0 \\ 0 & 0 & 1 & 0 \end{bmatrix} \begin{bmatrix} X_s \\ Y_s \\ Z_s \\ 1 \end{bmatrix}
\tag{15.5}
$$

where u, v, w are the homogenous coordinates of the image point
We must note that

$$
x_i = \frac{u}{w}
\tag{15.6}
$$

$$
y_i = \frac{v}{w}
$$

An autonomous vehicle (using a computer vision system) could evaluate its distance from incoming cars at night partly from a model of cars. Thus, knowing the distance between car headlights (Figure 15.10), the distance to the car can be calculated with the equation

$$
Z = \frac{fD}{d}
\tag{15.7}
$$

The scene coordinates (measured in meters) are scaled to fit onto an image sensor (ultimately measured in pixels). Considering Equations 15.4, these coordinates (x_{pi}, y_{pi}) are calculated as follows (Figure 15.11):

$$
x_{pi} = k_x x_i + x_0 = f k_x \frac{X_s + Z_s x_0}{Z_s}
\tag{15.8}
$$

$$
y_{pi} = k_y y_i + y_0 = f k_y \frac{Y_s + Z_s y_0}{Z_s}
\tag{15.9}
$$

where k_x and k_y are scaling factors on the x and y axes of the image.

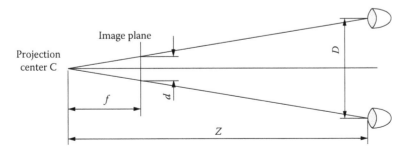

FIGURE 15.10
Parameters required to calculate the distance from a car.

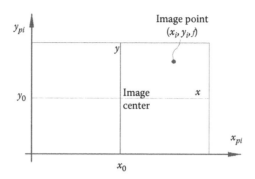

FIGURE 15.11
Image pixel coordinates after length transformation.

Using homogeneous coordinates, we can write

$$
\begin{bmatrix} u' \\ v' \\ w' \end{bmatrix} = \begin{bmatrix} \alpha_x & 0 & x_0 & 0 \\ 0 & \alpha_y & y_0 & 0 \\ 0 & 0 & 1 & 0 \end{bmatrix} \begin{bmatrix} X_s \\ Y_s \\ Z_s \\ 1 \end{bmatrix}
\tag{15.10}
$$

with x_0 and y_0 being the coordinates of the image center, expressed in pixels, and the focal lengths α_x and α_y given by the following equations:

$$
\alpha_x = fk_x
$$
$$
\alpha_y = fk_y
\tag{15.11}
$$

By similitude with Equations 15.6, we can write

$$
x_{pi} = \frac{u'}{w'}
$$
$$
y_{pi} = \frac{v'}{w'}
\tag{15.12}
$$

Adding the skew parameter s, the 4×3 matrix from Equation 15.10 can be written as a product of *camera calibration matrix* **K** and unitary matrix **I** as follows:

$$
\begin{bmatrix} \alpha_x & s & x_0 & 0 \\ 0 & \alpha_y & y_0 & 0 \\ 0 & 0 & 1 & 0 \end{bmatrix} = \begin{bmatrix} \alpha_x & s & x_0 \\ 0 & a_y & y_0 \\ 0 & 0 & 1 \end{bmatrix} \begin{bmatrix} 1 & 0 & 0 & 0 \\ 0 & 1 & 0 & 0 \\ 0 & 0 & 1 & 0 \end{bmatrix} = K\begin{bmatrix} I_3 & 0_3 \end{bmatrix}
\tag{15.13}
$$

K is an upper-triangular matrix with five degrees of freedom. When calibrating a camera based on 3D points, we end up estimating the intrinsic matrix **K**.

When acquiring an image, although 3D geometric features of a scene are projected into 2D features, the image is also made up of discrete color or brightness intensity values. For radiometric modeling of the pixel brightness and scene brightness, we take into account that

- The intensity of a light source falls off with the square of the distance between the light source and the object being lit because the same light is spread over a larger area.
- Pixel brightness is proportional to the radiance of the corresponding scene patch that is independent of the viewpoint.
- Radiance of the scene patch is proportional to the cosine of the angle between the normal to the patch and the direction of the illumination source.

Consequently, the pixel brightness is proportional to the cosine of the angle between the normal to the patch and the direction of the illumination source.

The incident illumination on an object sitting in an outdoor environment implies a more complex light distribution. It can be represented using an environment map or a so-called reflection map.

15.5.2 Object Recognition

The recognition of an object in a given image is the most important aspect of visual perception. The main goal of object recognition is retrieving information that is not apparent in the image. Recognition of common objects is way beyond the capability of the artificial systems proposed so far because visual object recognition is not a single or simple mechanism. For visual object recognition, multiple mechanisms are considered, such as characteristic shape, color pattern and texture, branching patterns, reflection-based material, object location relative to other objects, characteristic motion, prior knowledge-based anticipation (expectation), and reasoning.

The recognition methods can be divided into three classes: alignment methods, invariant property methods, and part decomposition methods. A vision system may combine methods from all three classes.

The *alignment methods* use point features (e.g., triangles) to find a sparse set of corresponding locations in different images. Such correspondences can also be used to align different images (e.g., rounded objects with smooth contours), or to perform object instance and category recognition. The best advantage of using key points is that they allow matching even in the presence of clusters. Two main approaches are used to find feature points and their correspondences. One approach is to identify features in an image that can be accurately tracked using a local searching technique, such as least squares sum. The other approach is to independently detect features in all images at hand and then match features based on their local appearance.

The *invariant property methods* are based on color indexing, salient features (points), and geometric hashing. Woebbeck et al. (1995) was among the first researchers to test vegetation color indices for differentiating green plants from soil, corn residue, and wheat straw residue in images. The color indices have been derived from color chromatic coordinates r, g, and b as follows: $r - g$, $g - b$, $(g - b)/(r - g)$, and $2g - r - b$.

The chromatic coordinates are defined as

$$r = \frac{R^*}{R^* + G^* + B^*}$$

$$g = \frac{G^*}{R^* + G^* + B^*} \qquad (15.14)$$

$$b = \frac{B^*}{R^* + G^* + B^*}$$

where the values R^*, G^*, and B^* are normalized RGB values from the interval $(0, 1)$, obtained as follows:

$$R^* = \frac{R}{R_m}$$

$$G^* = \frac{G}{G_m} \qquad (15.15)$$

$$R^* = \frac{R}{R_m}$$

R, G, and B are the actual pixel values obtained from color images, while the values R_m, G_m, and B_m take the value 255, which is the maximum tonal value for each primary color red, green, and blue.

The color indexing technique follows the steps given below:

- Build a color histogram **M** of the model to be identified in an image.
- Build a color histogram **I** of whole image.
- Build M/I; that is, each bin of color i is replaced by the ratio $\mathbf{M}_i/\mathbf{I}_i$.
- Replace each pixel of color i by its confidence value.
- Smooth the confidence image.
- Identify the model as peaks in the confidence image.

The geometric hashing uses the affine projection model. This involves using a special affine matrix transformation.

The invariant property methods are grounded on invariant features, such as color contiguities, shape moment (inertia of shape computed with respect to coordinate axes), and geometric features: elongation, perimeter length, and so forth. This method does not require the storage of a set of views. The main disadvantage is that the universal features are independent of the viewing position but dependent on the object nature.

The part decomposition method is based on an assumptions series, such as each object can be decomposed into a small set of generic elements, and these elements can be classified independently from the whole object to be identified in the image. The recognition process can be divided into the following subprocesses: object description in terms of constituent parts, part location, part classification as different types of generic components, checking the relationships between the parts, and selection of the objects for which the structure matches the detected relationship in the best way.

Computer vision is a distinct discipline whose theory and practice have made remarkable advances in recent years. The basic definitions given in this subsection should be viewed only as an invitation addressed to readers for a systematic deepening of knowledge in this area by consulting designated texts and research papers.

15.6 Machine Path Tracking and Control

The purpose of path tracking by a combine harvester is to follow a predefined path, that is, a series of waypoints represented by (x, y) coordinates that can be indicated using GPS through metric path planning. The main problem for an autonomous combine harvester is avoiding obstacles. Having a predefined path, the vehicle could be programmed to follow the path, and an algorithm should be used to search for the optimal path. This implies that if there is a need to avoid an obstacle, the machine will go around it, and have the path replanned to reach the desired location.

A proper path-tracking technique should allow the control of all subsystems that contribute to the vehicle performance, for example, engine, throttle, transmission, brakes, steering, tire rolling, and slippage.

In the following, let us solve the control problem of a combine harvester that moves in the field at a constant velocity v_c, and the objective is for it to be steered so that it will reach the destination in minimum time. We assume that there are no obstacles; therefore, the combine may travel anywhere without being confined to a roadway.

As an example of such a control algorithm, we start with the solution given in Application 2.2 (Figure 2.2) and the corresponding kinematics model, which are repeated here for convenience:

$$\dot{x} = v_c \cos \psi_c$$

$$\dot{y} = v_c \sin \psi_c$$

$$\dot{\psi}_c = -\frac{v_c}{L_c} \tan \psi_s \qquad (15.16)$$

$$\dot{\psi}_s = \omega_k$$

Hamilton's principle states that the actual motion of the vehicle along a path between two consecutive, generalized coordinates takes place through a stationary action. In regard to our application, we applied, adapted, and explained the optimal control method described by Cook (2011) for a vehicle steered in a different way.

Let us form the Hamiltonian H, as shown below:

$$H = 1 + \lambda_x v_c \cos \psi_c + \lambda_y v_c \sin \psi_c - \lambda_{\psi_c} \frac{v_c}{L_c} \tan \psi_s + \lambda_{\psi_s} \omega_k \qquad (15.17)$$

This means that for any small deviation of the path from the current one, keeping the initial and final configurations fixed, the action variation vanishes to first order in the derivation. Thus, to find out where a differentiable function has a stationary point, we differentiate and solve Equation 15.17 by setting the derivative to zero.

Consequently, the equations for the costates become

$$\lambda_x = \frac{\partial H}{\partial x} = 0 \tag{15.18}$$

That makes

$$\lambda_x = C_1 \tag{15.19}$$

$$\lambda_y = \frac{\partial H}{\partial y} = 0 \tag{15.20}$$

That makes

$$\lambda_y = C_2 \tag{15.21}$$

$$\lambda_{\psi_s} = \frac{\partial H}{\partial \omega} = 0 \tag{15.22}$$

or

$$\lambda_{\psi_s} = C_4 \tag{15.23}$$

and

$$\lambda_{\psi_c} = -\frac{\partial H}{\partial \psi_c} = -\left(-\lambda_x v_c \sin \psi_c + \lambda_y v_c \cos \psi_c\right) = \lambda_x \dot{y} - \lambda_y \dot{x} \tag{15.24}$$

That means

$$\lambda_{\psi_c}(t) = C_3 + C_1\left[y(t) - y(0)\right] - C_2\left[x(t) - x(0)\right] \tag{15.25}$$

From the geometry given in Figure 2.2, we get

$$\tan \psi_s = \frac{L_c}{R} \tag{15.26}$$

which, by solving, yields the instantaneous radius of curvature for the path of the center of the vehicle's front axle.

By examining the Hamiltonian (15.17), it can be minimized by choosing a value ψ_s of the steering angle that satisfies the following condition:

$$\psi_s = \psi_{s\max} \text{sign}\left(\lambda_{\psi_s}\right) \tag{15.27}$$

whenever

$$\lambda_{\psi_s} \neq 0$$

Thus, the optimal control action determines the vehicle moving on segments of circles with the radius

$$R = \frac{L_c}{\tan \psi_{s\max}} \tag{15.28}$$

An interesting control option is the singular control, that is, $\lambda_{\psi_s} = 0$. Consequently, over the nonzero interval, we should have $\dot{\lambda}_{\psi_s} = 0$. Thus, Equation 15.24 becomes

$$\lambda_x \dot{y} - \lambda_y \dot{x} = 0 \tag{15.29}$$

or

$$C_1 \dot{y} - C_2 \dot{x} = 0 \tag{15.30}$$

Interpreted geometrically, Equation 15.30 describes the movement of the vehicle on a linear path with the slope

$$m = \frac{C_2}{C_1} \tag{15.31}$$

When the vehicle moves on a straight path, $\psi_s = 0$. Thus, the optimal control obeys the law

$$\psi_s = \psi_{s\max} \text{sign}\left(\lambda_{\psi_s}\right)$$

$$\lambda_{\psi_s} \neq 0$$

and

$$\psi_s = 0$$
$$\lambda_{\psi_s} = 0 \tag{15.32}$$

Consequently, the conclusion is that the optimal control is seen to be a series of segments (clockwise or counterclockwise circle segments, $\psi_s = \pm\psi_{s\ max}$) or straight line segments $\psi_s = 0$.

EXERCISE

Apply the above-described control method when the vehicle moves on a defined path, starting from the kinematics model given in Application 2.2 for the vehicle kinematics shown in Figure 2.3.

15.7 Framework of an Autonomous Combine Harvester

The key elements of an autonomous combine harvester are guidance sensors (GPS receiver), process sensors, computational methods, navigation planning, and steering controller. In lieu of conclusions, the framework of such a combine harvester is given in Figure 15.12.

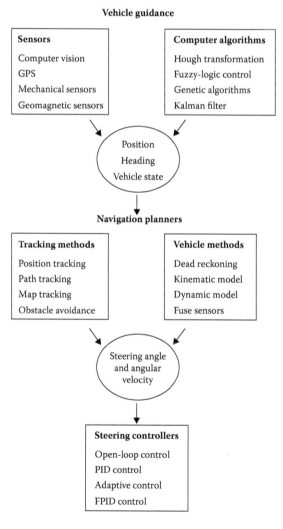

FIGURE 15.12
Framework of an autonomous combine harvester.

This topic was already partially discussed in this chapter; however, the whole subject requires extensive reading from other sources, such as those quoted in the references list of this chapter. The author also recommends reading the RAND publication "Autonomous Vehicle Technology: A Guide for Policymakers" (Anderson et al., 2014).

References

Anderson, J.M., N. Kalra, K.D. Stanley, et al. 2014. Autonomous vehicle technology: A guide for policymakers. RAND_RR443-1. Santa Monica, CA: RAND Corporation. www.rand.org.

Caruso, M.J. and L.S. Withanawasam. 2000. *Vehicle Detection and Compass Applications Using AMR Magnetic Sensor*. Plymouth, MN: Honeywell.

Cook, G. 2011. *Mobile Robots: Navigation, Control, and Remote Sensing*. Hoboken, NJ: IEEE Press, John Wiley & Sons.

Marr, D. 1982. *Vision*. San Francisco, CA: W.H. Freeman.

Neumann, B. 1993. Bildverstehen- ein Überblick. In *Einführung in die künstliche Intelligenz*, ed. G. Görz, 559–588. Bonn: Addison-Wesley.

Stephens, S.C. and V.P. Rasmussen. 2010. High-end DGPS and RTK systems. Utah geospatial extension program. USU/NASA space grant.

Woebbeck, D.M. et al. 1995. Color indices for weed identification under various soil, residue, and lighting conditions. *Transactions of the ASAE* 38:259–269.

Bibliography

Choi, Ji-W., R.E. Curry, and G.H. Elkaim. 2009. Smooth path generation based on Bézier curves for autonomous vehicles. http://users.soe.ucsc.edu/~elkaim/Documents/Choi_WCECS2009_ICIAR.pdf (accessed November 23, 2014).

Deelertpaiboon, C. and M. Parnichkun. 2008. Fusion of GPS, compass, and camera for localization of an intelligent vehicle. *International Journal of Advanced Robotic Systems* 5(4):315–326.

Langley, R.B. 2003. Getting your bearings. The magnetic compass and GPS. *GPS World* 9:70–79.

Li, M., K. Imou, K. Wakabayashi, and S. Yokohama. 2009. Review of research on agricultural vehicle autonomous guidance. *International Journal of Agricultural and Biological Engineering* 2(3):1–16.

Metternicht, G. 2006. Use of remote sensing and GNSS in precision agriculture. Presented at UN-Zambia-ESA Regional Workshop on the Applications of GNSS in Sub-Saharan Africa.

Meyer, G.E. 2011. Machine vision identification of plants. In *Recent Trends for Enhancing the Diversity and Quality of Soybean Products*, ed. D. Krezhova, 401–421. Rijeka, Croatia: InTech. http://www.intechopen.com/books/ (accessed November 18, 2014).

Miu, P. 1999. *Introducere in mecatronica*. Bucharest: Didactic and Pedagogic R.A.

Ringdahl, O. 2007. Techniques and algorithms for autonomous vehicles in forest environment. MSc. thesis, Umeå University, Sweden.

Schierwagen, A. 2001. Vision as computation, or: Does a computer vision really assign meaning to images? In *Integrative Systems Approaches to Natural and Social Dynamics*, ed. H. Matthies, H. Malchow, and J. Kriz, pp. 579–587. Berlin: Springer-Verlag.

Schmidhalter, U., F.-X. Maidl, H. Heuwinkel, et al. 2008. Precision farming: Adaptation of land use management to small scale heterogeneity. In *Perspectives for Agroecosystem Management*, ed. P. Schröder, J. Pfadenhauer, and J.C. Munch, pp. 121–199. Amsterdam: Elsevier.

Szelinski, R. 2010. *Computer Vision: Algorithms and Applications.* Berlin: Springer. http://szeliski.org/Book/.

Ting, K.C., T. Abdelzaher, A. Alleyne, and L. Rodriguez. 2011. Information technology and agriculture. In *The Bridge*, ed. C.R. Arenberg, 6–64. Vol. 41, no. 3. Washington, DC: National Academy of Engineering.

Ze, Z. 2014. Development of a robot combine harvester based on GNSS. PhD dissertation, Hokkaido University, Japan.

Appendix

MATLAB Application: Genetic Algorithm–Based Optimization of Threshing Process

```
% This is the BINMUT.m file
function y = binmut(x)
if x = =0
xloco = 1;
else
xloco = 0;
end

y = xloco;

% This is the cptpet.m file
h = hbar*lp;
lmbdh = lmbd0*(lp-h)/(lp-lmbd0*h);
epsh = epsb*(lb-h)/(lb-epsb*h);
gamh = 1-lmbdh;
lmbdhb = lmbdh+gamh*epsh;
qph = qp*lmbdh/lmbdhb;
p = eta*(s0-pi*de^2/4)/s0;
beta = kbeta*p*sqrt(v*up*sqrt(qph)/sqrt(rop)*exp(-qph/qp0-up/upmin));
lambda = klmbd*sqrt(rop*v*delta/eta/qph/sqrt(up))*exp(qph/qp0+up/upmax-v/
v0);
sectpp = diapp*sqrt(s0/2);
ppp = eta*(1-sectpp/s0);
aapp = ppp*caa*qph*v*delta*up/exp(qph^cab+cac*(v+delta+up));
bbpp = cbd*qph*delta*up/exp(qph^cbe+cbf*(delta+up));
ss = 100/(lambda-beta)*(lambda*(1-exp(-beta*L))-...
beta*(1-exp(-lambda*L)));
pts = dsp*v*v/(R*qph)+asp*up*exp(qph^0.5+csp*(v+delta-up))/...
(qph^bsp*v^csp*delta);
Ls = 100*lambda/(lambda-beta)*(exp(-beta*L)-exp(-lambda*L));

us = 5.4456*exp(0.02763*up);
sv = av*up*exp(bv*v+qph^(-0.5)-dv*up)/(qph*v*us);
pp = 100*aapp/bbpp/(bbpp-aapp)*(bbpp*(1-exp(-aapp*L))-...
aapp*(1-exp(-bbpp*L)));
perf(1) = pts;
perf(2) = Ls;
perf(3) = sv;
perf(4) = ss;
perf(5) = pp;
```

```
if iter = =genstart
if iind = =1
perf = perf';
end
end


% This is the evltpet.m file
limitperf = 1;
if performance(iind,1)<0
evalind = unstcoef*ones(1,perfnr);
end


if performance(iind,1)>-1

if performance(iind,1)<aover
evalind(1) = maxeval;
elseif performance(iind,1)<bover
evalind(1) = polyval([Aover,Bover,Cover,Dover],performance(iind,1));
elseif performance(iind,1)<cover
evalind(1) = mineval;
else
evalind(1) = mineval;
limitperf = 0;
end


if performance(iind,2)<asett
evalind(2) = maxeval;
elseif performance(iind,2)<bsett
evalind(2) = polyval([Asett,Bsett,Csett,Dsett],performance(iind,2));
elseif performance(iind,2)<csett
evalind(2) = mineval;
else
evalind(2) = mineval;
limitperf = 0;
end


if performance(iind,3)<atctr
evalind(3) = maxeval;
elseif performance(iind,3)<btctr
evalind(3) = polyval([Atctr,Btctr,Ctctr,Dtctr],performance(iind,3));
elseif performance(iind,3)<ctctr
evalind(3) = mineval;
else
evalind(3) = mineval;
limitperf = 0;
end


if abs(performance(iind,4))<cgminp
evalind(4) = mineval;
limitperf = 0;
elseif abs(performance(iind,4))<agminp
evalind(4) = mineval;
elseif abs(performance(iind,4))<bgminp
```

```
evalind(4) = polyval([Agminp,Bgminp,Cgminp,Dgminp],abs(performance(i
ind,4)));
else
evalind(4) = maxeval;
end

if performance(iind,5)<app
evalind(5) = maxeval;
elseif performance(iind,5)<bpp
evalind(5) = polyval([App,Bpp,Cpp,Dpp],performance(iind,5));
elseif performance(iind,5)<cpp
evalind(5) = mineval;
else
evalind(5) = mineval;
limitperf = 0;
end
end
perfvalue = weight*evalind';
if limitperf = =0
perfvalue = perfvalue/limitdim;
end
perfvaluevect(iind) = perfvalue;

% This is the genpet.m file for Axial Threshing Units

clear
path(path,'d:\matl\genp')
initpet
fileid = fopen('d:\matl\genrez\FUZrez.txt','w');
fileid3 = fopen('d:\matl\genrez\graf.txt','w');

for iter = genstart:gennr
disp('********************************')
disp('********************************')
disp('GENERATION ')
disp(iter)
disp('********************************')
disp('********************************')

fprintf(fileid,'\nGENERATION :%i\n',iter);
fprintf(fileid,'*************************** \n');

% Convert to decimal float gains

for i1 = 1:popsize
      for i2 = 1:gainsnr
      if i2 = =1
% First variable hbar
      kstar = 0;
      bitnr1 = bitnr(i2)-1;
      indprt = 0;
      for isum = 1:i2
      indprt = indprt+bitnr(isum);
      end
```

```
      for istar = 0:bitnr1
      kstar = kstar+parent(i1,indprt-istar)*2^istar;
      end
      parentfloat(i1,i2) = hlow+kstar*(hhigh-hlow)/(2^bitnr(i2)-1);

      elseif i2 = =2
% Second variable qp
      kstar = 0;
      bitnr1 = bitnr(i2)-1;
      indprt = 0;
      for isum = 1:i2
      indprt = indprt+bitnr(isum);
      end
      for istar = 0:bitnr1
      kstar = kstar+parent(i1,indprt-istar)*2^istar;
      end
      parentfloat(i1,i2) = qplow+kstar*(qphigh-qplow)/(2^bitnr(i2)-1);
      elseif i2 = =3
% Second bis variable up
      kstar = 0;
      bitnr1 = bitnr(i2)-1;
      indprt = 0;
      for isum = 1:i2
      indprt = indprt+bitnr(isum);
      end
      for istar = 0:bitnr1
      kstar = kstar+parent(i1,indprt-istar)*2^istar;
      end
      parentfloat(i1,i2) = uplow+kstar*(uphigh-uplow)/(2^bitnr(i2)-1);
      elseif i2 = =4
% Third variable v
      kstar = 0;
      bitnr1 = bitnr(i2)-1;
      indprt = 0;
      for isum = 1:i2
      indprt = indprt+bitnr(isum);
      end
      for istar = 0:bitnr1
      kstar = kstar+parent(i1,indprt-istar)*2^istar;
      end
      parentfloat(i1,i2) = vlow+kstar*...
            (vhigh-vlow)/(2^bitnr(i2)-1);
      elseif i2 = =5
% Third bis variable v0
      kstar = 0;
      bitnr1 = bitnr(i2)-1;
      indprt = 0;
      for isum = 1:i2
      indprt = indprt+bitnr(isum);
      end
      for istar = 0:bitnr1
      kstar = kstar+parent(i1,indprt-istar)*2^istar;
      end
      parentfloat(i1,i2) = vlow+kstar*...
```

```
                    (vhigh-vlow)/(2^bitnr(i2)-1);
        elseif i2 = =6
% Fourth variable delta
        kstar = 0;
        bitnr1 = bitnr(i2)-1;
        indprt = 0;
        for isum = 1:i2
        indprt = indprt+bitnr(isum);
        end
        for istar = 0:bitnr1
        kstar = kstar+parent(i1,indprt-istar)*2^istar;
        end
        parentfloat(i1,i2) = dellow+kstar*(delhigh-dellow)/
(2^bitnr(i2)-1);
        elseif i2 = =7
% Fifth variable L
        kstar = 0;
        bitnr1 = bitnr(i2)-1;
        indprt = 0;
        for isum = 1:i2
        indprt = indprt+bitnr(isum);
        end
        for istar = 0:bitnr1
        kstar = kstar+parent(i1,indprt-istar)*2^istar;
        end
        parentfloat(i1,i2) = Llow+kstar*(Lhigh-Llow)/(2^bitnr(i2)-1);
        elseif i2 = =8
% Sixth variable s0
        kstar = 0;
        bitnr1 = bitnr(i2)-1;
        indprt = 0;
        for isum = 1:i2
        indprt = indprt+bitnr(isum);
        end
        for istar = 0:bitnr1
        kstar = kstar+parent(i1,indprt-istar)*2^istar;
        end
        parentfloat(i1,i2) = s0low+kstar*(s0high-s0low)/(2^bitnr(i2)-1);
        elseif i2 = =9
% Seventh variable eta
        kstar = 0;
        bitnr1 = bitnr(i2)-1;
        indprt = 0;
        for isum = 1:i2
        indprt = indprt+bitnr(isum);
        end
        for istar = 0:bitnr1
        kstar = kstar+parent(i1,indprt-istar)*2^istar;
        end
        parentfloat(i1,i2) = etalow+kstar*(etahigh-etalow)/
(2^bitnr(i2)-1);
        else
% Eighth variable R
        kstar = 0;
```

```
            bitnr1 = bitnr(i2)-1;
            indprt = 0;
            for isum = 1:i2
            indprt = indprt+bitnr(isum);
            end
            for istar = 0:bitnr1
            kstar = kstar+parent(i1,indprt-istar)*2^istar;
            end
            parentfloat(i1,i2) = Rlow+kstar*(Rhigh-Rlow)/(2^bitnr(i2)-1);
            end
            end
end


for iind = 1:popsize
disp(' ')
disp('*******************************')
disp('Individual ')
disp(iind)
disp('*******************************')


if parentfloat(iind,1)<0.05/lp
parentfloat(iind,1) = 0.05/lp;
end
if parentfloat(iind,1)>1-0.25/lp
parentfloat(iind,1) = 1-0.25/lp;
end
if parentfloat(iind,6)<ls
parentfloat(iind,6) = ls;
end
if parentfloat(iind,8)<1.1*ls^2
parentfloat(iind,8) = 1.1*ls^2;
end


%parentfloat(iind,6) = 30;
vax = 2*parentfloat(iind,4)/pi/tan(gama);
h = parentfloat(iind,1)*lp;
lmbdh = lmbd0*(lp-h)/(lp-lmbd0*h);
sect = parentfloat(iind,2)/(lmbdh*vax*rotr);
%parentfloat(iind,10) = sect/(2*pi*delti/1000)-delti/2000;
%if parentfloat(iind,10)<lp/3.14
% parentfloat(iind,10) = lp/3.14;
%end


hbar = parentfloat(iind,1);
qp = parentfloat(iind,2);
      up = parentfloat(iind,3);
v = parentfloat(iind,4);
      v0 = parentfloat(iind,5);
delta = parentfloat(iind,6);
L = parentfloat(iind,7);
s0 = parentfloat(iind,8);
eta = parentfloat(iind,9);
R = parentfloat(iind,10);
```

```
cptpet
      for i3 = 1:size(perf)
      performance(iind,i3) = perf(i3);
      end

evltpet
      for i3 = 1:size(perf)
      evalvect(iind,i3) = evalind(i3);
      end

if 0
disp(' hbar qp      up v    v0')
disp(parentfloat(iind,1:5))
disp(' delta L      s0     eta R')
disp(parentfloat(iind,6:10))
disp('Performance Indeces:')
disp(' pts Ls sv ss pp')
disp(performance(iind,:))
disp('Performance Evaluation Vector')
disp(evalind)
disp('Performance Coefficient')
disp(perfvalue)
%pause
end
end

orderpet

genaverage = sum(pvvord)/popsize;
%genaverage3 = (pvvord(1)+pvvord(2)+pvvord(3))/3;

fprintf(fileid,'\nBest individual *** Performance coefficient
:%f\n',pvvord(1));
fprintf(fileid,'Performance vector : \n');
fprintf(fileid,'%6.2f%6.2f%6.2f%6.2f%6.2f\n',performance(ordervect(1),:));
fprintf(fileid,'Performance index : \n');
fprintf(fileid,'%6.2f%6.2f%6.2f%6.2f%6.2f\n',evalvect(ordervect(1),:));
fprintf(fileid,'\nhbar =%6.3f qp =%6.3f up =%6.3f v =%6.3f v0 =%6.3f\n',p
arentfloat(ordervect(1),1:5));
fprintf(fileid,'\ndelta =%6.3f L =%6.3f s0 =%6.3f eta =%6.3f R =%6.3f\n',
parentfloat(ordervect(1),6:10));
fprintf(fileid,'\nGeneration average :%5.3f\n',genaverage);
fprintf(fileid3,'%3i%5.3f%6.3f\n',iter,genaverage,pvvord(1));

grf(iter,1) = iter;
grf(iter,2) = genaverage;
grf(iter,3) = pvvord(1);

nxtpet

parent = child;
end
fclose('all')
save d:\matl\genrez\in_prnt parent
```

```
% This is init.m file for threshing units design

clear

lp = 0.8;      % plant length [m]
lmbd0 = 0.6;   % coef
rop = 30;      % [kg/m3]
epsb = 0.03;   %%
de = 4.2;      % [mm] Equivalent diameter of grain
delti = 50;    % [mm] Concave clearance at entrance
ls = 7.25;     % [mm] Max. diameter of grain
upmin = 12;    %%
upmax = 42;    %%
lb = 0.3;      % [m]
kbeta = 1.15;  % coef
klmbd = 0.25;  % coef


% Coef for MOG separation
diapp = 4.22;
caa = 0.009224;
cab = 0.6177;
cac = 0.04;
cbd = 0.1;
cbe = 0.5627;
cbf = 0.028;


dsp = 0.001251;
asp = 1.29;
bsp = 0.41;
csp = 0.021;


av = 0.0374;
bv = 0.3246;
cv = 0.2;
dv = 0.1263;


gama = 70*pi/180; % [rad]
rotr = 35; % [kg/m3]


qp0 = 4.5;

% Genetic Algorithm parameters

popsize = 100; % population size
gennr = 100; % number of generations (= number of iterations)
bitnr = [7 8 8 8 8 7 9 10 7 6]; % number of bits for binary
representation
gainsnr = 10;% number of variables to be determined

hlow = 0.05; % lower limit
hhigh = 0.5; % upper limit
qplow = 2; % lower limit
qphigh = 7; % upper limit
```

```
uplow = 15; % lower limit
uphigh = 50; % upper limit
vlow = 25; % lower limit
vhigh = 35; % upper limit
dellow = 5; % lower limit
delhigh = 25; % upper limit
Llow = 1.2; % lower limit
Lhigh = 2; % upper limit
s0low = 200; % lower limit
s0high = 900; % upper limit
etalow = 0.1; % lower limit
etahigh = 0.6; % upper limit
Rlow = 0.2; % lower limit
Rhigh = 0.4; % upper limit

perfnr = 5; % number of performance parameters
limitdim = 4;
probcross = 0.50;
probmut = 0.03;

performance = zeros(popsize,perfnr);

% Initial population
%
disp('Iteration starting at generation number 1 ? YES = 1, NO = 0')
gen_start_flag = input('gen_start_flag = ');

if gen_start_flag = =1
genstart = 1;
for i = 1:popsize
for j = 1:sum(bitnr)
if rand<0.5
parent(i,j) = 0;
else
parent(i,j) = 1;
end
end
end
else
genstart = input('Starting Generation Number = ');
gennr = input('Ending Generation Number = ');
load d:\matl\genrez\in_prnt
end

% Evaluation function parameters

mineval = 0; % minimum value of the detailed evaluation functions
maxeval = 1; % maximum value of the detailed evaluation functions

weight = [2/10 2/10 2/10 2/10 2/10]; %

aover = 6; % overshoot = pts
bover = 10;
cover = 12;
```

```
asett = 0.5; % settling time = Ls
bsett = 2.5;
csett = 4;

atctr = 0.8; % time constant = sv (Wpar right)
btctr = 2;
ctctr = 3;
%atctl = 0.10; % time constant = Wpar left
%btctl = 0.25;
%ctctl = 0.05;

agminp = 97.2; % gain margin at input = ss
bgminp = 99.4;
cgminp = 95;

app = 18;      % overshoot = pp
bpp = 23;
cpp = 28;

[Aover,Bover,Cover,Dover] = polinom3(aover,bover,mineval,maxeval);
[Asett,Bsett,Csett,Dsett] = polinom3(asett,bsett,mineval,maxeval);
[Atctr,Btctr,Ctctr,Dtctr] = polinom3(atctr,btctr,mineval,maxeval);
%[Atctl,Btctl,Ctctl,Dtctl] = polinom3(atctl,btctl,maxeval,mineval);
[Agminp,Bgminp,Cgminp,Dgminp] = polinom3(agminp,bgminp,maxeval,mine
val);
[App,Bpp,Cpp,Dpp] = polinom3(app,bpp,mineval,maxeval);

%end
% This is the nxtpet.m file

best = parent(ordervect(1),:);

totfit = sum(perfvaluevect);

for inext1 = 1:popsize
probsel(inext1) = perfvaluevect(inext1)/totfit;
end

for inext1 = 1:popsize
cumprob(inext1) = 0;
for inext2 = 1:inext1
cumprob(inext1) = cumprob(inext1)+probsel(inext2);
end
end

for inext1 = 1:popsize
aleator = rand;
icnext = 0;
for inext3 = 1:popsize
if icnext = =0
      if aleator<cumprob(inext3)
      parentselect(inext1,:) = parent(inext3,:);
      icnext = 1;
      end
```

```
end
end
end

parentselect(1,:) = best;

sumbit = sum(bitnr);
child = zeros(popsize,sumbit);
crosscount = 0;
crossnumber = 0;

for icross1 = 1:popsize
aleacross = rand;
if aleacross>probcross
child(icross1,:) = parentselect(icross1,:);
elseif crosscount = =0
crossind1 = parentselect(icross1,:);
crossindex1 = icross1;
crosscount = 1;
else
aleacross1 = rand;
border = 1/(sumbit-1);
poscross = fix(aleacross1/border);
for icross2 = 1:poscross
child(crossindex1,icross2) = parentselect(crossindex1,icross2);
child(icross1,icross2) = parentselect(icross1,icross2);
end
poscross1 = poscross+1;
for icross3 = poscross1:sumbit
child(crossindex1,icross3) = parentselect(icross1,icross3);
child(icross1,icross3) = parentselect(crossindex1,icross3);
end
crosscount = 0;
crossnumber = crossnumber+1;
end
end

if crosscount = =1
child(crossindex1,:) = parentselect(crossindex1,:);
end
crossnumber

mutnumber = 0;
for imut1 = 1:popsize
for imut2 = 1:sumbit
aleamut = rand;
if aleamut<probmut
child(imut1,imut2) = binmut(child(imut1,imut2));
mutnumber = mutnumber+1;
end
end
end
mutnumber
child(1,:) = best;
```

```
% This is the orderpet.m file
pvvinter = perfvaluevect;

for iord1 = 1:popsize
pvvord(iord1) = pvvinter(1);
iord2max = 0;
for iord2 = 1:popsize
if pvvord(iord1)<pvvinter(iord2)
pvvord(iord1) = pvvinter(iord2);
iord2max = iord2;
end
end
if iord2max = =0
pvvinter(1) = -2;
ordervect(iord1) = 1;
else
pvvinter(iord2max) = -2;
ordervect(iord1) = iord2max;
end

end

%This is the polinom3.m file
function [A,B,C,D] = polinom3(a,b,min,max)
delta = [a^3 a^2 a 1
b^3 b^2 b 1
3*a^2 2*a 1 0
3*b^2 2*b 1 0];
unu = [max min 0 0]';
deltaA = [unu delta(:,2:4)];
deltaB = [delta(:,1) unu delta(:,3:4)];
deltaC = [delta(:,1:2) unu delta(:,4)];
deltaD = [delta(:,1:3) unu];

detdelt = det(delta);
A = det(deltaA)/det(delta);
B = det(deltaB)/det(delta);
C = det(deltaC)/det(delta);
D = det(deltaD)/det(delta);
```

Index

Milton Keynes UK
Ingram Content Group UK Ltd.
UKHW051537141024
449569UK00028B/1513